POLLUTION CONTROL AND RESOURCE RECOVERY: SEWAGE SLUDGE

POLLUTION CONTROL AND RESOURCE RECOVERY: SEWAGE SLUDGE

ZHEN GUANGYIN AND ZHAO YOUCAI*

**State Key Laboratory of Pollution Control and Resource Reuse, School of Environmental Science and Engineering, Tongji University, Shanghai, China*

ELSEVIER

Amsterdam • Boston • Heidelberg • London
New York • Oxford • Paris • San Diego
San Francisco • Singapore • Sydney • Tokyo

Butterworth-Heinemann is an imprint of Elsevier

BH

Butterworth-Heinemann is an imprint of Elsevier
The Boulevard, Langford Lane, Kidlington, Oxford OX5 1GB, United Kingdom
50 Hampshire Street, 5th Floor, Cambridge, MA 02139, United States

Library of Congress Cataloging-in-Publication Data
A catalog record for this book is available from the Library of Congress

British Library Cataloguing-in-Publication Data
A catalogue record for this book is available from the British Library

ISBN: 978-0-12-811639-5

For information on all Butterworth-Heinemann publications visit our website at https://www.elsevier.com/

Publisher: Cathleen Sether
Acquisition Editor: Ken McCombs
Editorial Project Manager: Peter Jardim
Production Project Manager: Kiruthika Govindaraju
Cover Designer: Matthew Limbert

CONTENTS

LIST OF CONTRIBUTORS

Zhen Guangyin
Tongji University, Shanghai, China
Tohoku University, Sendai, Miyagi, Japan

Zhou Haiyan
Shanghai Laogang Waste Disposal Company, Shanghai, China

Chen Haoquan
Shanghai Environmental Engineering Company, Shanghai, China

Zhang Hua
Tongji University, Shanghai, China

Ma Jianli
Tongji University, Shanghai, China

Tai Jun
Shanghai Design Institute of Environmental Sanitation Engineering, Shanghai, China

Su Lianghu
Tongji University, Shanghai, China

Zhang Meilan
Shanghai Laogang Waste Disposal Company, Shanghai, China

Tang Ping
Tongji University, Shanghai, China

Huang Renhua
Shanghai Laogang Waste Disposal Company, Shanghai, China

Chen Shanping
Shanghai Design Institute of Environmental Sanitation Engineering, Shanghai, China

Chai Xiaoli
Tongji University, Shanghai, China

Lu Xueqin
Tohoku University, Sendai, Miyagi, Japan

Zhu Ying
Tongji University, Shanghai, China

Zhao Youcai
Tongji University, Shanghai, China

ABOUT THE AUTHORS

Zhen Guangyin is currently JSPS research fellow at the National Institute for Environmental Studies (NIES), Japan. He received his BE degree in Environmental Engineering from Hunan University, China in 2008 and his PhD from Tongji University, China in 2014. Prior to joining in the NIES, he worked as a visiting researcher at the Tohoku University, Japan. He has authored or coauthored 30 peer-reviewed international papers, 5 Chinese papers, and 4 invited book chapters. His principal research interests lie in sewage sludge treatment and renewable energy conversion (anaerobic digestion, microbial electrolysis cell, etc.).

Zhao Youcai is currently a professor of environmental engineering at School of Environmental Science and Engineering, Tongji University. He got bachelor degree from Sichuan University (1984) and PhD from Institute of Chemical Metallurgy (now Institute of Process Engineering), Chinese Academy of Sciences, Beijing (1989). After finishing his postdoctoral research work at Fudan University, Shanghai, he joined in Tongji University in 1991. Meanwhile, he had also worked at Aristotle University, Greece of Thessaloniki, National University of Singapore, Tulane University, United States, and Paul Scherrer Institute, Switzerland, for 4 years as research fellow or visiting professor. He had authored or coauthored 138 publications published in the peer-reviewed internationally recognized journals, 420 publications in China journals, and authored or coauthored 75 books (as an author or Editor-in-chief). Currently, his research interests include treatment of municipal solid wastes, sewage sludge, hazardous wastes, polluted construction wastes, and industrial wastes.

PREFACE

The increase in wastewater treatment activities has been confronted with a dramatically increased amount of sewage sludge, which accounts for up to 50–60% of the total operational costs of a wastewater treatment plant. Sewage sludge floc is a multiphase medium and constitutes the vast majority of components that hamper its dewatering and subsequent disposal or reuse. The treatment and disposal of sludge represents a bottleneck in water industries, which has caused an urgent call for the development of cost-efficient approaches to reduce and minimize sludge production.

During the last decennia, the authors have conducted a series of scientific research on developing various sustainable technologies for sewage sludge treatment and disposal, including advanced dewatering through chemical conditioning, solidification/stabilization, reuse for the development of construction and building materials, anaerobic bioenergy recovery, as well as sanitary landfill and odor control. Apart from the contribution to the promotion of fundamental science, the proposed technologies in our scientific work are of great promise in practice as well. Some of the technologies have been demonstrated by the great success of their full-scale implementations in the real-world scenarios, which made great contribution to sustainable sludge management in many cities in China. The positive outcomes encourage us to introduce our currently available scientific advances and well-proven experience to the world, especially to the countries that are the lack of proper sludge management strategies, and well-trained professionals.

This book covers contents of sewage sludge generation and characteristics, chemical and physical conditioning and deep dewatering, sludge solidification/stabilization, drying and incineration, sewage sludge-derived controlled low-strength materials, harvest of bioenergy by anaerobic digestion, and pollution control and recycling of sewage sludge in sanitary landfill. It will therefore provide environmentally friendly alternatives as well as valuable guidelines for engineers and researchers to address the sludge issues faced. In addition, it is convinced that the book also has the great potential to enhance promote the international exchange and cooperation of environmental researchers worldwide.

SUMMARY

Continuous sewage sludge production is becoming one of the crucial issues in the field of water industries. In practice, the majority of sewage sludge has been sent directly to landfill sites without proper pretreatment while this practice is not environmental friendly. In this book, several environmentally friendly technologies for sewage sludge treatment and disposal have been introduced, including sewage sludge generation and characteristics, chemical and physical conditioning and deep dewatering, sludge solidification/stabilization, drying and incineration, sewage sludge—derived controlled low-strength materials, harvest of bioenergy by anaerobic digestion, and pollution control and recycling of sewage sludge in sanitary landfill, with the purpose of providing proper alternatives and valuable guidelines for engineers and researchers to address the sludge issues faced.

ABBREVIATIONS

AS Calcined aluminum salts
ANC Acid neutralization capacity
AVS-S Acid-volatile sulfide
AD Anaerobic digestion
AR Aged refuse
CST Capillary suction time
CSA Calcium sulfoaluminate cement
CLSM Controlled low-strength materials
CRS-S Cr(II)-reducible sulfide
DS Dewatered sludge
3D-EEM Three-dimensional excitation-emission matrix fluorescence spectroscopy
DD_{SCOD} Disintegration degree
EPS Extracellular polymeric substance
ES-S Elemental sulfur
FT-IR Fourier transforms infrared spectra
FA Fulvic acid
FH Ferric hydroxide
HA Humic acid
LB-EPS Loosely bound EPS
MSWI BA Municipal solid waste incineration bottom ash
MOC Magnesium oxychlorides cement
PS Polysaccharides
PN Proteins
PC Portland cement
SCOD Soluble chemical oxygen demand
S-EPS Soluble/slime EPS
S/S Solidification/stabilization
SEM-EDX Scanning electron microscopy combined with an energy-dispersive X-ray spectroscopy
SV Sludge volume
TSS Total suspended solids
TCOD Total chemical oxygen demand
TB-EPS Tightly bound EPS
TG-DSC Thermogravimetry-differential scanning calorimetry

TOC Total organic carbon
TCLP Toxicity characteristic leaching procedure
UCS Unconfined compressive strength
VSS Volatile suspended solids
WWTP Wastewater treatment plant
WAS Waste activated sludge
XRD X-ray powder diffraction
ZVSI Zero valent scrap iron
ZVI Zero valent iron

CHAPTER ONE

Sewage Sludge Generation and Characteristics

1.1 SEWAGE SLUDGE PRODUCTION

In biological wastewater treatment process, the part of chemical oxygen demands (CODs) removed is converted into biosolids, which makes up sewage sludge. Sewage sludge usually represents 1–2% of the treated wastewater volume (Fig. 1.1). As per UN-Habitat's statistics, the existing wastewater treatment plants (WWTPs) in the United States, for instance, generate over 6.5 million tons of dry solids annually; it is estimated to be around 3.0 and 2.0 Mt per year produced in China and Japan, respectively (Fig. 1.2). The figures are naturally anticipated to increase in the near future when considering the growing applications of WWTPs in developing countries. The main disposal routes and rates are different in different countries, heavily depending upon the economic development level. As illustrated in Fig. 1.2, in developed countries such as the United States, the reuse and disposal rate reaches up to 94%, and it is roughly 97% in Japan, where more than half (52%) of sewage sludge is being recycled to produce building materials and 12% anaerobically digested for bioenergy recovery. Comparatively, the situation of sewage sludge use and disposal in developing countries is far beyond optimism. Note that for a sanitary landfill, the threshold of sludge water content is 60% from the view of safety regulations. The improper

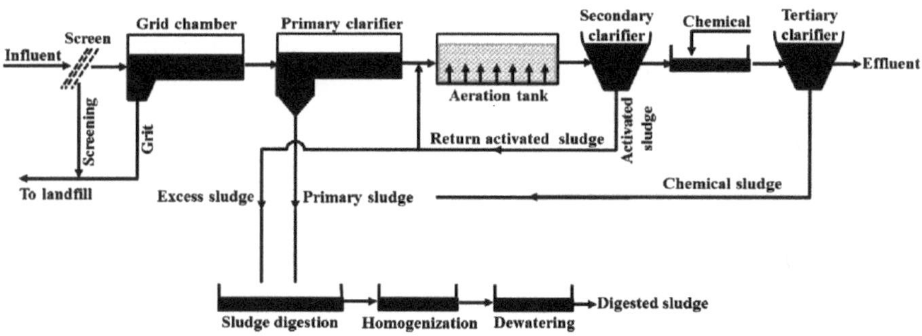

Figure 1.1 Sources and types of sewage sludge generated in wastewater treatment plants (WWTPs).

Pollution Control and Resource Recovery: Sewage Sludge
ISBN: 978-0-12-811639-5
http://dx.doi.org/10.1016/B978-0-12-811639-5.00001-2

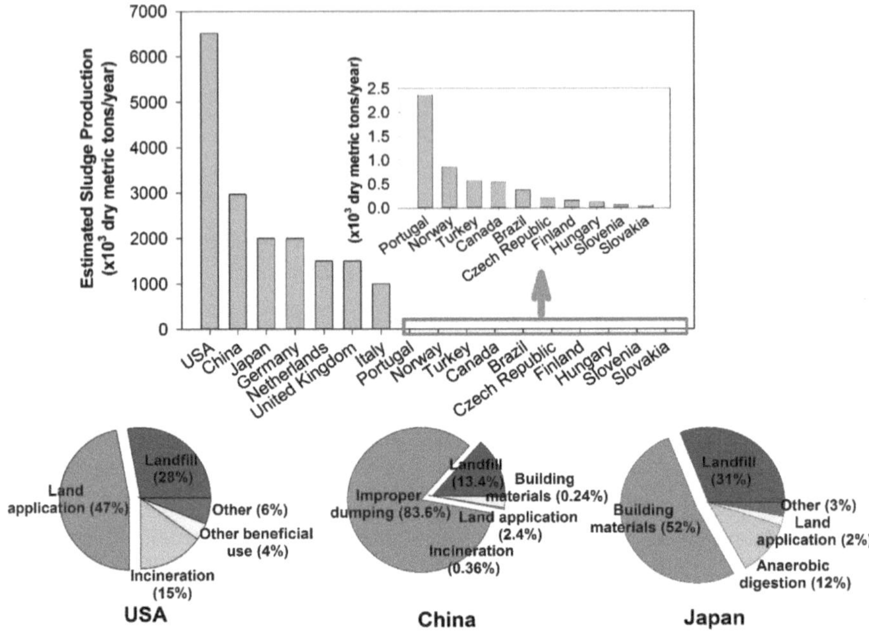

Figure 1.2 Estimated sewage sludge production as per the statistics of UN-Habitat and the situation of sewage sludge use and disposal. *United States: From "US Wastewater Treatment Fact sheet" Pub. No. CSS04-14, 2015. China, Japan: Ministry of Land, Infrastructure, Transport and Tourism, Japan.*

disposal not only causes the wasting of resources but also brings about a series of secondary disasters (e.g., landslide, environmental pollution, etc.). Sewage sludge management is highly complex and costly, representing a stern global challenge.

1.2 SPECIAL FEATURES OF SEWAGE SLUDGE

Sewage sludge floc is a multiphase medium and constitutes the vast majority of components including microbial aggregates, filamentous bacterial strains, organic and inorganic particles, extracellular polymeric substances (EPSs), and large amount of water. The composition of sewage sludge varies, depending upon the type and original components of the raw wastewater. EPSs, originating from the microbial activity (secretion and cell lysis) and/or from the wastewater itself, that is, from the adsorption of organic matter (e.g., cellulose, humic acids, etc.), are the major constituent of sludge organic fractions, mainly composed of proteins, polysaccharides, nucleic acids, humic substances, lipids, etc. (Fig. 1.3A). Depending on the spatial distribution

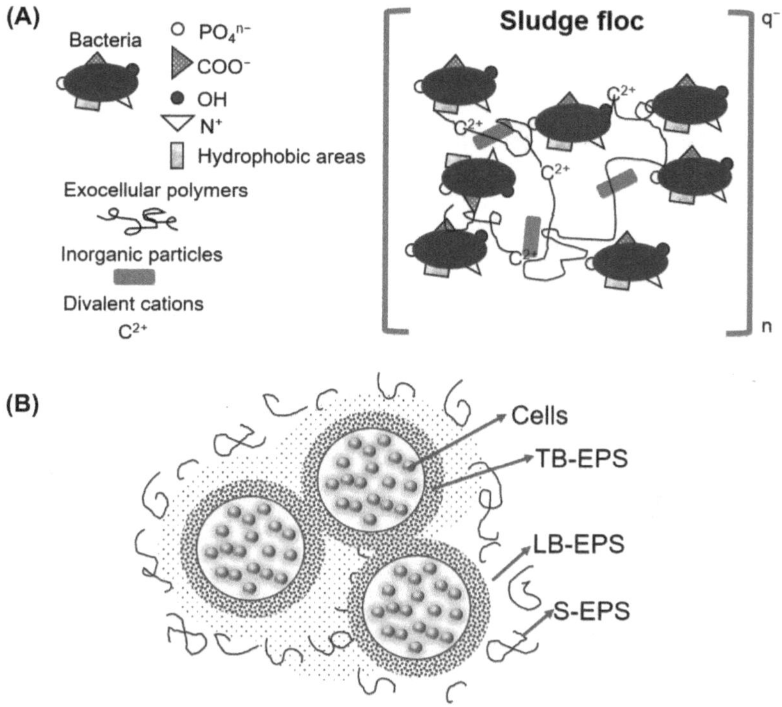

Figure 1.3 Schematic representation of (A) the sewage sludge floc and (B) the sketch of extracellular polymeric substance (EPS) structure. *(A) Modified from Urbain, V., Block, J.C., Manem, J., 1992. Bioflocculation in activated-sludge, an analytic approach. Water Sci. Technol. 25, 441–443.*

within sludge floc matrixes, EPSs are usually divided into three categories: slime EPS (S-EPS), loosely bound EPS (LB-EPS), and tightly bound EPS (TB-EPS). S-EPS are evenly distributed in the aqueous phase and LB-EPS extend from TB-EPS and are characterized by a highly porous and dispersible structure; comparatively, TB-EPS adhere to the surface of the bacterial cells inside the sludge flocs (Fig. 1.3B). The presence of these three-dimensional, gel-like, and negatively charged biopolymers governs the surface physicochemical properties of sludge matrixes. EPSs provide the protective shielding and prevent the cell rupture and lysis, thereby influencing sludge functional integrity, strength, flocculation, dewaterability, and even biodegradability. Besides the protection from EPS, microbial cells themselves possess a hard cell envelope composed of glycan strands cross-linked by peptide that presents physical and chemical barriers to direct anaerobic digestion. In consequence, the sewage sludge with high EPS and cells-content has the stiff structure and will be more difficult to hydrolyze and digest.

1.3 GENERAL PROCESSES OF POLLUTION CONTROL AND RESOURCE RECOVERY FOR SEWAGE SLUDGE

The term sludge treatment encompasses all alternatives that facilitate transport, storage, reuse, or final disposal of sewage sludge. The measures to avoid discharge into the environment need to be adopted compulsorily to protect human and environmental health from the risks linked to sludge (Fig. 1.4). The methods used most frequently for sludge handling and disposal include dewatering, composting, anaerobic digestion, drying–incineration, and sanitary landfill. An overview of various alternatives for treatment and disposal is presented in Figs. 1.5 and 1.6.

The main purpose of sludge dewatering is to reduce sludge volume by removing as much water as possible from the sludge. The water in sludge flocs is generally classified into four categories (Fig. 1.7): free water (65–85% of the total water), interstitial water (15–25%), vicinal water (7%), and water of hydration (3%). Free water is removed by gravity or flotation thickening. The removal of interstitial and vicinal water requires the addition of flocculants or strong mechanical forces. Water of hydration, retained in sludge flocs, can only be removed after harsh pretreatment such as thermal drying or freezing–thawing processes.

The composting is a biothermal aerobic process that decomposes the organic portion of sewage sludge and degrades the organic material by approximately 25%. Meanwhile, the heat can be generated during this process, which reduces the moisture content of the sludge. The process performance strictly relies upon temperature, dry matter, and volatile solids (VS). In general, the sludge used for composting should meet the requirements of below 60% moisture, over 40% organic matters (dry basis), and

Figure 1.4 Illegal sewage sludge discharge directly into the environment.

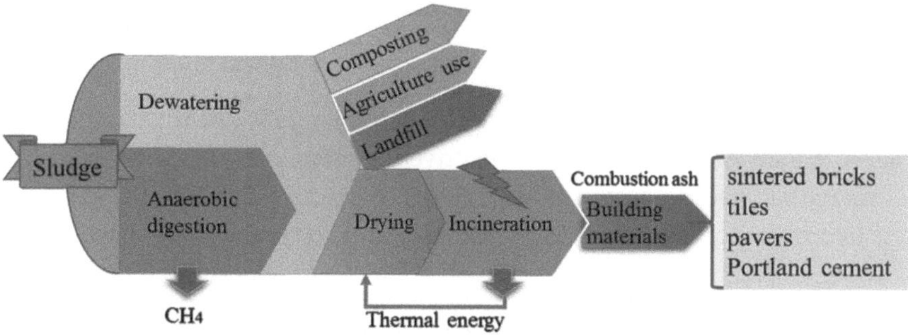

Figure 1.5 Treatment and disposal routes of sewage sludge.

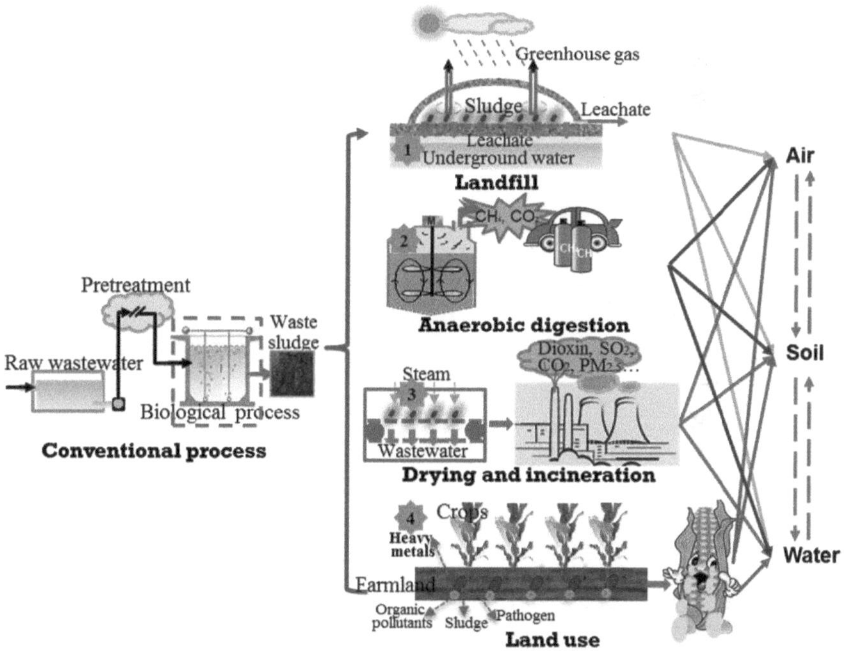

Figure 1.6 Treatment and disposal of sewage sludge and corresponding environmental impacts.

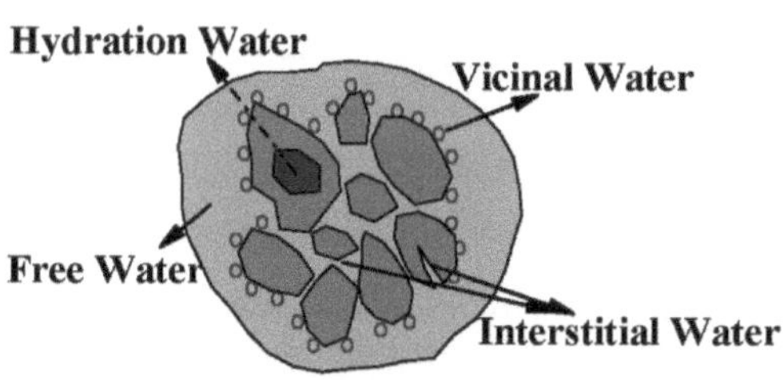

Figure 1.7 Water distribution of sewage sludge.

acceptable level of heavy metals and POPs as regulated. When the 80% moisture sludge is used for composting, lower moisture organic matters, such as straw, sawdust, should be added to adjust the water contents to around 55–65%. Moreover, strong odor will generate during composting process, which is very difficult to effectively control.

Anaerobic digestion comprises several successive biological processes (i.e., hydrolysis, acidogenesis, acetogenesis, and methanogenesis) in which microorganisms break up biodegradable matter and produce methane-rich biogas (60–70 vol% CH_4); it also reduces the amount of final residual solids for disposal while destroying the pathogens present in the sludge and eliminating offensive odors. The digested residual is disposed of in agriculture fields as soil conditioner/fertilizer, when it meets the environmental constraints reflected by the local regulations and standards. Anaerobic digestion is thus regarded as a major and essential part of a modern WWTP. In some cases, the heavy metals in the residues may be too high to be used in agriculture and should be landfilled. To date, three types of anaerobic processes (or digesters) have been proposed and commercially applied in the real world (Fig. 1.8), such as single-stage process in which the bioconversion of sludge is completed in a single chamber, double-stage process where the acidogenic and methanogenic stages are separated into two chambers, and temperature-phased anaerobic digestion (TPAD) that combines a thermophilic pretreatment unit prior to mesophilic anaerobic digestion.

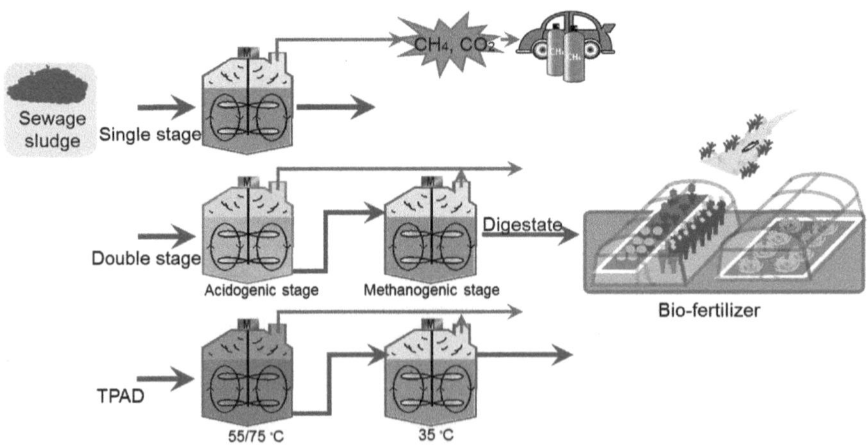

Figure 1.8 Schematic diagrams of single-stage process, double-stage process, and temperature-phased anaerobic digesters (TPAD).

The obstacles for anaerobic digestion of sludge include the relatively lower organic matters and high inorganic matters. Silt, sand, gravel, and wood may be present in the sludge, which makes the digestion very difficult and low output of biogas, resulting in a high operation cost and low weight reduction. The residue generated in the digestion has no way to go as it still contains biodegradable matters and should be composted before using in agriculture and city green plantation. Sanitary landfill seems to be the ultimate way for these residues.

Drying and incineration can totally carbonize organic constituents in sludge and minimize the volume of sludge (90% reduction). The chemical energy in the dried sludge can be harvested during combustion and subsequently employed for thermal drying. Nonetheless, thermal drying is not a cheap option because of its high energy demand to remove biologically bound water from sludge. The economics of the process depend upon dry matter content, VS/TS (total solids) ratio, and calorific value (Fig. 1.9). Meanwhile, the resulting combustion ash can be used as a raw material for the manufacture of different building materials such as sintered bricks, tiles, pavers, or Portland cement if it is justified as a general waste other than a hazardous waste.

Fig. 1.10 shows the typical schematic diagram of a typical sludge drying and incineration process. In the conventional drying and incineration process, the drying and incineration operations are separated. The sludge is first dried in one equipment and then incinerated in another incinerator. The

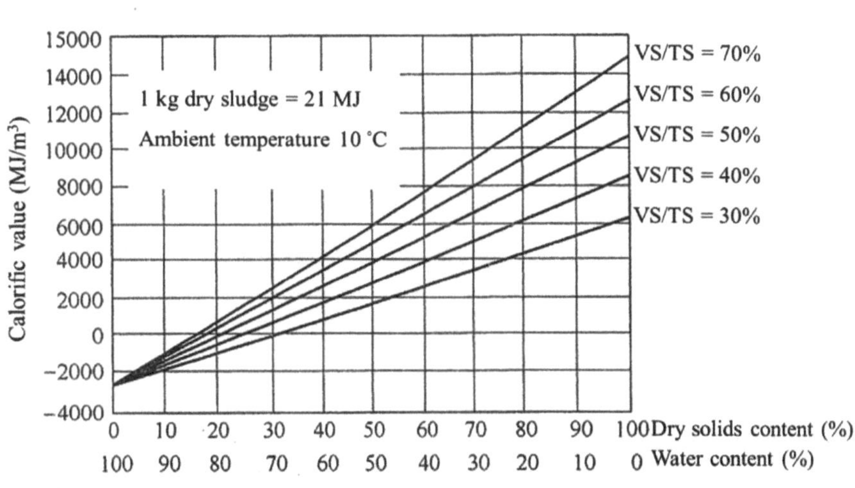

Figure 1.9 Sludge calorific value versus water content (or dry solids content).

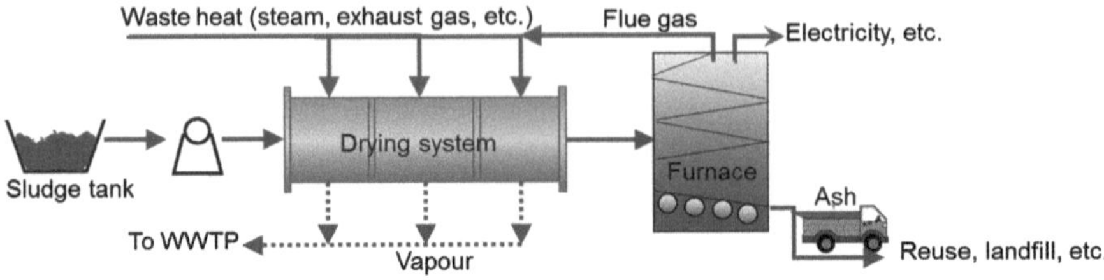

Figure 1.10 Flow sheet of a typical sludge drying and incineration process.

heat for sludge drying may be from burning of oil, natural gas, coal, and even electricity. The heat from the dry sludge is also used for the sludge drying. To raise the heat utilization efficiency, some equipment makes the hot flue gas from coal combustion direct contact with sludge. In this case, the oxygen content in the drying system must be controlled very carefully, otherwise explosion may occur. Indirect drying process will lead to a lower heat utilization efficiency and thus will raise the cost.

1.4 SANITARY LANDFILL OF SLUDGE

Sanitary landfill has been viewed as an ultimate disposal site for sludge. One of the critical factors affecting sludge landfill is its rheological properties. Dewatered sludge with around 80% moisture and a low compressive strength makes the landfill operation impossible because of construction difficulties and instability of the landfill slopes. According to the practical experiences gained in the real landfill for sludge landfill, the moisture of sludge should be below 60% when being disposed of in a landfill to support the weight of operation equipment and to prevent the deposited sludge from the sliding disaster. Therefore, different solidifying additives, including aged refuse, magnesium salts, coal ash, demolition wastes, lime, and soil, have been adopted to modify sewage sludge so as to transform the geotechnical properties. Taking Shanghai Refuse Landfill, the biggest landfill in China, as an example, all of sewage sludge has been landfilled after modified with aged refuse and magnesium salts, conducted by the authors (Fig. 1.11). According to the authors' research and experiences, the common modification methods and mixing ratios are provided in Table 1.1, which can be referred as references for sewage sludge sanitary landfill.

Sludge could meet the requirement of shear stress and bearing capacity of not less than 50 and 25 kPa, respectively when its moisture content decreased to 64%. Side slope of 12.6 degrees may be constructed without

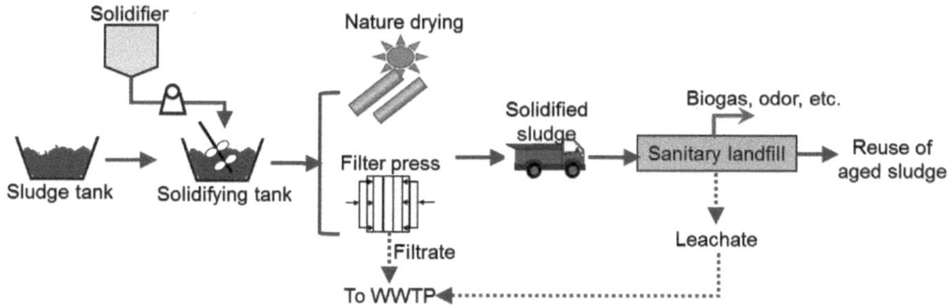

Figure 1.11 Flow sheet of a typical sludge solidification and sanitary landfill process.

Table 1.1 Common Modifying Additives and Mixing Ratios of Sewage Sludge

Modifying Additives	Requirements	Recommended Ratios (Modifiers: Wet Sludge)	Time for Curing
Magnesium or aluminum salts	–	5–10%	7 days
Aged refuse	Diameter < 4 cm; moisture content: 30–40%	>3:2 >1:2	Landfilled directly 7 days
Refuse incineration bottom ash	Moisture content < 20%	>1:1	Landfilled directly
Municipal refuse	–	Sludge content ≤ 8%	Landfilled directly
Composting residual	Moisture content < 20%	>1:1	Landfilled directly
Demolition wastes	Moisture content < 20%; shredded and screened through sieve of 15 mm	>1:1	Landfilled directly
Lime	–	>1:3	Landfilled directly
Coal ash	Moisture content < 20%	>1:1	Landfilled directly

causing sliding problems during landfill operation. The malodor could decrease to third grade. Consolidation coefficient under 50 kPa was calculated to be 0.010726 cm^2/s, and the compressive coefficient under 50–100 kPa was 1.16.

It was found that the geotechnical properties of sludge for landfill operation should meet the requirements of the following specifications: bearing capacity over 50 kPa, vane shear strength over 25 kPa, permeability of over the magnitude of $10^{-6} \sim 10^{-5}$ cm/s, and moderate odor. In this case, the sludge should be mixed with the additives, and the ratios of additives to sludge should be over 6:10, 9:10, 7:10, and 9:10 for coal fly ash, aged refuse, demolition wastes, and soil, respectively.

The annual production rate of leachate from fresh sludge with a moisture of 80% was about 44 L/t. The biodegradability of leachate decreased and annual production of leachate increased to 287 L/t in the presence of 33% aged refuse in the mixtures. Meanwhile, it was observed that the presence of the additives can accelerate the stabilization of sludge. Pollutants' concentrations in leachate from mixtures with additives were always lower than those of the fresh sludge.

A field lysimeter with 36 t of fresh sludge was established and its long-term stabilization process was monitored. It was found that pollutants' concentrations in the leachate from the field-test lysimeter were significantly lower than those of the laboratory lysimeters. Moreover, it seemed that the degradation in the field lysimeter was faster. After 498 days' biodegradation, the organic matters decreased by 67.1%, 61.6%, 30.5%, and 51.4% for the laboratory-scale lysimeters of biosludge, biosludge + aged refuse, chemical sludge, and field-test lysimeter of chemical sludge, respectively. The biodegradation of the biosludge was observed to be faster than the chemical sludge. It was found that humus, humic percentum, and humic index slightly increased during the stabilization process. Molecular weight of humic acid decreased in early days revealing its mineralization process and increased with time in metaphase and anaphase showing its maturation process. Fulvic acid decomposed in prophase and composed in metaphase and anaphase. Humic percentum decreased in the sequence of the laboratory-scale lysimeters of biosludge + aged refuse, biosludge, field-test lysimeter of chemical sludge, and chemical sludge, whereas humic index decreased in the sequence of the laboratory-scale lysimeters of biosludge, biosludge + aged refuse, field-test lysimeter of chemical sludge, and biosludge.

Biological toxicity faded during the stabilization of sludge, slow in prophase and rapid in metaphase and anaphase, according to the logarithm formula. The germinating index of biosuldge, aged refuse + biosludge, chemical sludge and field-test chemical sludge can reach upto 80% at 612, 626, 650, and 670 days of placement, respectively. Mixing of aged refuse could reduce the toxicity of the biosludge and increase the germinating index.

Settlement of simulated landfills could be classified into the initial stress settlement occuring mainly in the first year and secondary degradation settlement in the following years. The predicted ultimate settlement of the chemical sludge in field-test lysimeter, chemical sludge, biosludge, and aged refuse + biosludge in the laboratory-scale lysimeters were evaluated to be 12.33%, 20.54%, 13.77%, and 26.57% of the initial sludge height, and the corresponding settlements in first year were 98.9%, 94.4%, 97.4%, and 98.9% of the ultimate settlement, respectively. 99.99% of the ultimate settlement may reach at 870, 1170, 995, and 762 days of the placement, respectively. Consolidation coefficient of field-test lysimeter was calculated to be 0.00249 cm^2/s.

The equations among the settlements, pollutants' concentrations in leachate (COD, NH_3-N, etc.), chemical compositions in the sludge (organic matters, BDM, VS, TS, moisture, etc.), and the placement time were established. Taking the leachate discharge standard and the BDM in soil as reference, it was estimated that the resultant biodegraded sludge can be regarded as the stabilized sludge or aged sludge that can be excavated and recycled after around 5.82-year placement, when the BDM in the sludge decreased from around 30–60% in sludge to 4.76% in soil and COD to 100 mg/L in the leachate. If the aged sludge was to be used for planting, 1.68–1.84 years at least would be needed.

CHAPTER TWO

Enhanced Sewage Sludge Dewaterability by Chemical Conditioning

2.1 ENHANCED DEWATERING CHARACTERISTICS WITH FENTON PRETREATMENT

Advanced oxidation processes (AOPs) have attracted great attention as alternative methods for sludge conditioning. AOPs, through the generation of very reactive species such as hydroxyl radicals can rapidly degrade extracellular polymeric substances (EPSs) and disrupt sludge flocs, improving sludge settleability and dewaterability. Among AOPs, Fenton pretreatment is regarded as the most promising technology for sludge-dewatering in view of its ease of implementation and negligible environmental impact. Effectiveness of Fenton pretreatment is strongly dependent upon the reaction conditions, that is, H_2O_2 concentrations, Fe^{2+} concentrations, and initial pH values. Inappropriate operating conditions do not effectively improve sludge dewaterability and sometimes even deteriorate it, attributable to low rates of reaction. Optimizing such governing parameters therefore is very important for Fenton conditioning to maximize the dewatering efficiency.

The aim of this section is to evaluate the effectiveness of Fenton pretreatment in improving the waste-activated sludge-dewatering characteristics, and to determine the optimum conditions (H_2O_2, Fe^{2+} and initial pH) for sludge-dewatering by Fenton pretreatment through response surface methodology (RSM) and central composite design (CDD).

2.1.1 Sewage Sludge Sampling and Analytical Procedures

Waste-activated sludge was taken from the end of aeration basin of a local sewage treatment plant in Shanghai, China (Table 2.1). Fresh Fe^{2+} solution of 0.5 mol/L in Fenton's reagent was prepared from $FeSO_4 \cdot 7H_2O$ prior to each experiment. Hydrogen peroxide (30 wt%) was utilized for Fenton reaction in the experiments. The pH was adjusted to the desired values using 1 mol/L H_2SO_4 or NaOH solutions.

Pollution Control and Resource Recovery: Sewage Sludge
ISBN: 978-0-12-811639-5
http://dx.doi.org/10.1016/B978-0-12-811639-5.00002-4

Table 2.1 Characteristics of Sewage Sludge Used in This Study

Water Content (%)	pH	Viscosity (mPa·s)	CST (s)	TSS (g/L)	VSS (g/L)
96.93 ± 0.06	6.94 ± 0.20	179.80 ± 6.44	1360 ± 141	13.70 ± 0.13	8.42 ± 0.07

Dewaterability was evaluated using a capillary suction time (CST) apparatus (Model 440, Fann UK) with a standard Whatman No. 17 chromatography grade paper (Fig. 2.1). CST reduction efficiency (E) was calculated to compare the dewaterability efficiencies under various experimental conditions.

$$\mathrm{E}(\%) = \frac{CST_0 - CST}{CST_0} \times 100\% \tag{2.1}$$

where CST_0 and CST are, respectively, the CST of sludge before and after Fenton conditioning.

Three-level-three-factor CCD was chosen to evaluate the interactive effects of the three independent variables, that is, H_2O_2 concentrations, Fe^{2+} concentrations, and initial pH value as χ_1, χ_2, and χ_3 on the response variable E (%), respectively, during Fenton conditioning. The ranges of the experimental variables were chosen according to the preliminary signal factor tests. The levels of the factors used for optimization are presented in Table 2.2. A total number of 20 experiments were augmented by six axial points coded ($\pm\alpha$) and six replications of center points (all factors at level zero), and carried out in a randomized order as required in many design procedures. The center runs and axial points provided an evaluation of the experimental errors and curvature of the model, respectively. The distance from the center point was calculated by $\alpha = 2^{n/4}$ (for the three factors $n = 3$, $\alpha = 1.682$). The levels of H_2O_2 concentrations (χ_1), Fe^{2+}

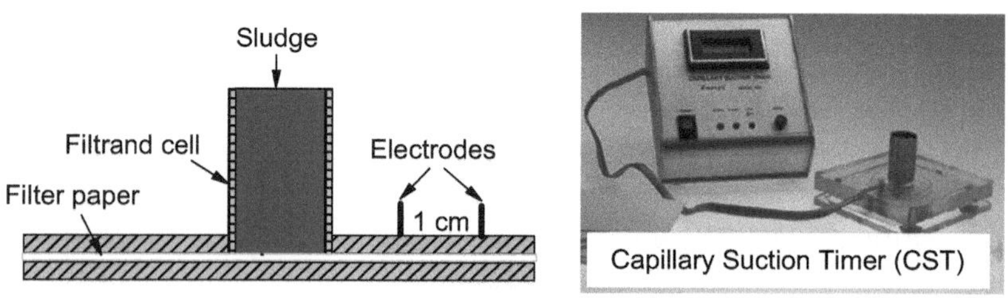

Figure 2.1 Standard capillary suction time (CST) apparatus.

Table 2.2 Range and Levels of Natural and Corresponding Coded Variables for RSM of Sewage Sludge Dewaterability

	Symbols		Levels				
Variable	**Natural**	**Coded**	**−1.682 (−α)**	**−1**	**0**	**1**	**1.682 (+α)**
H_2O_2 (mg/g-VSS)	χ_1	X_1	72.72	100	140	180	207.28
Fe^{2+} (mg/g-VSS)	χ_2	X_2	110.49	133	166	199	221.51
pH	χ_3	X_3	2.32	3	4	5	5.68

concentrations (χ_2), and initial pH values (χ_3) were identified from the following equations:

$$X_1 = \frac{\chi_1 - 140}{40} \tag{2.2}$$

$$X_2 = \frac{\chi_2 - 166}{33} \tag{2.3}$$

$$X_3 = (\chi_3 - 4) \tag{2.4}$$

The corresponding matrix is displayed in Table 2.3. An empirical second order polynomial equation was fitted to find out the true functional relationship between the response variable and the independent variables. The general form of the empirical second-order polynomial equation is expressed as follows:

$$E = A_0 + \sum_{i=1}^{k} A_i\chi_i + \sum_{i=1}^{k} A_{ii}\chi_i^2 + \sum_{i=1}^{k}\sum_{j=1}^{k} A_{ij}\chi_i\chi_j \tag{2.5}$$

where E is the predicated response (CST reduction efficiency, %); A_0, A_i, A_{ii}, and A_{ij} ($i = 1, 2, 3$ and $j = 1, 2, 3$) are the model regression coefficients; χ_i and χ_j ($i = 1, 2, 3$ and $j = 1, 2, 3$) are the input variables, affecting the response variable E. The analysis of variance (ANOVA) was used for analyses of experimental data. The goodness of fit of the model was expressed by the coefficient of determination R^2 and R^2_{adj}, and its statistical significance was determined by Fisher's F-test. The significance of regression coefficients was checked by Student's t-test.

2.1.2 Regression Model and Statistical Testing for Sludge Conditioning

The results based on a CCD are given in Table 2.3. The following polynomial quadratic model for CST reduction efficiency was then fitted by applying the response surface regression (RSREG).

Table 2.3 Experimental and Predicted Response Results of Fenton's Reagent to Condition Sewage Sludge

Run No.	Coded Values			Natural Values			Response (E, %)	
	X_1	X_2	X_3	χ_1	χ_2	χ_3	Experimental	Predicted
1	−1	−1	−1	100	133	3	95.42	95.37
2	1	−1	−1	180	133	3	96.65	96.47
3	−1	1	−1	100	199	3	96.98	96.68
4	1	1	−1	180	199	3	97.29	97.49
5	−1	−1	1	100	133	5	95.60	95.36
6	1	−1	1	180	133	5	96.13	96.38
7	−1	1	1	100	199	5	96.46	96.59
8	1	1	1	180	199	5	97.32	97.31
9	−1.682	0	0	72.72	166	4	95.56	95.80
10	1.682	0	0	207.28	166	4	97.51	97.33
11	0	−1.682	0	140	110.49	4	96.13	96.35
12	0	1.682	0	140	221.51	4	97.54	97.53
13	0	0	−1.682	140	166	2.32	96.51	96.68
14	0	0	1.682	140	166	5.68	96.31	96.25
15	0	0	0	140	166	4	97.08	96.76
16	0	0	0	140	166	4	97.17	97.09
17	0	0	0	140	166	4	97.29	97.09
18	0	0	0	140	166	4	96.92	97.09
19	0	0	0	140	166	4	97.30	97.09
20	0	0	0	140	166	4	97.14	97.09

$$\begin{aligned} E(\%) = {} & 79.8287 + 0.055\chi_1 + 0.0813\chi_2 + 2.2775\chi_3 - 0.0001\chi_1^2 \\ & - 0.0002\chi_2^2 - 0.2683\chi_3^2 - 0.0001\chi_1\chi_2 - 0.0005\chi_1\chi_3 \\ & - 0.0006\chi_2\chi_3 \end{aligned} \quad (2.6)$$

The ANOVA were conducted for the response variable and presented in Table 2.4. The model *F*-value of 12.59 indicated that the second-order polynomial model was statistically significant for CST reduction efficiency and there was only a 0.01% chance that model *F*-value of this large could occur due to noise. The high R^2 coefficient of 0.9189 ensured a satisfactory adjustment of the quadratic model to the experimental data and simultaneously implied that only 8.11% of CST reduction efficiency variability could not be explained by the proposed model. The adjusted *R*-square (${R_{adj}}^2 = 0.8459$) was also high to advocate for high significance of the model. Furthermore, the "Lack of Fit *F*-value" of 5.60 revealed that the *Lack of Fit* was of insignificance. This was a 4.1% chance that a "Lack of

Table 2.4 ANOVA for the Polynomial Model for CST Reduction Efficiency (%)

Source	Degrees of Freedom (DF)	Sum of Squares (SS)	Mean Square (MS)	*F*-Value	*p*
Regression	9	7.83695	0.87077	12.59	<0.0001
Linear	3	6.23455	0.37041	5.36	0.019
Quadratic	3	1.55109	0.51703	7.48	0.007
Interactive	3	0.05131	0.01710	0.25	0.861
Residual	10	0.69166	0.06917		
Lack of fit	5	0.58682	0.11736	5.60	0.041
Pure error	5	0.10485	0.02097		
Cor. Total	19	8.52861			
R^2		0.9189			
R^2_{adj}		0.8559			

Fit *F*-value" of this large could occur due to noise. Thus, it is reasonable to believe that the proposed empirical model (Eq. (2.6)) is reliable to predicate CST reduction efficiency and reveal the Fenton pretreatment behavior in sludge conditioning.

The significance test results for the regression coefficients in terms of the uncoded factors χ_i determined by Student's *t*-test and *p*-values are listed in Table 2.5. The larger the magnitude of *t*-value and smaller the *p*-value, the more significant is the corresponding coefficient. The coefficient estimates and the corresponding *p*-values showed that χ_1, χ_2, χ_3, $\chi_1{}^2$, and $\chi_3{}^2$ were significant model terms (*p*-value = 0.017, 0.028, 0.035, 0.023, and 0.013, respectively). H_2O_2 (*t*-value = 2.854, *p*-value = .017) had the largest effect on CST reduction efficiency, followed by Fe^{2+} and initial pH. It is

Table 2.5 Significance Test of Regression Coefficients

Terms	Coefficients	Standardized Residual	*T*-Value	*p*
Constant	79.8287	3.90213	20.458	0.000
χ_1	0.0550	0.01926	2.854	0.017
χ_2	0.0813	0.03157	2.574	0.028
χ_3	2.2775	0.93131	2.445	0.035
$\chi_1{}^2$	−0.0001	0.00004	−2.679	0.023
$\chi_2{}^2$	−0.0002	0.00008	−1.999	0.074
$\chi_3{}^2$	−0.2683	0.08840	−3.035	0.013
$\chi_1\chi_2$	−0.0001	0.00007	−0.802	0.441
$\chi_1\chi_3$	−0.0005	0.00232	−0.213	0.835
$\chi_2\chi_3$	−0.0006	0.00282	−0.230	0.823

clear that the interactive terms ($\chi_1\ \chi_2$, $\chi_1\ \chi_3$, $\chi_2\ \chi_3$) of H_2O_2, Fe^{2+} and initial pH were the least significant (p-value $>$ 0.05).

The checking of model adequacy is an important part in optimizing a response surface, as the model would give poor or misleading outcome if it was not a good fit. The normal probability plot indicates whether the residuals follow a normal distribution, in which case the points will follow a straight line. Through constructing the plots of standardized residuals versus the normal % of probability (Fig. 2.2A), a check was made for the normality

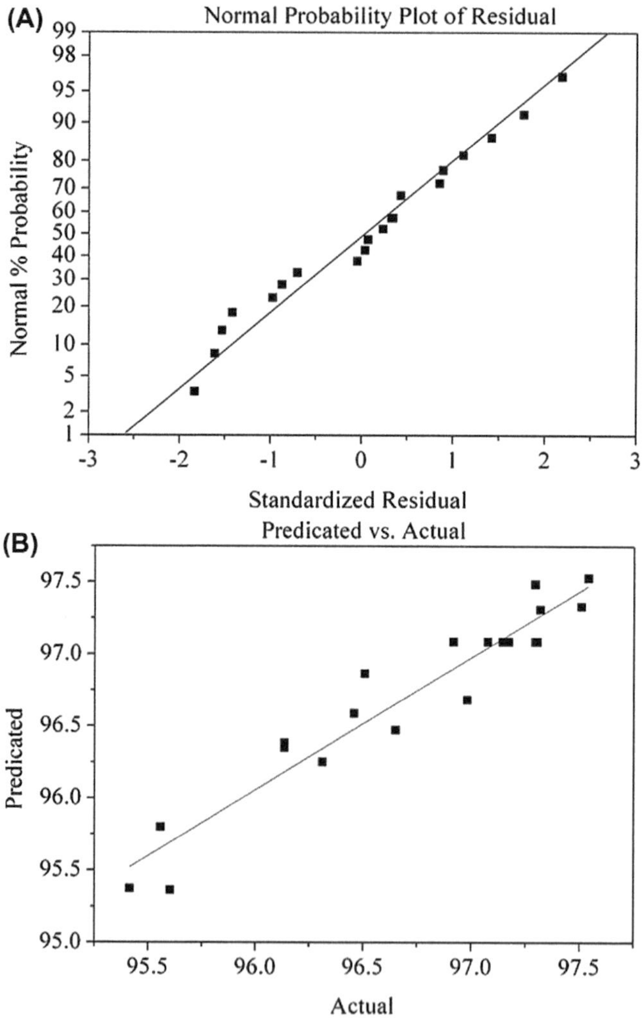

Figure 2.2 Plots of standardized residuals vs. normal % of probability (A) and predicted vs. actual values (B) for CST reduction efficiency.

assumption. The normal probability plots approximated well a straight line, revealing that neither the transformation of the response variable was required nor there was any apparent problem with abnormality. The random scatter of the predicted E (%) versus standardized residual plots also indicated no need for a transformation.

Fig. 2.2 shows the actual and predicted values of the responses from the model. It was found that the predicted values accorded well with the actual values. The distribution of data plots relatively close to a straight line suggested a reasonably adequate agreement between the actual and predicted values, which confirmed that the model could be used to navigate the space defined by the CCD. The outlier *t* for the batch runs of CST reduction efficiency (%) was further examined, since it is a measure of how many standard deviations the actual value deviates from the predicted value. Any observation with a standardized residual outside the interval of ± 3.0 can be rejected as outlier. All outlier *t* values below the interval of ± 3.0 demonstrated an adequate precision and reliability of the designed experiments.

2.1.3 Optimization of Fenton Conditioning Operating Variables

The 3D response surface and the corresponding contour plot (E, %), generated by Minitab 11.0, with one variable maintained at its central level, are utilized to visualize the relations between two independent factors and response. Fig. 2.3 illustrates the effects of H_2O_2 and Fe^{2+} concentrations on CST reduction efficiency at initial pH 4.0. It can be seen that a significant enhancement of CST reduction was observed with increasing H_2O_2 and

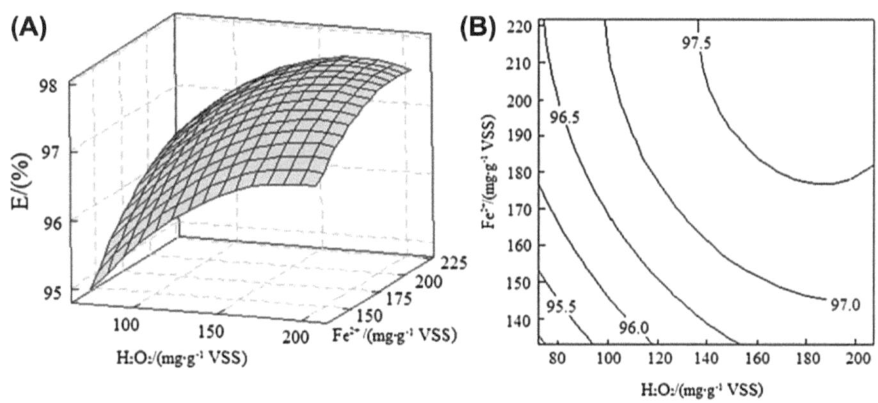

Figure 2.3 Response surface (A) and contour lines (B) for the interactive effects of H_2O_2 and Fe^{2+} concentrations on CST reduction efficiency (%).

Fe^{2+} concentrations; however, the CST reduction efficiency was negatively affected when an increase in their dosages was beyond a certain limit. It is evident that the overloading of H_2O_2 and Fe^{2+} did not mean a continuous increase in dewatering efficiency, attributed to the fact that excess H_2O_2 or Fe^{2+} had a scavenging effect on hydroxyl radicals (Eqs. (2.7) and (2.8)), thus yielding a smaller concentration of OH• radicals as

$$H_2O_2 + OH^\bullet \rightarrow HO_2^\bullet + H_2O \tag{2.7}$$

$$Fe^{2+} + OH^\bullet \rightarrow OH^- + Fe^{3+} \tag{2.8}$$

The effects of H_2O_2 concentrations and initial pH on CST reduction efficiency at the Fe^{2+} concentration of 166 mg/g-VSS are plotted in Fig. 2.4. The combination of H_2O_2 and initial pH had a crucial effect on the reduction of CST. Increasing H_2O_2 concentration with a decrease in initial pH raised the CST reduction efficiency greatly. At an initial pH higher than 5.0, however, CST reduction efficiency was rapidly decreased. Similarly, the effects of Fe^{2+} concentrations and initial pH values at the H_2O_2 of 140 mg/g-VSS (Fig. 2.5) also showed that acidic pH levels were usually optimum for sludge-dewatering.

In Fenton process, pH plays a key role in dewatering efficiency since it controls the generation of OH• radicals and stability of Fe^{2+}. The optimum pH value in present study was determined to be 3.5—4.0, where CST reduction efficiency reached up roughly 98%. When pH was outside this interval, CST reduction efficiency was reduced greatly. The findings might be due to the inhibition of complexation of Fe(III) and H_2O_2 at the extremely acidic

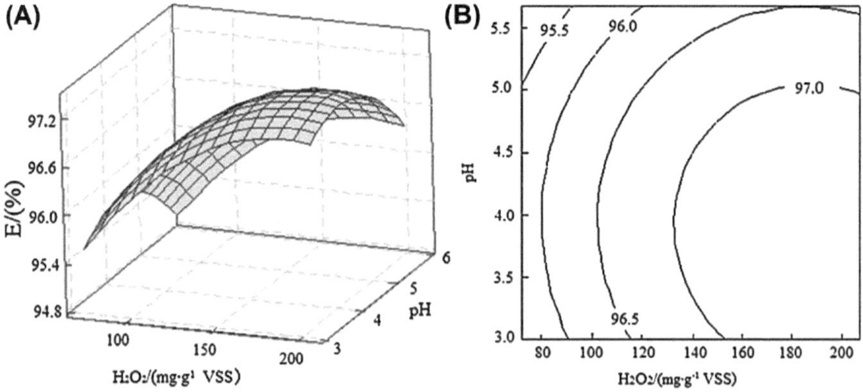

Figure 2.4 Response surface (A) and contour lines (B) for the interactive effects of H_2O_2 concentration and initial pH on CST reduction efficiency (%).

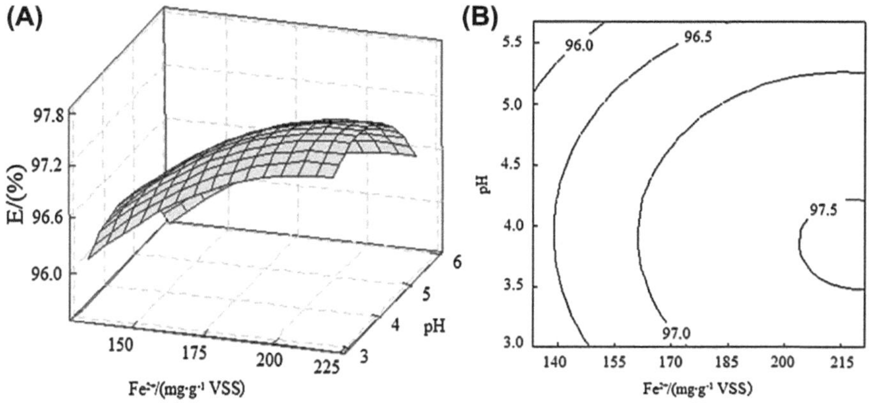

Figure 2.5 Response surface (A) and contour lines (B) for the interactive effects of Fe^{2+} concentration and initial pH on CST reduction efficiency (%).

condition, and the precipitation of Fe(II) and Fe(III) at the alkaline environment. The molecular weight (MW) distribution of S-EPS in sludge filtrates before and after conditioning was analyzed using gel permeation chromatography (GPC) (LC-10ADVP, Shimadzu, Japan) equipped with a differential detector (RID-10A, Shimadzu, Japan) and a TSKgel column (G4000PWXL, TOSOH Co., Japan), and results are shown in Fig. 2.6.

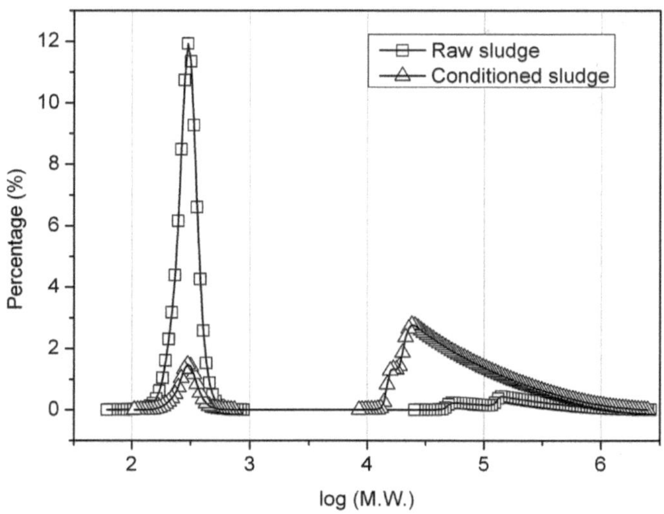

Figure 2.6 Molecular weight distribution of organic maters in sludge filtrates before and after Fenton conditioning (the mobile phase is Milli-Q water. Polyethylene glycol (MW at 1169, 771, 128, 12, 4, 0.62 and 0.194 kDa) is used as molecular standard for the calculation).

It is noted that the both sludge samples had heterogeneous MW distribution patterns, with one peak at MW of 0.16–0.63 kDa presented in raw sludge EPS, however, two peaks at MW of 0.16–0.63 and 16–100 kDa detected in the EPS from the conditioned sludge. The remarkable difference was the latter possessing the high content of large MW organics, probably attributable to the disintegration of sludge micro-floc aggregates via highly active OH• radicals.

It is well-known that sludge is a multiphase medium, containing water, mineral and organic substances, microorganism cells and EPS, etc. EPS are considered important for supporting the structural stability of sludge granules and play a critical part in the retention of water and cell attachment; however, they can be solubilized and destroyed by some oxidants. Microscopic examination in this study suggested that Fenton conditioning could effectively damage the EPS biopolymers that surrounded the sludge micro-floc aggregates, which eventually weakened the attachment between cells and disrupted the structure of the sludge flocs (Fig. 2.7). The raw sludge flocs, as displayed in Fig. 2.7A, presented a tight pack state. The outer portion of a thick EPS layer that extended from the floc's surface to the water solution could be clearly observed. Excessive EPS biopolymers on the floc's surface, such as proteins, polysaccharides and humic-like substances, were likely to cause steric stabilization and poor performance in sludge-dewatering. After conditioned with Fenton oxidation, EPS were degraded and damaged rapidly, causing a break-up of sludge matrix and partial cells exposed. The floc size therefore appeared small, and the flocs were uniformly dispersed (Fig. 2.7B). The sludge floc breakage might be more beneficial than

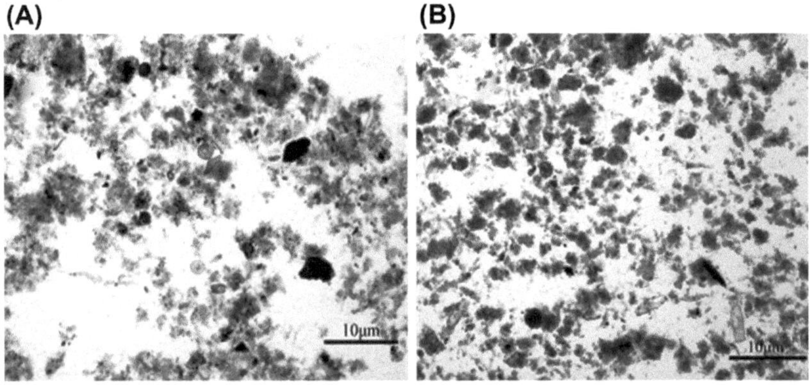

Figure 2.7 Micrographs of raw sludge (A) and sludge conditioned with Fenton's reagent (B).

detrimental to dewaterability, as it could provide more passages for the release of water. Furthermore, associated with the exposed microorganism cells, Fenton pretreatment could crack cell membranes, which further resulted in the release of bound water and intracellular large molecular organics.

$$RH + OH\bullet \rightarrow H_2O + R\bullet \quad (2.9)$$

Therefore, it is reasonable to address that the enhancement of sludge dewaterability by Fenton's reagent lies in the migration of sludge bound water due to the disintegration of sludge flocs and microbial cells lysis. The more detailed dewatering mechanisms were summarized and depicted in Fig. 2.8.

2.1.4 Conditions for Optimum Response and Validation Experiment

Numerical optimization was used to determine the optimum conditions of the operating variables in the Fenton process. Based on the response surface and desirability function, optimized conditions for CST reduction efficiency were obtained at H_2O_2 178 mg/g-VSS, Fe^{2+} 211 mg/g-VSS, and initial pH 3.8. Under this condition, a 97.70% CST reduction efficiency was predicated based on desirability function of 0.85. To confirm the accuracy of the predicated model and the reliability of the optimum combination, a confirmation experiment with a triplicate set was conducted at optimum operating conditions. The replicate experiments yielded a CST reduction efficiency of 98.25%, which was close to the predicated value (97.70%). These results clearly demonstrate that RSM is a powerful tool to determine and optimize the operational conditions of Fenton process for sludge conditioning.

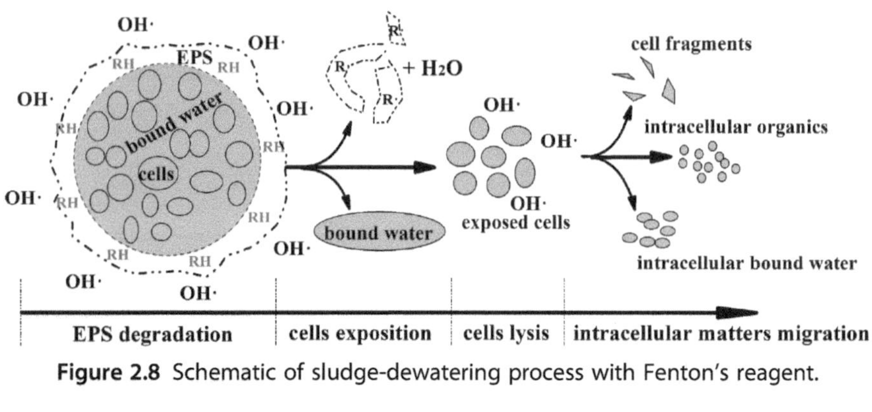

Figure 2.8 Schematic of sludge-dewatering process with Fenton's reagent.

2.2 ENHANCED DEWATERABILITY USING FE(II)-ACTIVATED PERSULFATE OXIDATION

Persulfate is a strong and nonselective oxidant with a high redox potential of 2.01 V. It can be effectively activated by initiators, including heat, UV light, or transition metals (Me^{n+}) to generate sulfate-free radicals ($SO_4^{-}\bullet$), which has an even higher redox potential estimated to be 2.60 V, similar to that of the hydroxyl radical (2.70 V). Some of the reactions during the activation of $S_2O_8^{2-}$ may be expressed as follows:

$$S_2O_8^{2-} + 2e^- \rightarrow 2SO_4^{2-} \quad E^o = 2.01\ \mathrm{V} \tag{2.10}$$

$$S_2O_8^{2-} + \mathrm{heat/UV} \rightarrow 2SO_4^{-}\bullet \tag{2.11}$$

$$S_2O_8^{2-} + Me^{n+} \rightarrow Me^{(n+1)+} + SO_4^{-}\bullet + SO_4^{2-} \tag{2.12}$$

$$SO_4^{-}\bullet + e^- \rightarrow SO_4^{2-} \quad E^o = 2.60\ \mathrm{V} \tag{2.13}$$

The sulfate radicals have higher reduction potential at neutral pH and are more selective for oxidation at acidic pH as compared to hydroxyl radical. Moreover, it is relatively more stable in water, and thus may be more effective for the degradation of recalcitrant organic compounds. With the advantages of high solubility, strongly oxidizing potential, wide operative pH range, and slow rate of consumption (due to its stability), sulfate radical-based advanced oxidation processes (SR-AOPs) have been successfully used in situ chemical oxidation (ISCO) remediation of a variety of contaminants in aqueous and sediment systems. In this section, potassium persulfate ($K_2S_2O_8$) activated with ferrous ion (Fe(II)) was investigated as a potentially alternative oxidant. The performance of the Fe(II)-persulfate system to improve sludge dewaterability was evaluated through the examination of various operational parameters, including applied persulfate and Fe(II) concentrations, way of adding Fe(II), and initial solution pH, by which the optimal experimental conditions were identified.

2.2.1 Test Materials and Analytical Procedures

Chemicals including potassium persulfate ($K_2S_2O_8 > 99.5\%$) and ferrous sulfate heptahydrate ($FeSO_4 \cdot 7H_2O > 99.0\%$) were purchased from Sinopharm Chemical Reagents Co., China. A series of bench-scale experiments were conducted in 500-mL beakers with the sludge sample of 300 mL at inherent pH of 6.95. The reaction was initiated by adding an appropriate amount of persulfate once a specified constant Fe(II) was applied. At predetermined time intervals, a 5.0-mL sample was collected and analyzed for its dewaterability. The effect of initial pH on the dewaterability was evaluated by varying the pH from 1.5 to 10.0.

2.2.2 Effect of Persulfate and Ferrous Ion Concentrations on Sludge Dewaterability

The effect of the concentrations of persulfate and ferrous ion on sludge dewaterability was investigated at pH 6.95 at 25°C, and the results are shown in Fig. 2.9. It can be clearly seen that all the possible reductions of CST under all the tested conditions were almost completed within 1 min; hence, the average CST reduction efficiency ($1 - CST/CST_0$, %) over the period from 0 to 1 min was used to elucidate the influence of any parameters on sludge dewaterability.

As shown in the inset of Fig. 2.9A, the reduction of CST within 1 min increased from 58.7% to 61.8% and then to 88.8% as the concentration of $S_2O_8^{2-}$ increased from 0.1 to 0.3 and to 1.2 mmol/g-VSS in the presence of 1.5 mmol/g-VSS Fe(II). However, the CST reduction did not proceed apparently as the concentration of $S_2O_8^{2-}$ further increased. The initial efficiency increment at lower persulfate concentrations was attributed to the elevated sulfate radical concentration as the $S_2O_8^{2-}$ concentration increased. When the $S_2O_8^{2-}$ concentration accumulated to certain level in the solution, however, the influence of $S_2O_8^{2-}$ concentration would be less for a given Fe(II) concentration due to the insufficiency of initiator. Therefore, it is observed that the increment of CST reduction was relatively little when the $S_2O_8^{2-}$ was added over 1.2-mmol/g-VSS.

As an initiator, Fe(II) activates persulfate, yields sulfate radicals, and consequently, accelerates the reduction of sludge CST. Hence, it is reasonable that increasing the dosage of Fe(II) is favorable to improving the sludge dewaterability. Interestingly, the CST reduction increased very fast initially with applied Fe(II) concentration increasing from 0.3 to 1.5 mmol/g-VSS (i.e., optimized at 1.5 mmol/g-VSS), then decreased at higher Fe(II) concentration (see inset of Fig. 2.9B). The significant jump of initial CST reduction from low to mid Fe(II) dosage was ascribed to the higher initiator concentration in the solution, contributing a lot to a faster production of sulfate radicals and, therefore, the better CST reduction efficiency. However, as Fe(II) was added in excess, and the excess Fe(II) may directly quench $SO_4^{-}\bullet$ through Eq. (2.14), which resulted in the unfavorable consumption of $SO_4^{-}\bullet$ by the excess Fe(II), and ultimately, lower CST reduction

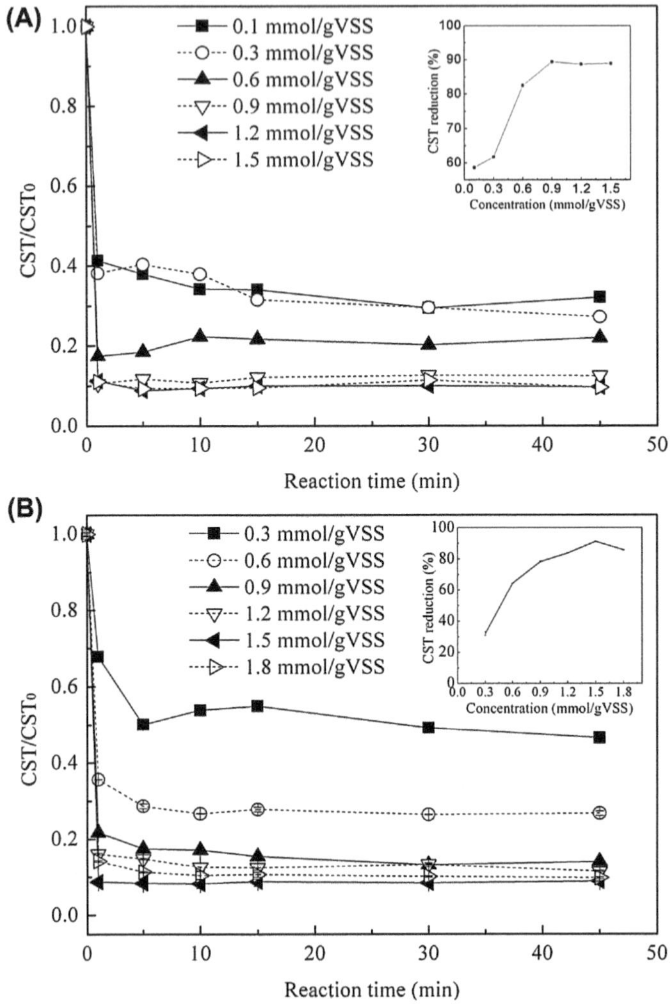

Figure 2.9 Variation of sludge dewaterability in Fe(II)-$K_2S_2O_8$ system at pH = 6.95: (A) the initial concentration of $S_2O_8^{2-}$ was varied in the presence of Fe(II) (1.5 mmol/g-VSS), and (B) the initial concentration of Fe(II) was varied in the presence of $S_2O_8^{2-}$ (1.2 mmol/g-VSS).

efficiency. To minimize the possible quenching effect and to achieve the better dewatering efficiency, the concentrations of $S_2O_8^{2-}$ and Fe(II) expected during sludge-dewatering in this study were 1.2 mmol/g-VSS and 1.5 mmol/g-VSS, respectively, which corresponded to $S_2O_8^{2-}/Fe(II)$ molar ratio of 1: 1.25 as

$$Fe^{2+} + SO_4^{-}\bullet \rightarrow Fe^{3+} + SO_4^{2-} \quad k = 4.6 \times 10^9 \text{ mol/Ls} \tag{2.14}$$

2.2.3 Effect of the Way of Adding Ferrous Ion on Sludge Dewaterability

The effect of the way of adding ferrous ion on enhancing the sludge-dewatering efficiency in the presence of persulfate was evaluated by following tests: (1) adding all the Fe(II) in one time, and (2) adding Fe(II) in three times. The results were compared in Fig. 2.10. The way of adding Fe(II) had no appreciable impact on the efficiency of CST reduction. In test (2), a relatively small increase was observed in the CST reduction (87.9%) in comparison with that (83.9%) obtained in test (1). The divergence was possibly caused by the different reaction processes. Persulfate activation with Fe(II) was a fast reaction because of its requiring a low activation energy of 14.8 kcal/mol, and a considerable amount of powerful sulfate radicals could be yielded transiently once the Fe(II) was attached with persulfate. Thus, for a specified Fe(II) dosage, the way of adding Fe(II) hardly influenced the formation pathway of sulfate radicals.

In contrast, persulfate activation using Fe_3O_4 MNs as the initiator appeared more complicated. The surface of the Fe_3O_4 MNs may be inactivated due to its high reactivity, as a result of which, there was a strong competition between Fe_3O_4 MNs inactivating and reacting with persulfate after injected into the reaction system, while it was not the case for persulfate

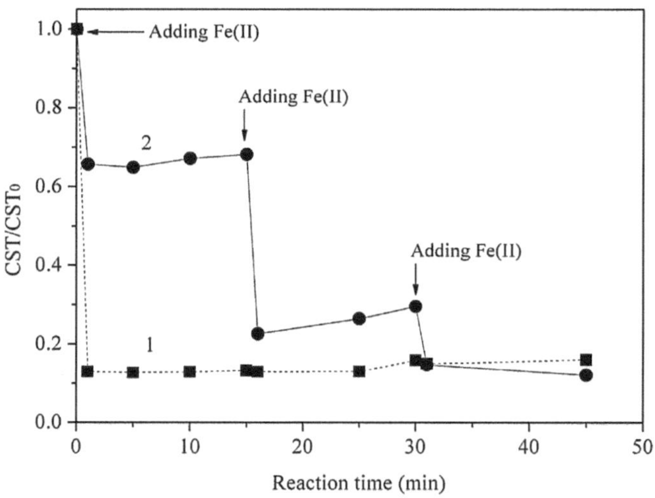

Figure 2.10 Effect of the way of adding Fe(II) on sludge dewaterability: (1) all the Fe(II) were added in one time, and (2) the Fe(II) were added in three times after the Fe(II) were divided into three equal portions. Other experimental conditions: total amount of Fe(II) 0.9 mmol/g-VSS; initial concentration of $S_2O_8^{2-}$ 1.2 mmol/g-VSS; no pH adjustment.

oxidation driven by Fe(II). Therefore, when Fe_3O_4 MNs were divided into several parts and added at a fixed time interval, the inactivation would be noticeably controlled, thanks to the strong competition of excess persulfate in the solution for the activator, which partially resulted in the observed increase in SMM degradation. Besides, this pronounced discrepancy may possibly arise from the inherent differences in the mechanisms of sludge-dewatering and SMM degradation.

2.2.4 Effect of Initial pH on Sludge Dewaterability

The effect of initial pH value on enhancing the sludge dewaterability was also explored by conducting experiments using six different initial pH levels (i.e., 1.5, 3, 5.5, 7.0, 8.5, and 10.0) under certain loading of oxidizing system (1.2 mmol/g-VSS $S_2O_8^{2-}$ + 1.5 mmol/g-VSS Fe(II)). Fig. 2.11 illustrates the corresponding CST reduction efficiency with the change of solution pH level. As exhibited in the inset of Fig. 2.11, reduction of CST was highly efficient in a wide pH range from 3.0 to 8.5; in this range, the corresponding reduction was determined to be 90.0% or so after 15 min of treatment. Hence, it could be considered that the more aggressive oxidation induced by both $SO_4^{-}\bullet$ and •HO radicals may be potentially responsible for the large CST reduction.

In the case of initial pH 1.5, however, a slight decrease of around 20% in CST reduction was observed, most likely because of the scavenging

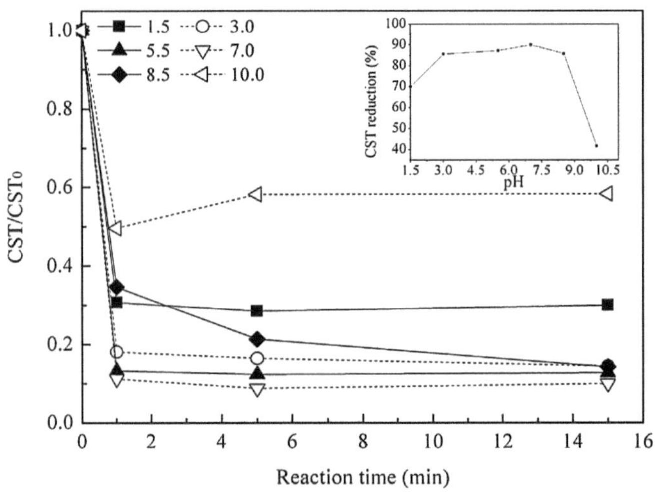

Figure 2.11 Effect of initial pH on sludge dewaterability. Experimental conditions: $[S_2O_8^{2-}]$ = 1.2 mmol/g-VSS; [Fe(II)] = 1.5 mmol/g-VSS.

reactions of $SO_4^{-}\bullet$. Strongly acidic conditions (i.e., pH 1.5) could cause acceleration of $SO_4^{-}\bullet$ formation due to acid-catalyzation through Eqs. (2.15) and (2.16). Higher radical concentrations that favored a scavenging $SO_4^{-}\bullet$ by itself or $SO_4^{-}\bullet/S_2O_8^{2-}$ reactions in accordance with Eqs. (2.17) and (2.18) retarded the performance of persulfate oxidation and yielding a lower CST reduction over $SO_4^{-}\bullet$/sludge floc's interaction. Nevertheless, the CST reduction efficiency achieved at pH 1.5 was still as high as 70.0% comparable to that at near neutral pH, suggesting that persulfate oxidation could perform well at the extremely acidic environment. At first glance, it looked similar to the conventional Fenton's process, which was generally regarded to have better performance at acidic condition. However, the Fenton's process was greatly hindered at the extremely acidic condition due to formation of $(Fe(II)(H_2O_2))^{2+}$ and inhibition of the complexation of Fe(III) with hydrogen peroxide. The former reduced the availability of free-Fe(II), while the latter limited the regeneration of Fe(II) from Fe(III). In the present study, the negative effects were not observed in the present study, as evidenced by a higher CST reduction efficiency.

$$S_2O_8^{2-} + H^+ \rightarrow HS_2O_8^{-} \tag{2.15}$$

$$HS_2O_8^{-} \rightarrow H^+ + SO_4^{-}\bullet + SO_4^{2-} \tag{2.16}$$

$$SO_4^{-}\bullet + SO_4^{-}\bullet \rightarrow S_2O_8^{2-} \quad k = 4.0 \times 10^8 \text{ mol/L} \tag{2.17}$$

$$SO_4^{-}\bullet + S_2O_8^{2-} \rightarrow SO_4^{2-} + S_2O_8\bullet \quad k = 6.1 \times 10^5 \text{ mol/L} \tag{2.18}$$

When compared to the acidic and neutral conditions, alkaline pH caused more pronounced inhibition of the persulfate effect, especially at pH 10.0. As can be seen, the reduction of CST at pH 10.0 was only 42.0%, approximately 28.0% and 48.0% lower than those obtained at pH 1.5 and 3.0–8.5, respectively. The apparently quenched CST reduction was mainly related to the precipitation of iron. At higher pH levels, both Fe(II) and Fe(III) precipitated to generate $Fe(OH)_{2(s)}$ and $Fe(OH)_{3(s)}$, respectively. It not only led to the decrease of dissolved Fe(II) and Fe(III) in the Fe(II)-persulfate system, but also retarded the regeneration of soluble ferrous catalyst, which in turn significantly reduced sludge-dewatering efficiency by stopping the decomposition of persulfate—another reason possibly resulted from $SO_4^{-}\bullet$ scavenging by OH^{-} through Eq. (2.19). Although the scavenging reaction may yield $\bullet HO$ radicals, this strong nonspecific oxidant was also vulnerable to scavenging by various reactants including SO_4^{2-}, an abundant species in persulfate oxidation system. The strong competition for •HO radicals between SO_4^{2-} and sludge flocs reduced the available free radicals, consequently resulting in the more remarkable decrease in CST reduction efficiency.

$$SO_4^{-}\bullet + OH^{-} \rightarrow SO_4^{2-} + \bullet OH \quad k = 6.5 \times 10^{7}\ \text{mol/L} \qquad (2.19)$$

Overall, persulfate oxidation driven by Fe(II) could exhibit a perfect performance in a wide operative pH rang of 3.0–8.5, into which the pH of raw sewage sludge usually falls, despite the appreciable drops in CST reduction efficiency under initial pH 1.5 and 10.0 as earlier discussed. These previous observations implied that pH value was not a key factor for Fe(II)-activated persulfate oxidation during sludge-dewatering. The persulfate oxidation was, therefore, more beneficial and cost-effective in enhancing the sludge dewaterability, owing to the lower requirement for pH adjustment.

2.2.5 Possible Mechanisms of Sludge Dewaterability Improvement

2.2.5.1 Radical Quenching Studies

The addition of 50 mmol/g-VSS EtOH significantly inhibited the system performance with CST reduction efficiency decreasing from 88.9% to 80.1%, whereas the addition of 50 mmol/g-VSS TBA only resulted in around 5.8% reduction efficiency drop (Fig. 2.12). Such difference in CST reduction-efficiency drop by the two radical scavengers implied that the presence of both $SO_4^{-}\bullet$ and •OH species at near neutral pH. This was because EtOH

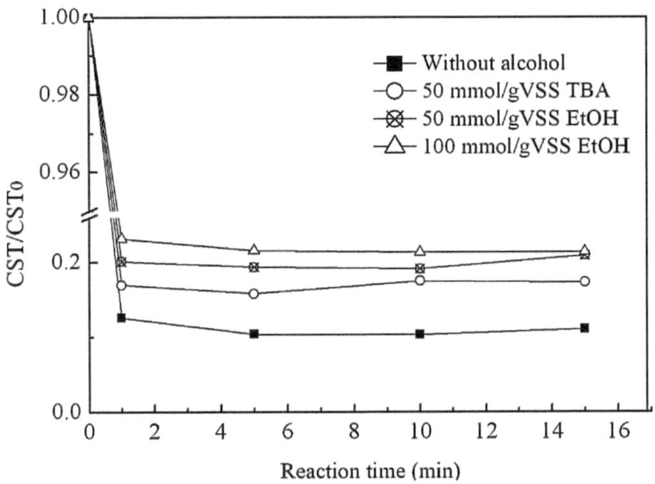

Figure 2.12 Effect of EtOH and TBA on sludge dewaterability. Experimental conditions: $[S_2O_8^{2-}] = 2.4$ mmol/g-VSS; [Fe(II)] = 3.0 mmol/g-VSS; No pH adjustment.

contains α-hydrogen that could rapidly quench hydroxyl and sulfate radicals. The TBA, however, as a weaker quencher for sulfate radicals, selectively quenched hydroxyl radicals and thus resulted in around 3.0% higher CST reduction than EtOH. Nevertheless, CST reduction was not fully quenched indicating the possibility of insufficient EtOH or nonradical enhancement pathway. To further verify its contribution, EtOH concentration was increased to 100 mmol/g-VSS and approximately 10.9% drop in CST reduction was observed compared with that of 8.8% by the addition of 50 mmol/g-VSS EtOH. The obvious decrease in CST reduction efficiency with increasing amount of EtOH confirmed that the radical-based pathway was the dominating dewatering mechanism in Fe(II)-persulfate system.

2.2.5.2 Effect of EPS on Sludge Dewaterability

Microbial EPSs are major components of the sludge floc matrix. They are regarded as one of the most important factors that influence the dewatering characteristics of sludge. Information regarding EPS is thus relatively favorable for understanding the exact roles of EPS in controlling dewaterability and for revealing the mechanisms enhancing dewatering.

Three-dimensional EEM spectroscopy was applied to characterize the EPS. Fig. 2.13 illustrates the EEM fluorescence spectra of EPS samples extracted from two types of sludge with obviously distinct dewaterability. The CST for raw sludge was about 210 s; after treated with Fe(II)-persulfate oxidation for 15 min, CST decreased to approximately18 s. It is evident from the results of Fig. 2.13 that persulfate oxidation resulted in more considerable changes in the fluorescent characteristics of soluble EPS compared to bound EPS. After treated, only two peaks with much lower fluorescence intensity were identified at the excitation/emission (Ex/Em) of 275–280/305–310 and 310–315/380–390 in soluble EPS in comparison with three peaks located at the Ex/Em of 275–280/295–310, 325–220/365–370 and 310–315/380–390 in the raw sludge, suggesting that the partial soluble EPS were degraded oxidatively by the reactive free radicals from the decomposition of persulfate. In contrast, fluorescence EEMs of bound EPS were not greatly affected, in which the fluorescence peak was always located at the Ex/Em of 315–325/380–390. Notably, sludge dewaterability (i.e., CST) was much more closely related to soluble EPS, rather than the bound EPS. Soluble EPS aid in the retention of water especially responsible for the strong water-binding capacity of sludge, and thus an increased amount of soluble EPS may bring a greater amount of bound water, causing poor performance of sludge-dewatering.

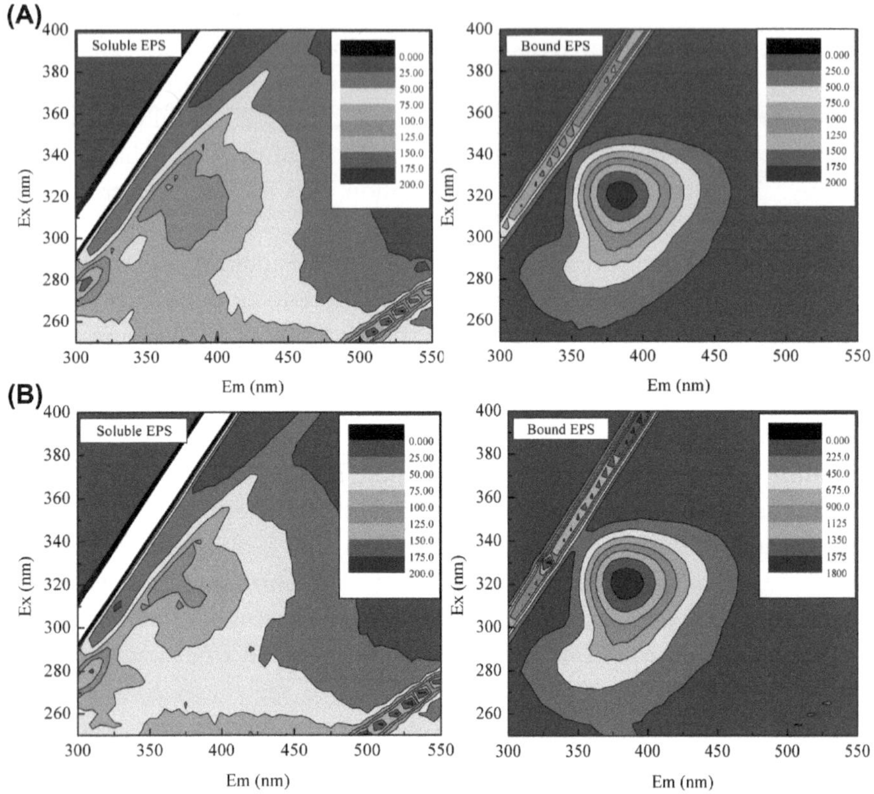

Figure 2.13 EEM fluorescence spectra of different EPS fractions of (A) the raw sludge and (B) the sludge after a 15-min of treatment by Fe(II)-persulfate oxidation process (Experimental conditions: $[S_2O_8^{2-}] = 1.2$ mmol/g-VSS; $[Fe(II)] = 1.5$ mmol/g-VSS; No pH adjustment).

Considering the high hydratability of soluble EPS, together with the finding that the mass of total EPS represents up to 80% of the mass of activated sludge, it is therefore reasonable to suppose that the profound improvement in sludge dewaterability following Fe(II)-persulfate oxidation was mainly brought about by the decreased amount of soluble EPS. Soluble EPS removal reduced sludge viscosity, as a result of which, sludge dewaterability was improved. Good dewaterability in the presence of low levels of EPS was presumably due to the enhancement of sludge flocculation, which resulted in the increase of flocs' size and decrease of its fractal dimension, improving the sludge dewaterability. Note that the effect of Fe(II)-persulfate oxidation on the dewaterability of sludge were that the powerful oxidizing radices not only degraded EPS but also disrupted floc structure and cell membranes and reduced the floc size. The interstitial and inner water of sludge would consequently be released into the solution, and as a result, sludge dewaterability was enhanced significantly.

2.2.6 Effect of Viscosity on Sludge Dewaterability

The viscosity of sludge was another parameter affecting the sludge dewaterability. The CST value showed a significant correlation ($R = 0.883, p = 0.00$) with viscosity as exhibited in Fig. 2.14, and the sludge samples with a higher viscosity had more difficulties with dewatering. Otherwise, some facts seem to show that the viscosity of sludge was positively correlated with its EPS content. This is mostly because soluble EPS located in the outer parts of cells and flocs had high affinity for water and were thus highly hydrated, as a result of which, an increase in EPS content induced the increasing of sludge viscosity. Sludge that is characterized by higher viscosity would therefore be more difficult to dewater if it has a higher CST.

2.3 NOVEL INSIGHTS INTO ENHANCED DEWATERABILITY BY FE(II)-ACTIVATED PERSULFATE OXIDATION

Effective dewatering requires the attack of EPSs and microorganism cells in sludge flocs. The specific objectives of this section are (1) to evaluate the potential of $\left(Fe(II) - S_2O_8{}^{2-}\right)$ oxidation as a novel technique for improving the dewaterability; (2) to determine the exact role of EPS in sludge-dewatering; and (3) to explore the responsible influencing mechanism involved in the dewaterability of sludge.

2.3.1 General Sludge Characteristics

Sewage sludge samples were taken from three full-scale WWTPs in Shanghai, China. The characteristics for each sludge sample are listed in Table 2.6. The dewaterability of raw sludge from sewage WWTPs (sample 1), measured as normalized CST, was similar to that of industrial WWTPs (sample 2), but obviously better than that of special WWTPs (sample 3). On the contrary, the viscosity of sample three was the lowest followed by samples 1 and 2.

Microscopic examination suggested that the sludge flocs from the 3 full-scale WWTPs were relatively small and loose and contained low numbers of filaments and some pin-pointed flocs. The sludge flocs in samples 2 and 3 appeared to be slightly more compact in comparison with sample 1. In spite

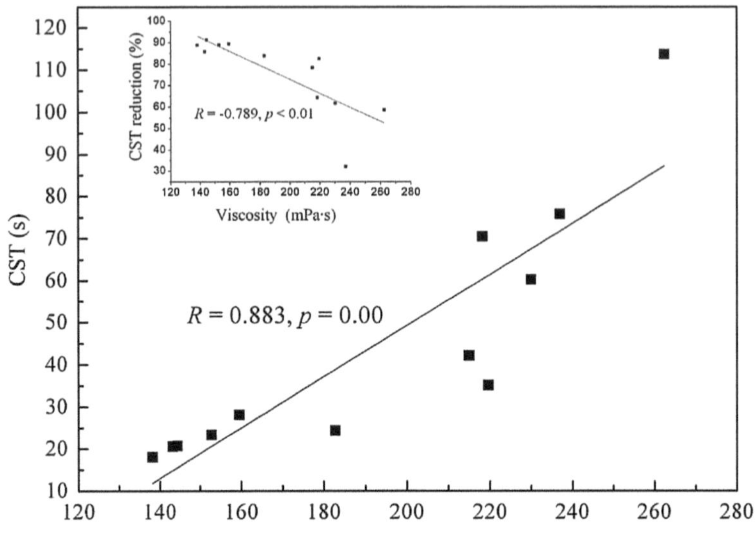

Figure 2.14 Correlation between the viscosity of sludge and its dewaterability as measured by CST.

of the tight structure, the porous zoogloea flocs could be drastically disintegrated when subjected to the $(Fe(II) - S_2O_8^{2-})$ conditioning, which would result in high numbers of small flocs. The volume-based median size of the flocs was approximately 45 μm for sample 1, 15 μm for sample 2, and 4 μm for sample 3 and decreased to 6, 5 and 2 μm for samples 1, 2, and 3 respectively, after 30 min of conditioning. The small floc fragments were uniformly dispersed and some single bacterial cells could be identified distinctly between the larger flocs. The observation of a remarkable reduction in particle size following the conditioning indicated that the oxidation by $(Fe(II) - S_2O_8^{2-})$ system strongly destroyed Zeta-potential and structure of surface electric double layers, and consequently contributed to the disintegration of the sludge flocs and complete breakage of microbial cells that make up sludge.

2.3.2 Enhanced Dewaterability of Sludge by $(Fe(II) - S_2O_8^{2-})$ Oxidation

$K_2S_2O_8$ and $FeSO_4 \cdot 7H_2O$ were applied as the sludge conditioner. The optimum condition, that is, $S_2O_8^{2-}$ 1.2 mmol/g-VSS and Fe(II) 1.5 mmol/g-VSS, based on Section 2.2, was adopted to improve the dewaterability of sludge in this study. The profiles of normalized CST of the

Table 2.6 Fundamental Characteristics of Waste-Activated Sludge Samples From Different WWTPs[a]

Sources	Symbol	Process[b]	Characteristics of the Sludge Samples[c]							
			TSS (g/L)	VSS (g/L)	Water Content (%)	pH	Viscosity (mPa·s)	Normalized CST (s L/g-TSS)	Conductivity (μs/cm)	Redox Potential (mV)
Sewage WWTP	Sample 1	A^2O	16.83 ± 0.35	11.32 ± 0.89	98.33 ± 0.12	6.93 ± 0.01	92.33 ± 0.35	13.79 ± 0.52	1422 ± 9.85	1.0± 0
Industrial WWTP	Sample 2	SBR	42.22 ± 0.14	16.67 ± 0.05	95.83 ± 0.01	6.68 ± 0.57	259.50 ± 11.30	11.90 ± 0.01	1582 ± 9.00	15 ± 0.10
Special WWTP	Sample 3	AO	25.67 ± 0.02	13.41 ± 0.35	97.47 ± 0.03	6.86 ± 0.01	8.00 ± 0.50	23.45 ± 0.46	699 ± 2.12	5.0 ± 0.50

[a] *WWTP*, wastewater treatment plants.
[b] *A^2O*, anaerobic-anoxic-oxic process; *SBR*, sequencing batch reactor; *AO*, anaerobic-oxic process.
[c] *TSS*, total suspended solids; *VSS*, volatile suspended solids; *CST*, capillary suction time.

waste-activated sludge samples before and after $(Fe(II) - S_2O_8^{2-})$ conditioning are depicted in Fig. 2.15. It is apparent that normalized CST of all the sludge samples remained relatively unaltered before exposed to the conditioning. When $(Fe(II) - S_2O_8^{2-})$ conditioner was injected, however, the normalized CST sharply decreased from 13.79 to 2.73, 11.90 to 1.62, and 23.45 to 4.39 s L/g-TSS for sludge samples 1, 2, and 3 within the first 1 min, drops of roughly 80.2%, 86.4% and 81.3%, respectively. Obviously, although the wastewater sources and treatment processes were definitely different, the performance of $(Fe(II) - S_2O_8^{2-})$ system on enhancing the dewaterability was not affected, in which all the possible reduction of normalized CST under all the tested conditions was essentially completed within 1 min. It is evident that the $(Fe(II) - S_2O_8^{2-})$ oxidation is capable of improving the sludge dewaterability as compared with other dewatering technologies.

The radical-based pathway was the dominating dewatering mechanisms of sludge through $(Fe(II) - S_2O_8^{2-})$ conditioning. The sulfate radicals $(SO_4^{-}\bullet)$ from the decomposition of $(Fe(II) - S_2O_8^{2-})$ system oxidatively degraded the EPS, which were responsible for the water-binding capacity and structural stability of sludge flocs, causing a great reduction in both EPS contents (Fig. 2.16) and floc's size. A decreased amount of EPS lowered the EPS-bound water content, reduced sludge viscosity, and as a

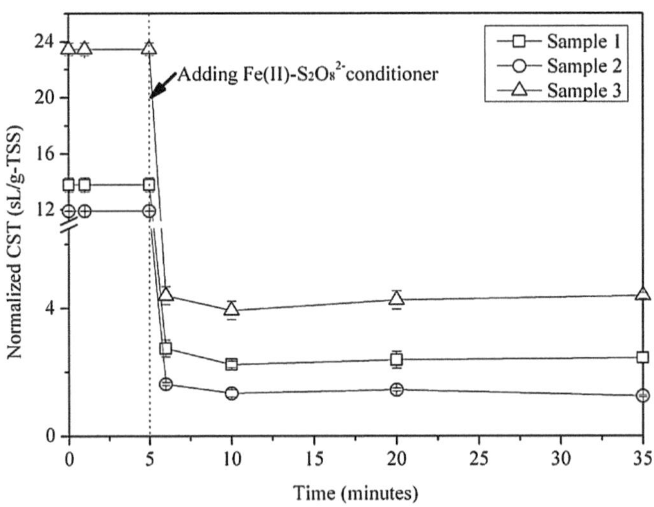

Figure 2.15 Normalized CST of activated sludge samples with $(Fe(II) - S_2O_8^{2-})$ conditioning (Experimental conditions: $[S_2O_8^{2-}] = 1.2$ mmol/g-VSS; $[Fe(II)] = 1.5$ mmol/g-VSS).

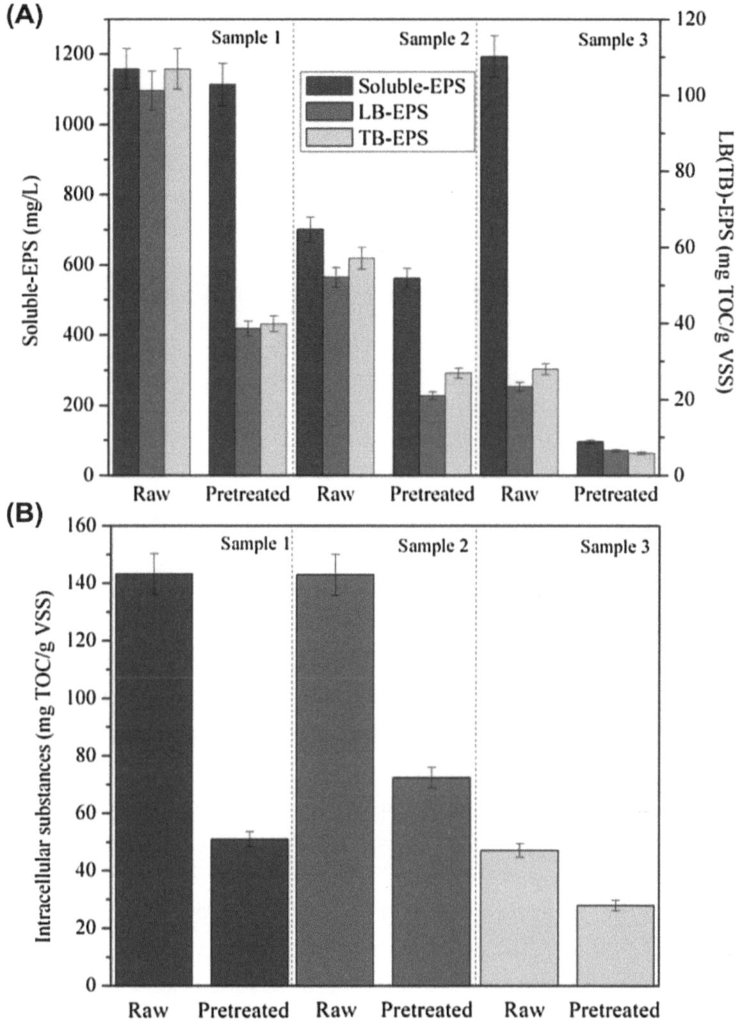

Figure 2.16 Different EPS fractions (A) and intracellular substances (B) contents of sludge flocs before and after pretreatment.

consequence, significantly improved sludge dewaterability in terms of normalized CST. Also, the sludge flocs after conditioning could yield a relatively lower surface charge than the unconditioned sludge flocs due to the removal of EPS. The decrease in surface charge declined repulsive electrostatic interactions and enhanced reflocculation of fine flocs by the production of the coagulant, Fe(III), during the oxidation reaction of $\left(Fe(II) - S_2O_8{}^{2-}\right)$, eventually contributing a lot to the better sludge dewaterability. As for the floc's size, the "supracolloidal" flocs in the range of 1−100 μm had the greatest adverse impact on sludge dewaterability, and that the dewaterability decreased as the quantities of flocs in this size range increased. In fact, the floc's breakage, as observed in present study, appeared to be more beneficial than detrimental to the dewatering characteristics of sludge, as it could provide more passages for the removal of water in the inner of sludge flocs.

2.3.3 Contribution of Different EPS Fractions to Sludge Dewaterability

The variations of different EPS fractions in sludge flocs with or without $\left(Fe(II) - S_2O_8^{2-}\right)$ pretreatment are presented in Fig. 2.16A. Waste-activated sludge has a diverse floc structure depending on both wastewater source and operational conditions in the WWTP. However, it is found that the changes of EPS contents of sludge flocs from different plants were surprisingly similar when subjected to $\left(Fe(II) - S_2O_8^{2-}\right)$ oxidation. Through the $\left(Fe(II) - S_2O_8^{2-}\right)$ oxidation process, different EPS fractions decreased apparently for all the sludge samples.

The EPS located in the outer parts of sludge flocs, which are very important to the colloidal stability of the flocs, were disintegrated during oxidation process. As observed in the figures, after 30 min of conditioning, soluble-EPS and LB-EPS in sample 1 declined to approximately 1110 mg/L and 38.6 mg TOC/g-VSS, respectively, in comparison with the unconditioned sludge. Similar trends were observed for samples 2 and 3 as well, and the corresponding values decreased to around 562 mg/L and 21.0 mg TOC/g-VSS, as well as 95 mg/L and 6.5 mg TOC/g-VSS, respectively. Decreased normalized CST (Fig. 2.15) accompanied by reduced soluble-EPS and LB-EPS contents (Fig. 2.16A) was observed, revealing that the disruption of both soluble- and LB-EPS was considerably favorable for the enhancement of sludge dewaterability. EPS have a high affinity for water and thus are highly hydrated, as a result of which, the viscosity of sludge was increased with the increasing of EPS contents. Sludge that carries a large amount of EPS has a higher CST value and will be thereby more difficult to dewater. When the EPS were attacked by $SO_4^{-}\bullet$ from $\left(Fe(II) - S_2O_8^{2-}\right)$ oxidation, sludge flocs were dissolved immediately, bringing about the release of interstitial and vicinal water inside the spaces of flocs into free water. Therefore, the dewaterability of sludge was promoted greatly.

Otherwise, TB-EPS, as compared with soluble- and LB-EPS, were often considered to be less crucial to dewaterability as they had a lower bound water. However, it is worth pointing out that, through the conventional dewatering pretreatments, such as electrolysis, potassium ferrate oxidation, etc., the contents of TB-EPS did not vary basically ascribed to the weaker oxidation of the methods to the thick soluble- and LB-EPS layers that surrounded the sludge micro-flocs. Accordingly, the real contribution of TB-EPS to dewaterability was not disclosed well so far. TB-EPS located in the inner flocs represents an essential structure and contributes significantly to cell adhesion and attachment in the inner floc structure through strong interactions. The presence of microorganism cells trapped in sludge zoogloea flocs, bounding a large quantity of water of hydration, could cause "difficult to dewater" of sludge. The removal of TB-EPS was therefore of a great importance to the sludge dewaterability. In thepresent study, TB-EPS (Fig. 2.16A) presented a considerable reduction from approximately 106.9 to 39.9, 57.1 to 26.9, and 27.9 to 5.8 mg TOC/g-VSS for samples 1, 2, and 3, after 30 min of $\left(Fe(II) - S_2O_8{}^{2-}\right)$ oxidation, respectively.

It is obvious that the oxidation applied in this study could effectively decompose the soluble- and LB-EPS layers, and further destroy the TB-EPS located in inner parts of sludge flocs. The collapse of TB-EPS led to the break-up of sludge matrix and also made more microorganism cells available for oxidation attack. Based on the above results, it is relatively reasonable to suppose that the powerful $SO_4{}^{-}\bullet$ from $\left(Fe(II) - S_2O_8{}^{2-}\right)$ system could subsequently penetrate the microorganism walls, increase the osmosis of cell membranes, and crack the uniformity of the cell walls.

In order to verify the hypothesis, the changes of metabolic activity of microorganisms in terms of intracellular substances contents were measured (Fig. 2.16B). It was clear that the sludge metabolic activity decreased remarkably through $\left(Fe(II) - S_2O_8{}^{2-}\right)$ oxidation, suggesting microorganism cells lysis. The rapid decrease in the metabolic activity confirmed that $\left(Fe(II) - S_2O_8{}^{2-}\right)$ oxidation improved sludge dewaterability principally via the degradation of EPS and rupture of cells. This resulted in the release of intracellular materials in sludge cells, most of which could be degraded by excess $SO_4{}^{-}\bullet$, also large quantities of water of hydration bound inside the sludge cells was released into the free water, which promoted sludge-dewatering characteristics.

2.3.4 EEM Spectra of Different EPS Fractions

Three-dimensional excitation-emission matrix (EEM) fluorescence spectroscopy is a rapid and sensitive technique. EEM fluorescence spectra can be regarded as an overall "fingerprint" of EPS mixtures in sludge flocs.

Therefore, it can be a potentially useful technique for rapidly determining the fluorescence compounds present in EPS and exploring the responsible mechanisms enhancing dewaterability.

Typical EEM fluorescence spectra of EPS and its corresponding fractions extracted from both raw and pretreated sludge flocs from the three full-scale WWTPs are illustrated in Fig. 2.17. Each EEM gave spectral information regarding the chemical compositions of EPS mixtures. Four main peaks could be identified from fluorescence spectra of EPS in raw sludge samples 1 and 2 (Fig. 2.17A and B) collected from sewage and industrial WWTPs, respectively. The first peak located at the excitation/emission wavelengths (Ex/Em) of 225/335−355 nm (Peak A) was attributed to aromatic protein-like substances. A second peak at the Ex/Em of around 280/335−340 nm (Peak B) was referred to as tryptophan protein-like substances. The third peak, identified at the Ex/Em of 275/425−435 nm (Peak C), and the fourth peak observed at the Ex/Em of around 340/420−435 nm (Peak D) were regarded as visible humic- and fulvic-like fluorescence, respectively. For the raw sludge sample 3 from the special WWTP, the Peaks C and D located at the Ex/Em of 275/425−435 nm and 340/420−435 nm were absent in the EEM spectra. Two main peaks could be readily identified from Fig. 2.17C. The first peak was located at the Ex/Em of 225/345−355 nm (Peak A) while the second peak was found at the Ex/Em of 280/335 nm (Peak B). The differences might be mainly due to the fact that the EPS were harvested from the different sludge origins and thus the structure and components in EPS were chemically different.

The fluorescence parameters of EEM spectra including peak location and maximum fluorescence intensity were summarized in Table 2.7, which could be used for quantitative analysis. $\left(Fe(II) - S_2O_8^{2-}\right)$ oxidation had a substantial effect on the fluorescence characteristics of the EEM spectra of the sludge EPS. Slight shifts were observed for the peak locations of sludge EPS following the conditioning. For the samples 1 and 2, the locations of Peaks A, B, C, and D of different EPS fractions (e.g., soluble-, LB- and TB-EPS) after pretreatment were all red-shifted by 5−10, 5−15, 5−15 and 0−15 nm, respectively, than those of the raw sludge EPS. For the sample 3, however, the locations of Peaks A and B demonstrated a blue shift of 5−10 and 5−15 nm, respectively, in comparison with those of the raw sludge EPS. Such a shift suggested the changes in the conformations of the fluorescence components in the sludge EPS during the oxidation process. A red shift is attributed to the increase of carbonyl, hydroxyl, alkoxyl, amino, and carbonyl groups in fluorophores, while a blue shift is ascribed to

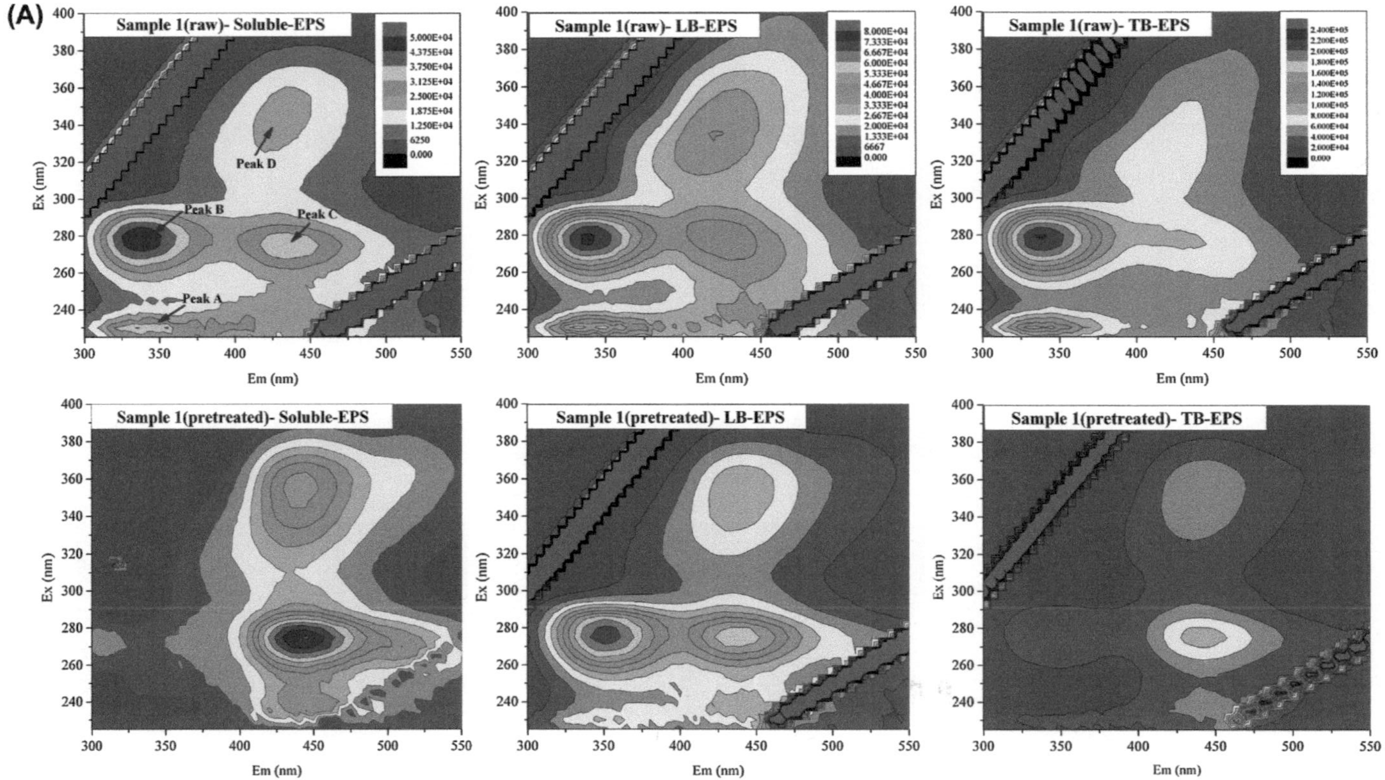

Figure 2.17 Typical EEM fluorescence spectra of different EPS fractions in the waste-activated sludge flocs achieved from sewage (A), industrial (B), and special (C) WWTPs, respectively.

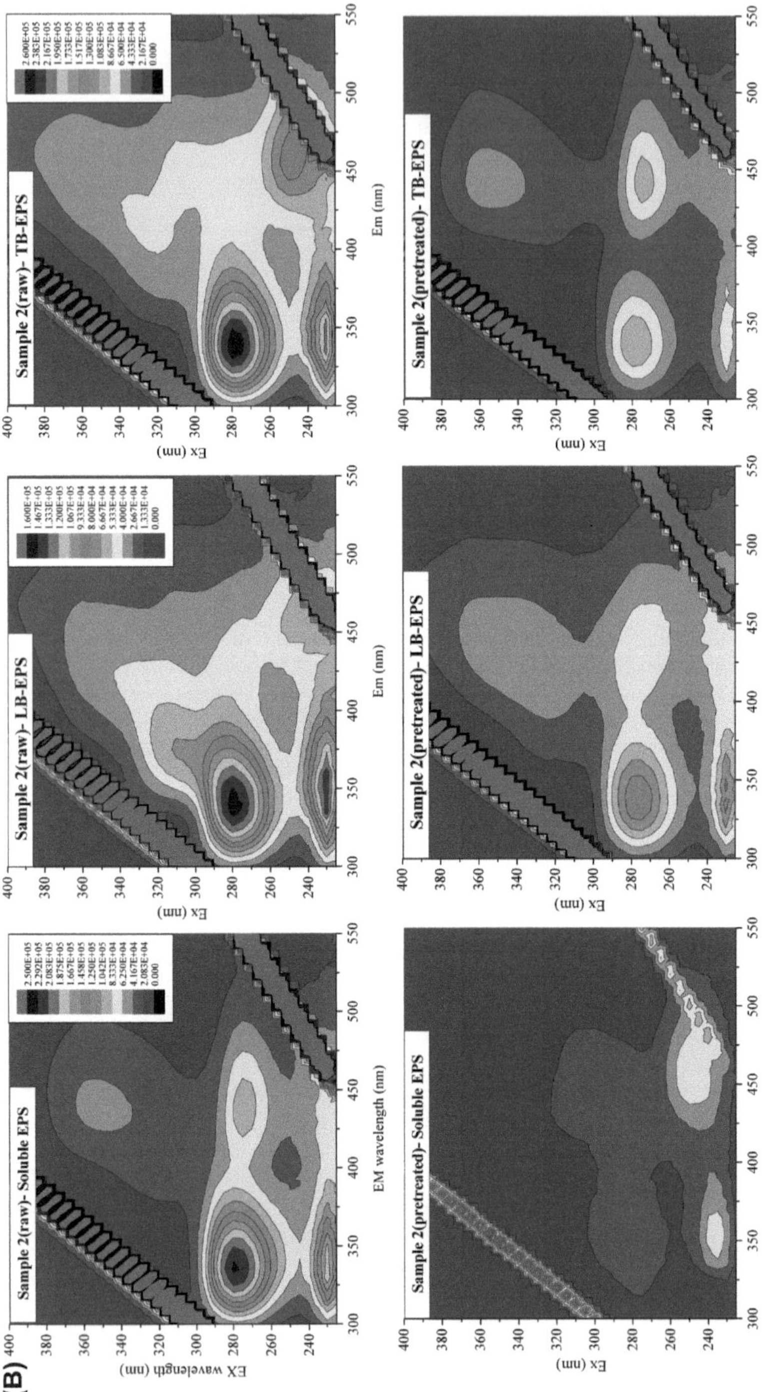

Figure 2.17 *(continued)*.

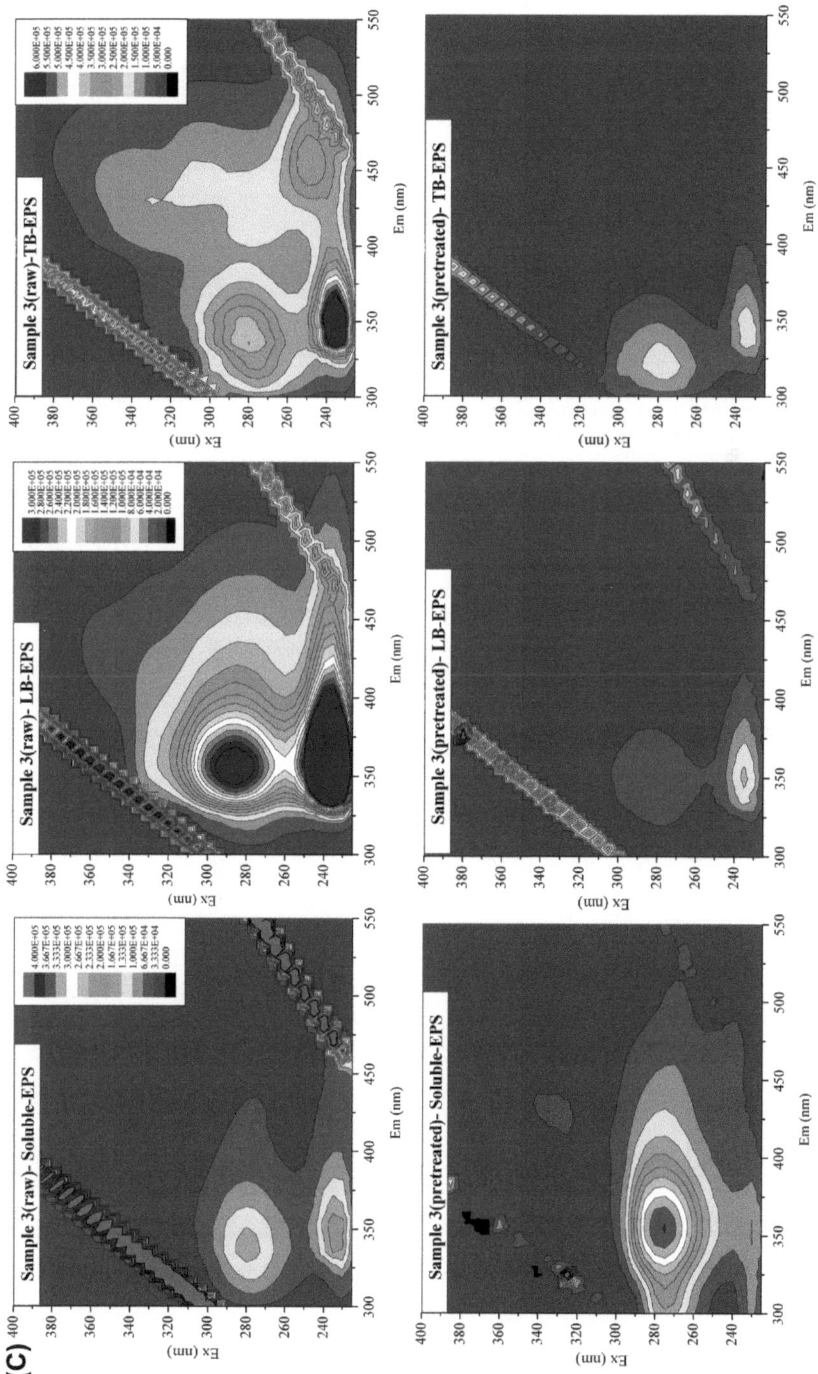

Figure 2.17 *(continued)*.

Table 2.7 Fluorescence Spectral Parameters of Different EPS Fractions in Sludge Flocs Before and After Pretreatment

			Peak A		Peak B		Peak C		Peak D	
Types of Sludge	Conditioning	EPS Fractions	Ex/Em	Intensity ($\times 10^4$)	Ex/Em	Intensity ($\times 10^4$)	Ex/Em	Intensity ($\times 10^4$)	Ex/Em	Intensity ($\times 10^4$)
Sample 1	Raw	Soluble-EPS	225/340	2.8	280/335	5.0	275/430	3.7	340/425	2.1
		LB-EPS	225/350	3.9	280/340	7.5	275/425	4.5	340/420	3.9
		TB-EPS	225/335	9.3	280/340	2.2	275/430	8.3	340/430	6.7
	Pretreated	Soluble-EPS	225/345	0.1	280/335	0.5	275/445	5.1	340/440	2.8
		LB-EPS	225/355	2.1	280/350	6.9	275/435	5.9	340/435	3.0
		TB-EPS	225/345	2.2	280/345	3.9	275/440	8.9	340/440	5.0
Sample 2	Raw	Soluble-EPS	225/340	12.7	280/335	23.2	275/435	9.4	340/435	4.4
		LB-EPS	225/350	10.3	280/340	15.5	275/430	4.8	340/420	4.1
		TB-EPS	225/335	12.7	280/335	25.2	275/435	7.7	340/430	6.5
	Pretreated	Soluble-EPS	225/350	1.9	280/350	3.7	275/430	2.6	340/435	1.9
		LB-EPS	225/340	7.1	280/340	8.9	275/435	5.1	340/430	3.7
		TB-EPS	225/345	5.9	280/335	10.3	275/445	9.6	340/440	4.7
Sample 3	Raw	Soluble-EPS	225/345	8.2	280/335	14.8	275/430	3.5	340/415	2.7
		LB-EPS	225/355	18.9	280/355	32.6	275/430	8.3	340/430	4.1
		TB-EPS	225/350	19.9	280/335	40.2	275/430	17.9	340/435	14.2
	Pretreated	Soluble-EPS	225/340	7.8	280/355	35.9	275/430	10.2	340/435	3.4
		LB-EPS	225/350	2.1	280/350	3.9	275/430	0.9	340/410	0.5
		TB-EPS	225/340	4.1	280/320	14.9	275/425	2.1	340/415	0.9

an elimination of particular functional groups such as carbonyl, hydroxyl and amine, a reduction in the degree of π-electron systems, a decrease in the number of aromatic rings and conjugated bonds in a chain structure, or a conversion of a linear ring system to a nonlinear system.

Generally, intensity reduction of the fluorescence peak before and after pretreatment is an indication for oxidation of fluorescing materials. It can be seen from Table 2.7 that the fluorescence intensities of the peaks in EEM spectra of EPS were decreased significantly resulted from $\left(Fe(II) - S_2O_8^{2-}\right)$ oxidation, which was in line with the tendency of sludge dewaterability indicated by normalized CST. It revealed that in soluble- and LB- as well as TB-EPS fractions, aromatic protein-, tryptophan protein-, humic- and fulvic-like substances together affected sludge dewaterability. These investigations obviously implied that the enhanced dewaterability of sludge flocs in this study was closely related to the degradation and removal of the fluorescence components in sludge EPS. The powerful oxidation of $SO_4^{-}\bullet$ from $\left(Fe(II) - S_2O_8^{2-}\right)$ system actually destroyed the particular functional groups of fluorescing substances present in sludge EPS and caused cleavage of linkages in the polymeric backbone, thereby inducing the release of EPS-bound water, intercellular materials and water of hydration inside microbial cells, and subsequent enhancement of sludge dewaterability.

2.4 SYNERGETIC PRETREATMENT BY FE(II)-ACTIVATED PERSULFATE OXIDATION UNDER MILD TEMPERATURE

Thermal treatment generally consumes higher energy and even has a larger hazardous emission potential as the temperature higher than 180°C, making the process expensive and increasing safety concerns. The combined thermochemical treatment might act synergistically and give an advantage of lowering the temperature, thus providing an energy-efficient and cost-effective alternative for sludge-dewatering. In this section, the synergetic effectiveness of combined thermal and $\left(Fe(II) - S_2O_8^{2-}\right)$ oxidation pretreatments for sludge-dewatering were systematically elucidated. CST was characterized to evaluate the sludge dewaterability.

2.4.1 Sludge Dewaterability

The sludge dewaterability in terms of CST depicted a definite difference under various pretreatment conditions (Fig. 2.18A). It is evident that CST

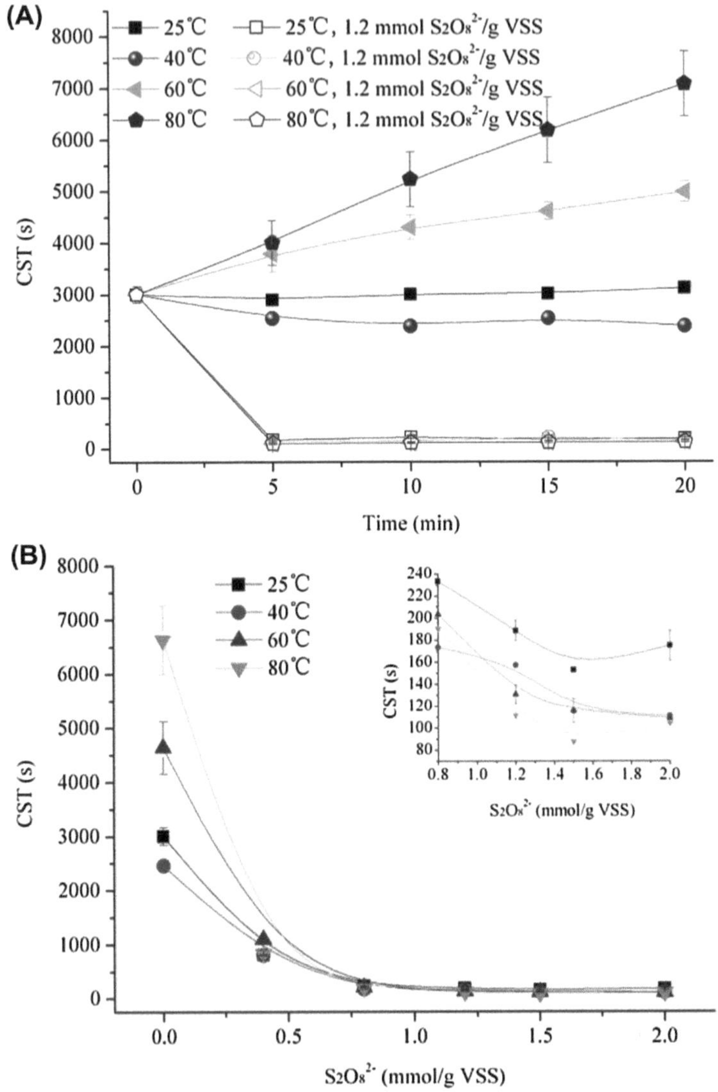

Figure 2.18 CST evolution of sludge as a function of pretreatment time (A) and $(Fe(II) - S_2O_8^{2-})$ dosages (B). (During the thermal-$(Fe(II) - S_2O_8^{2-})$ oxidation process, the $S_2O_8^{2-}/Fe(II)$ molar ratio was fixed at 0.8. In Figs. 2.18–2.22, only $S_2O_8^{2-}$ dosage is presented).

varied a little (from 3006.1 ± 160.0 to 3119.2 ± 92.5 s) within 20 min at 25°C, whereas as temperature elevated to 60 and 80°C, CST rapidly reached 4981.5 ± 202.7 and 7074.7 ± 631.9 s, increases of roughly 65.7% and 135.4% respectively, indicative of a significant deterioration in the dewaterability. This might be linked to the progressive cell lysis, releasing intracellular biopolymers, which decreased dewatering properties.

In contrast, when in the presence of $\left(Fe(II) - S_2O_8{}^{2-}\right)$ oxidation (i.e., $S_2O_8{}^{2-}$ 1.2 mmol/g-VSS and Fe(II) 1.5 mmol/g-VSS), CST apparently decreased from 3006.1 ± 160.0 to 174.2 ± 23.6, 136.9 ± 37.5, 103.3 ± 14.4 and 106.3 ± 12.8 s, decreases of 94.2%, 95.4%, 96.6% and 96.5% or so at 25, 40, 60 and 80°C respectively, within the first 5 min. Afterward, a stable plateau with less than 40.0-s variation was observed during the following 15 min (Fig. 2.18A), implying that the dewaterability was improved effectively by $\left(Fe(II) - S_2O_8{}^{2-}\right)$ oxidation within a relatively short time.

The effects of $\left(Fe(II) - S_2O_8{}^{2-}\right)$ dosages on the dewaterability at various temperatures were systematically investigated and main results are illustrated in Fig. 2.18B. The CST decreased with $S_2O_8{}^{2-}$ dosage increasing from 0 to 1.2 mmol/g-VSS in which the corresponding Fe(II) dosage was 1.5 mmol/g-VSS, then leveled off at higher dosages. Overdosed $\left(Fe(II) - S_2O_8{}^{2-}\right)$ produced less additional benefits to the dewatering. Therefore, the optimum $S_2O_8{}^{2-}$ and Fe(II) dosages in the current study were supposed to be 1.2 mmol/g-VSS and 1.5 mmol/g-VSS, respectively. For temperature, as expected, the higher the operational temperature, the better the dewaterability was at a certain $\left(Fe(II) - S_2O_8{}^{2-}\right)$ dosage. As depicted in the inset of Fig. 2.18B, when the temperature was raised from 25 to 40, 60 and 80°C, the corresponding CST jumped from 188.6 ± 9.4 to 157.3 ± 2.1, 131.0 ± 8.6, and 111.5 ± 2.7 s, drops of roughly 16.6%, 30.5% and 40.9%, respectively, at $S_2O_8{}^{2-}$ 1.2 mmol/g-VSS and Fe(II) 1.5 mmol/g-VSS. It may be attributable to the fact that the persulfate can be initiated via metal catalyzer as well as heat. The elevated temperatures, in current study, thereby increased the decomposition of persulfate into $SO_4{}^{-}\bullet$ radicals and enhanced $\left(Fe(II) - S_2O_8{}^{2-}\right)$ oxidation process, contributing greatly to the degradation of EPS and cell lysis. As a result of this, the dewaterability of sludge was improved substantially. Obviously, the mild thermal treatment coupled with $\left(Fe(II) - S_2O_8{}^{2-}\right)$ oxidation provides an alternative technique with a much lower operational temperature to promote the dewaterability, and therefore is more capable for sludge-dewatering in the view of efficiency and economy.

2.4.2 Influence of Zeta Potential on the Dewaterability

The zeta potential is a critical factor in influencing sludge dewaterability. Fig. 2.19A presents the zeta potential evolution of sludge flocs under differential operational temperatures. The surfaces of raw sludge were originally negatively charged with the zeta potential of -12.6 ± 1.4 mV at 25°C.

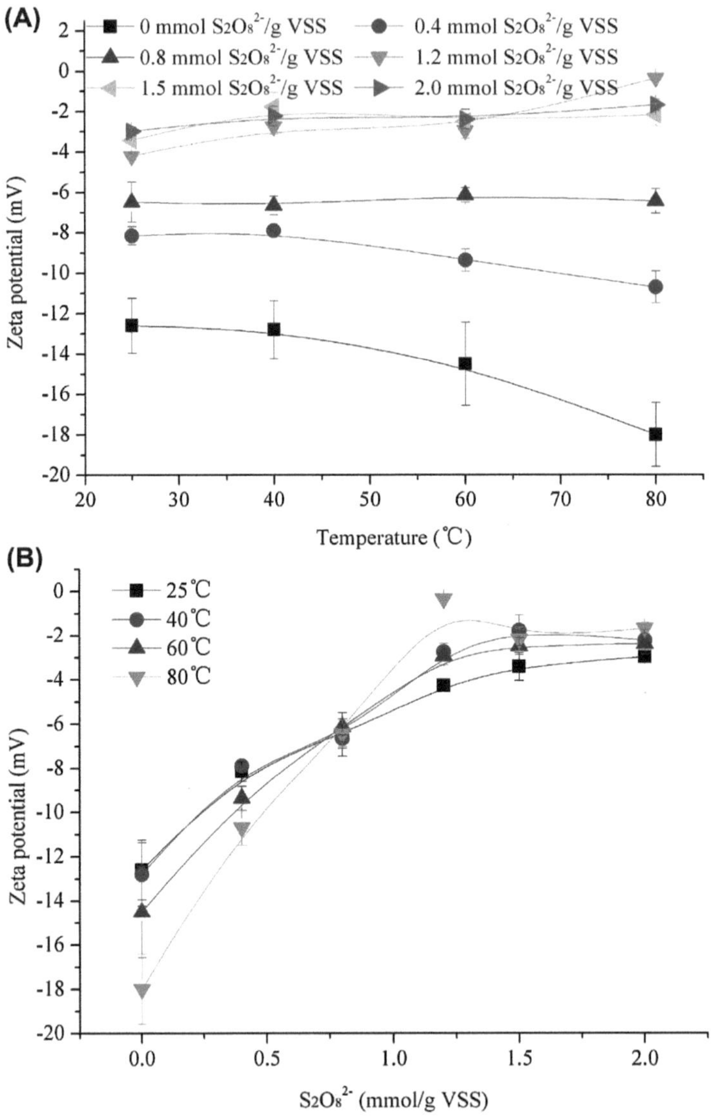

Figure 2.19 Zeta potential evolution of sludge as a function of operational temperatures (A) and $\left(Fe(II) - S_2O_8{}^{2-}\right)$ dosages (B) for 20 min.

As the temperature increased to 80°C, the zeta potential was found to sharply decrease to -18.0 ± 1.6 mV. The reduction in the zeta potential possibly originated from the cell disruption under thermal treatment, releasing biopolymers and some inorganic species. The rise in the biopolymers could cause the increasing negative surface charge. Interestingly, the decline trend with the temperature increasing could be basically controlled by $\left(Fe(II) - S_2O_8^{2-}\right)$ oxidation. The zeta potential showed no obvious shift with increasing temperature in the presence of $S_2O_8^{2-}$ 0.8 mmol/g-VSS and Fe(II) 1.0 mmol/g-VSS, while a marked rise was observed beyond $S_2O_8^{2-}$ 1.2 mmol/g-VSS and Fe(II) 1.5 mmol/g-VSS. The presence of $\left(Fe(II) - S_2O_8^{2-}\right)$ oxidation altered the zeta potential evolution of the flocs through EPS degradation by SO_4^{-}• radicals and charge neutralization of ferric and ferrous irons, and higher operational temperature could strengthen the impact.

The zeta potential of sludge was significantly affected by $\left(Fe(II) - S_2O_8^{2-}\right)$ dosages, and it shifted toward the less negative charge as $\left(Fe(II) - S_2O_8^{2-}\right)$ dosages rose (Fig. 2.19B). The zeta potential lifted drastically from -12.6 ± 1.4 to -4.2 ± 0.2 mV at 25°C, from -12.8 ± 1.4 to -2.8 ± 0.4 mV at 40°C, from -14.5 ± 2.1 to -2.9 ± 0.4 mV at 60°C and from -18.0 ± 1.6 to -0.4 ± 0.1 mV at 80°C as the $S_2O_8^{2-}$ dosage increased from 0 to 1.2 mmol/g-VSS, then elevated slightly and tended to the zero point of charge (0 mV) with further increasing $\left(Fe(II) - S_2O_8^{2-}\right)$ dosages. According to the DLVO theories, the aggregation of sludge flocs is primarily governed by the negative surface charge. Increasing the negative surface charge is expected to cause an increasing dispersion and, as a consequence, poor dewatering properties. Thus, with the decreasing surface charge related to elevating zeta potential, the sludge can aggregate, settle down quickly, and is dewatered more easily. This deduction is supported by our current findings, in which the best dewaterability occurred at the zeta potential of -0.4 ± 0.1 mV.

2.4.3 Influence of Particle Size Distribution on the Dewaterability

The variations of particle size distribution of sludge flocs are given in Fig. 2.20 based on cumulative volume distribution. It is noticeable that particle size distribution of sludge were not effected remarkably by $\left(Fe(II) - S_2O_8^{2-}\right)$ oxidation at 25°C and 50% of the volume of sludge consisted of particles larger than 110.4 ± 5.7 μm (Fig. 2.20A). For the sludge samples pretreated by single thermal treatment as shown in Fig. 2.20B–C,

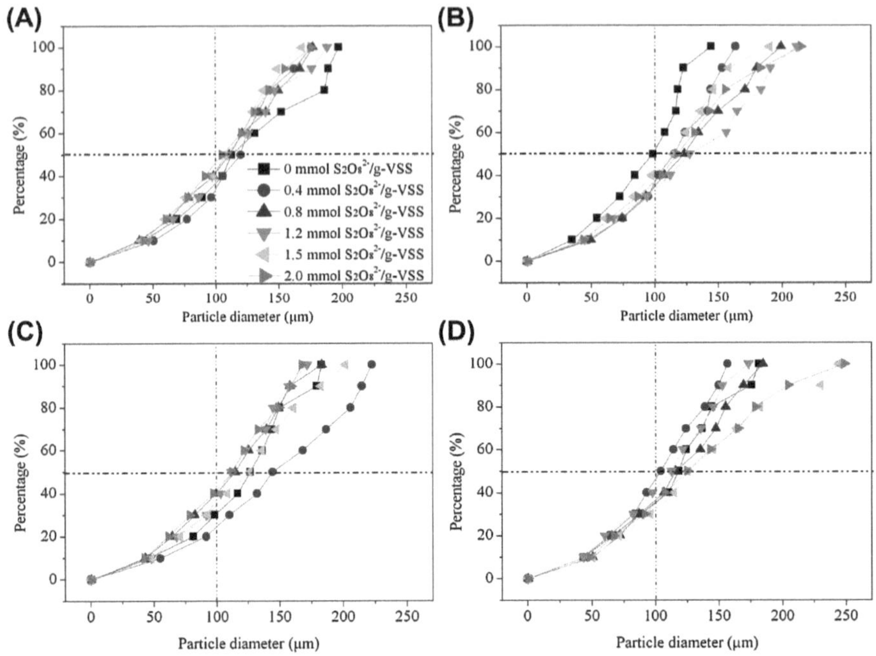

Figure 2.20 Cumulative particle size distribution (based on volume) of sludge as a function of $(Fe(II) - S_2O_8{}^{2-})$dosages at the operational temperatures of 25 (A), 40 (B), 60 (C), and 80°C (D) for 20 min.

particle size distribution remained more or less constant whatever be the applied operational temperatures. These results revealed that mild thermal treatment did not induce significant microflocs' breakage. When in the presence of $(Fe(II) - S_2O_8{}^{2-})$ oxidation, however, thermal treatment had a detectable effect on the granulometric distribution. The floc-size distribution curves were shifted to the large-size classes as $(Fe(II) - S_2O_8{}^{2-})$ dosages increased (Fig. 2.20B–C), with 50% sludge flocs being larger than 119.4 ± 5.4 μm at 40°C, than 124.2 ± 13.0 μm at 60°C and then 117.6 ± 8.6 μm at 80°C. These results could be explained by a reflocculation phenomenon. The reflocculation might occur after the combined pretreatment process, mainly due to the release of intracellular and extracellular substances that led to the creation of chemical bonds. In addition, this reflocculation was also strongly promoted by the coagulation effects of ferric and ferrous ions produced during $(Fe(II) - S_2O_8{}^{2-})$ oxidation process since they, acting as coagulants, could absorb onto the negatively charged sludge flocs and promoted agglomeration of the flocs through charge neutralization, absorption bridging, and precipitation catching.

Particle sizes have an effect on sludge-dewatering characteristics. The "supracolloidal" flocs in the range of 1−100 μm have the greatest adverse influence on sludge dewaterability, and the dewaterability decreases as the quantities of flocs in this size range increase because of the clog of filter-cake by smaller flocs during filtration. In this study, the combined thermal and $\left(Fe(II) - S_2O_8^{2-}\right)$ oxidation pretreatment increased the quantities of sludge flocs larger than 100 μm. Therefore, the dewaterability became better.

2.4.4 EEM Fluorescence Analysis of EPS

Three-dimensional EEM fluorescence spectroscopy is a potentially useful technique for rapidly determining the fluorescence compounds present in EPS mixtures in sludge flocs. Each EEM gave spectral information regarding the chemical compositions of EPS. Typical EEM fluorescence spectra of EPS and its corresponding fractions extracted from the raw and pretreated sludge are depicted in Fig. 2.21. Four main peaks could be readily identified from fluorescence spectra of EPS. The first peak (Peak A) was present at the

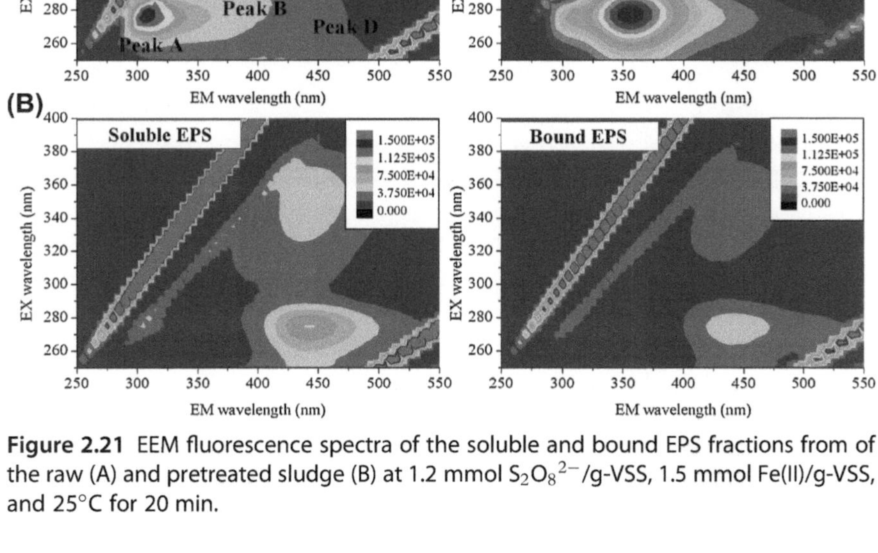

Figure 2.21 EEM fluorescence spectra of the soluble and bound EPS fractions from of the raw (A) and pretreated sludge (B) at 1.2 mmol $S_2O_8^{2-}$/g-VSS, 1.5 mmol Fe(II)/g-VSS, and 25°C for 20 min.

excitation/emission wavelengths (Ex/Em) of 270–280/305–310 nm in EEM spectra of EPS, which belonged to tyrosine protein-like substances. The second peak (Peak B) located at the Ex/Em of around 280/350–360 nm was described as tryptophan protein-like substances. The third peak (Peak C) was found at the Ex/Em of 345/435–445 nm, while the fourth peak (Peak D) was observed at the Ex/Em of around 275/435–445 nm. The fluorescence signals were associated with visible humic- and fulvic-like fluorescences.

The fluorescence parameters of EEM spectra including peak locations and maximum fluorescence intensity are displayed in Figs. 2.21 and 2.22, respectively, which could be used for quantitative analysis. After $(Fe(II) - S_2O_8^{2-})$ oxidation at 25°C for 20 min, the locations of Peaks C and D in soluble and bound EPS were both red-shifted (0–5 nm) to longer wavelength than those of raw sludge EPS, while Peak B was blue-shifted to shorter wavelength only in bound EPS by 5 nm compared to

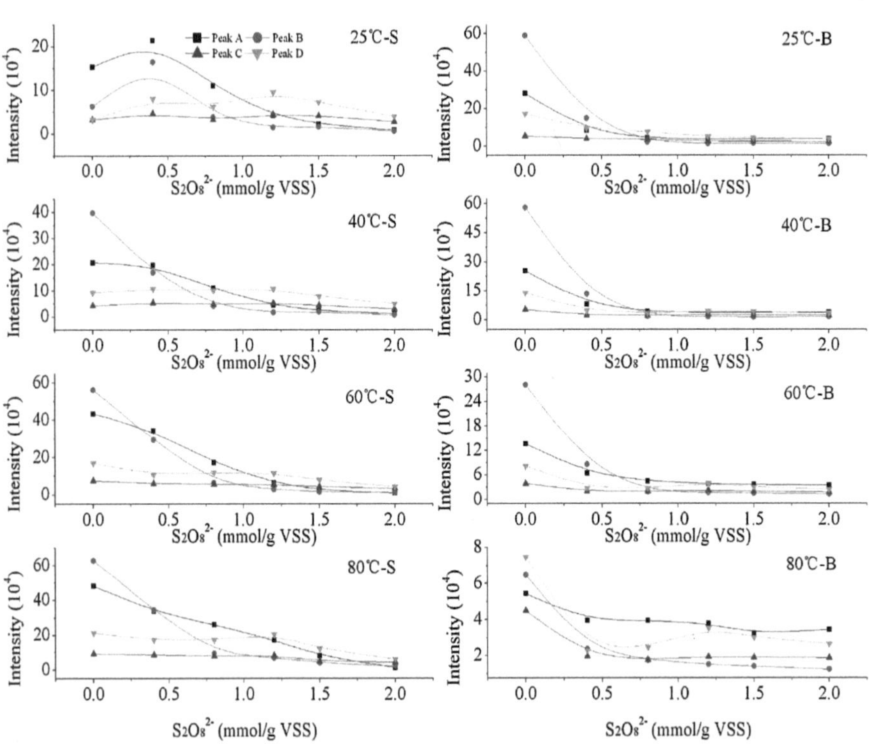

Figure 2.22 The fluorescence intensities of peaks of the soluble (S) and bound (B) EPS fractions under the various $(Fe(II) - S_2O_8^{2-})$ dosages and operational temperatures for 20 min.

that of raw sludge EPS (Fig. 2.21). The location shift of fluorescence peak provides spectral information on the chemical structure changes of the fluorescence components in EPS samples. A red shift is attributable to the increase of carbonyl, hydroxyl, alkoxyl, amino, and carbonyl groups in fluorophores, while a blue shift is ascribed to an elimination of particular functional groups such as carbonyl, hydroxyl, and amine, a reduction in the degree of π-electron systems, a decrease in the number of aromatic rings and conjugated bonds in a chain structure, or a conversion of a linear ring system to a nonlinear system.

Fig. 2.22 exhibits the evolution in the intensities of fluorescence peaks with $\left(Fe(II) - S_2O_8{}^{2-}\right)$ dosages under different temperatures. It was clear that as the temperature increased, more bound EPS, particularly tyrosine and tryptophan protein-like substances indicated, respectively, by Peaks A and B, were solubilized and converted into soluble EPS. For example, when pretreated by mild thermal treatment, the intensities of Peaks A and B in soluble EPS respectively increased about 215.0% and 896.8% as the temperature increased from 25 to 80°C, while the corresponding intensities in bound EPS reduced 80.7% and 88.9%, respectively. However, the intensities of Peaks C and D related to humic- and fulvic-like substances remained nearly constant, regardless of applied $\left(Fe(II) - S_2O_8{}^{2-}\right)$ dosages and temperatures. These observations suggested that thermal treatment caused flocs' disruption and cell lysis, releasing the extracellular and intracellular protein-like biopolymers. The rise in soluble EPS could often lead to the increasing surface charge and poorer dewaterability. However, in the presence of $\left(Fe(II) - S_2O_8{}^{2-}\right)$ oxidation, to the opposite, better dewaterability was observed as already described in Fig. 2.18, mainly attributed to the degradation of the fluorescence components in EPS by active $SO_4{}^{-}\bullet$. As depicted in Fig. 2.22, the intensities of Peaks A and B in soluble EPS declined approximately 71.9% and 76.2% at 25°C, 78.8% and 96.0% at 40°C, 85.3% and 95.0% at 60°C, 64.3% and 88.9% at 80°C, respectively as the $S_2O_8{}^{2-}$ dosage increased from 0 to 1.2 mmol/g-VSS. This result implied that the sludge dewaterability was mainly influenced by tyrosine and tryptophan protein-like fluorophores in EPS.

Based on the aforesaid results, it is thus reasonable to address that the enhanced dewaterability by the combined thermal and $\left(Fe(II) - S_2O_8{}^{2-}\right)$ oxidation pretreatment lies in the cleavage of linkages in the polymeric backbone (e.g., bound EPS) and microbial cells rupture, which induced the rise of tyrosine and tryptophan protein-like substances in soluble EPS, most of which could be degraded by $SO_4{}^{-}\bullet$. As a result of that, large quantities of EPS-bound water and interstitial water were released into the free water, and eventually sludge dewaterability was promoted greatly.

2.4.5 FT-IR Spectra of Sludge Flocs

Fig. 2.23 illustrates the FT-IR spectra of sludge flocs previously subjected to various pretreatment at 25°C. The strong band at 3299 cm^{-1} was attributed to the O—H stretching vibration. The bands at 2923 and 2852 cm^{-1} corresponded to the asymmetric and symmetric stretching vibrations of CH_2 of aliphatic structures and lipids and band at 1716 cm^{-1} to the C=O vibration of carboxylic acids, respectively. The typical bands located at 1654 and 1540 cm^{-1} were assigned to the stretching and deformation vibrations of C=O, C—N (amide I), and N—H (amide II) peptide bonds of proteins. The band at 1457 cm^{-1} was the CH_2 deformation vibration and band at 1418 cm^{-1} was related to the C=O stretching vibrations of carboxylates and OH deformation vibration of alcohols and phenols. The bands at 1037—1088 cm^{-1} was due to the C—O—C and C—O vibration of polysaccharides, while the bands at < 1000 cm^{-1} due to the "fingerprint" zone of phosphate or sulfur functional groups.

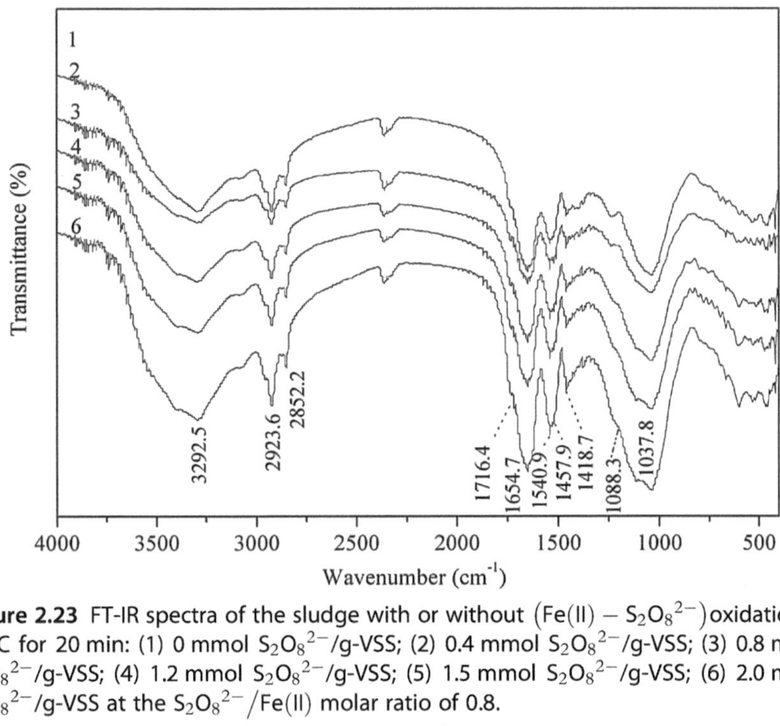

Figure 2.23 FT-IR spectra of the sludge with or without $(Fe(II) - S_2O_8^{2-})$ oxidation at 25°C for 20 min: (1) 0 mmol $S_2O_8^{2-}$/g-VSS; (2) 0.4 mmol $S_2O_8^{2-}$/g-VSS; (3) 0.8 mmol $S_2O_8^{2-}$/g-VSS; (4) 1.2 mmol $S_2O_8^{2-}$/g-VSS; (5) 1.5 mmol $S_2O_8^{2-}$/g-VSS; (6) 2.0 mmol $S_2O_8^{2-}$/g-VSS at the $S_2O_8^{2-}/Fe(II)$ molar ratio of 0.8.

EPS are one of the predominant components in sludge matrixes. The nature of functional groups present in the flocs may predominantly originate from the EPS compositions. The relative intensities of bands at 1654 and 1540 cm^{-1} in the six spectra gradually decreased following $\left(Fe(II) - S_2O_8{}^{2-}\right)$ oxidation, indicating the degradation and removal of protein-like substances in EPS as proved by the increased surface charge of flocs (Fig. 2.19). It was also supported by the analysis as shown for EEM spectra (Figs. 2.21 and 2.22), that is, the intensities of protein-like biopolymers decreased with the increase of $\left(Fe(II) - S_2O_8{}^{2-}\right)$ dosages.

2.4.6 Morphological (SEM-EDX) Analysis of Sludge Flocs

To get insight into the responsible dewatering mechanisms, SEM analysis of sludge with a different pretreatment was performed. The differences in sludge appearance were profoundly significant (Fig. 2.24). The surface of

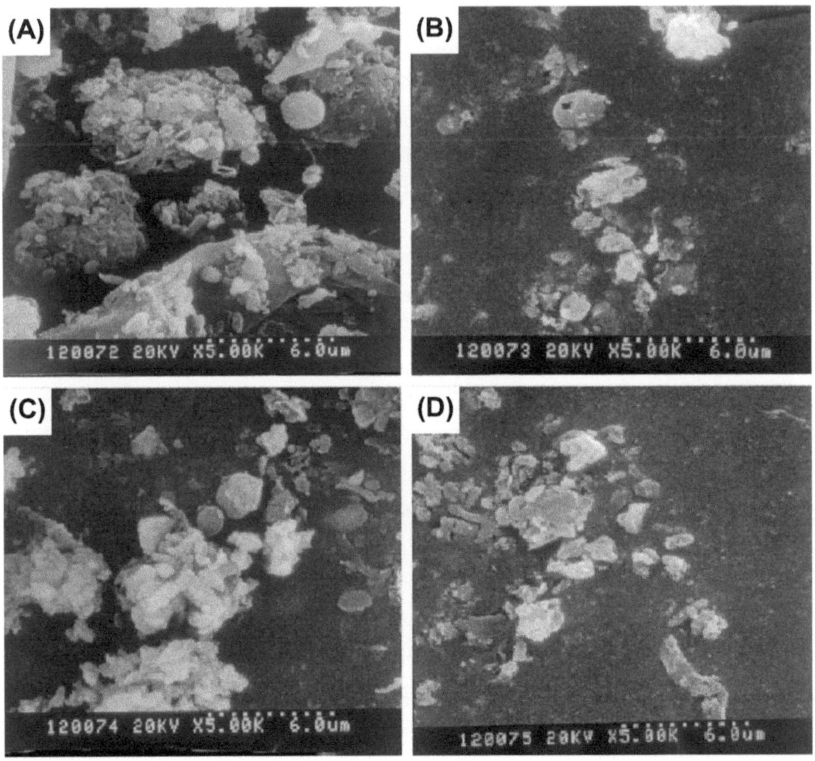

Figure 2.24 SEM of the sludge flocs under various operational conditions: (A) 25°C, (B) 25°C$\left(Fe(II) - S_2O_8{}^{2-}\right)$—$\left(Fe(II) - S_2O_8{}^{2-}\right)$ oxidation, (C) 80°C and (D) 80°C$\left(Fe(II) - S_2O_8{}^{2-}\right)$—$\left(Fe(II) - S_2O_8{}^{2-}\right)$ oxidation.

unpretreated sludge flocs was relatively round and smooth; while pretreated by $\left(Fe(II) - S_2O_8^{2-}\right)$ oxidation at 25°C, the flocs of waste-activated sludge appeared to be much less regular, forming a structure with many pores. When the temperature increased to 80°C, however, there were nearly no round sludge flocs to be detected, indicating the total collapse of the flocs. The results further corroborated that the combined thermal and $\left(Fe(II) - S_2O_8^{2-}\right)$ oxidation pretreatment resulted in the disruption of sludge flocs by degrading EPS, also EPS-bound water and interstitial water trapped between the flocs was released into the free water, which enhanced the dewaterability of waste-activated sludge.

2.5 COMBINATION OF ELECTROLYSIS AND FE(II)-ACTIVATED PERSULFATE OXIDATION FOR DEWATERABILITY

Electrolysis is considered an efficient tool not only because of its potential to remove interstitial water from sludge flocs but also because it has less adverse effects on the environment. However, the low corrosion resistance of the electrode and the high energy consumption of the process have been cited as problems. For these reasons, electrolysis is not widely used in the water industry. In the current research, the trend is to combine alternatives to enhance the dewatering ability of the conventional technique. The combined process does not effectively release the trapped water inside the EPS and cells and increase the dewatering rate.

In this section, the potential of combined electrolysis and $S_2O_8^{2-}/Fe(II)$ oxidation as a novel technique for improving sludge dewaterability is evaluated. Selected physicochemical properties, such as the sludge volume (SV), total suspended solids (TSS) and volatile suspended solids (VSS) and the contents of different EPS fractions (i.e., slime EPS, LB-EPS and TB-EPS) were investigated in order to ascertain the exact roles they play in sludge-dewatering.

2.5.1 Experimental Procedures

A series of bench-scale experiments were conducted in a 500-mL undivided single-compartment glass cell by using a pair of Ti/RuO_2 meshes as the anode and cathode. The electrodes were 7 cm × 10 cm in dimension and the distance between two electrodes was 4 cm in all experiments. The electrodes were connected to a regulated digital DC power supply (AD-8735, A&D Co., Japan; Fig. 2.25).

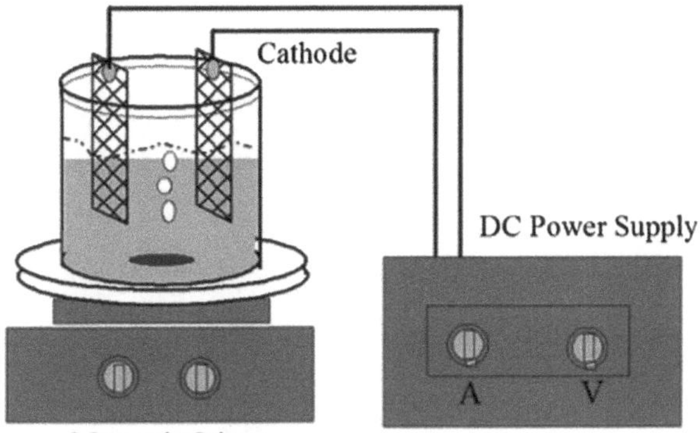

Figure 2.25 Schematic diagram of the electrolytic cell systems for sludge conditioning.

2.5.2 Optimal Electro-Dewatering Process

Fig. 2.26 illustrates the dewatering kinetics obtained for all sludge samples during the filtration phase. It shows: (I) the variation of dry-solids content as a function of time (Fig. 2.26A–D); (II) a column chart depicting the percentage of removed water over the filtration stage (Fig. 2.26E–H). It was evident from a perusal of the curve plots that the utilization of the voltage and $S_2O_8^{2-}/Fe(II)$ oxidation had a significant impact on the dewatering kinetics and the quantity of water removed.

Experiments carried out in the electro-dewatering stage showed that electrolysis alone led to a longer time of dewatering and lower final solids content. At the end of the filtration, the sludge treated by electrolysis gained a wetter filter-cake in comparison with the untreated sludge (0 V) except at 10 V, which produced the highest dry-solids content of 4.1 wt%. As presented in Fig. 2.26A, when the applied voltage increased from 0 to 15, 20, and 25 V, the final dry-solids contents progressively declined to around 2.3, 1.8, and 1.1 wt%, decreases of roughly 34.3%, 48.6%, and 68.6%, respectively, implying a remarkable deterioration in the sludge dewaterability. The adverse effect of higher voltages on the dewaterability might be attributable to a breakdown of the polymeric matrix because of EPS solubilization and cell lysis induced by ohmic heating. Besides the release of extracellular (i.e., EPS) and intracellular substances due to the polymeric matrix breakdown, which increased the viscosity of sludge and weakened the dewatering performance, the release of supracolloidal particles resulted in a clogged filter-cake and filter medium, which limited the filtrate flow.

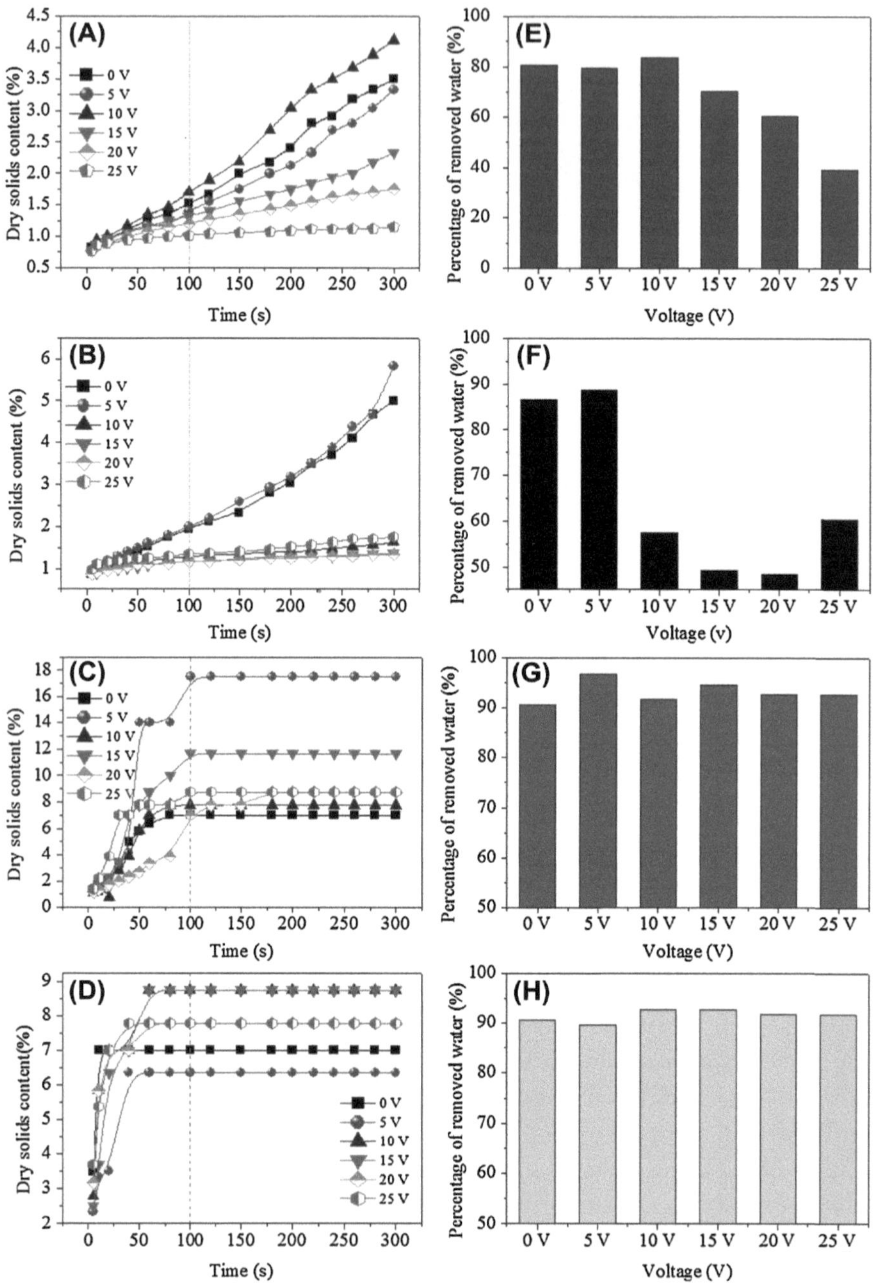

Figure 2.26 Left: dry-solids content versus filtration time for dewatering testing of pretreated sludge at an applied pressure of 1.5 kg/cm^2. Right: the percentage of removed water during filtration process: (A and E) electrolysis; (B and F) electrolysis and 0.2 mmol $S_2O_8^{2-}$/g-VSS + 0.25 mmol Fe^{2+}/g-VSS; (C and G) electrolysis and 0.4 mmol $S_2O_8^{2-}$/g-VSS + 0.5 mmol Fe^{2+}/g-VSS; and (D and H) electrolysis and 1.2 mmol $S_2O_8^{2-}$/g-VSS + 1.5 mmol Fe^{2+}/g-VSS.

The results of final dry-solids content of filter-cake during the combined electrolysis and $S_2O_8^{2-}/Fe(II)$ oxidation process are depicted in Fig. 2.26B—D. The sludge dewaterability was not always promoted during the combined pretreatment process. In the course of the electrolysis with low dosages of $S_2O_8^{2-}/Fe(II)$, better dewaterability was obtained when $S_2O_8^{2-}/Fe(II)$ oxidation was used alone or in combination at low voltage (5 V). Increases in the voltage resulted in a profound deterioration of dewaterability. The final dry-solids contents were found to be only 1.6, 1.4, 1.3, and 1.8 wt% for 10, 15, 20, and 25 V, about 54.3%, 60.0%, 62.9%, and 48.6% lower than that of the untreated sludge, respectively. The corresponding percentages of water removed from the inside of sludge flocs were around 57.4%, 49.3%, 48.3%, and 60.4%, respectively (Fig. 2.26F) while around 80.0% removal for the untreated sludge (Fig. 2.26E). This suggested that the combination of high voltage with low dosage of $S_2O_8^{2-}/Fe(II)$ tended to be detrimental rather than beneficial to the task of improving dewaterability.

In contrast, when $S_2O_8^{2-}$ and Fe(II) dosages were increased to 0.4 mmol/g-VSS and 0.5 mmol/g-VSS, respectively, the initial filtrate removal occurred relatively fast: more than 90% of the total water was removed within the first 100 s in all cases (Fig. 2.26G); afterward the dry-solids content stayed at a constant level for the rest of the filtration process, confirming the significant potential of the combined process for sludge-dewatering. Further data indicated that the overdosed $S_2O_8^{2-}/Fe(II)$ oxidation did not produce additional benefits to the sludge-dewatering (Fig. 2.26D and H). Hence, from the viewpoint of both effectiveness and economy, the optimal dewatering conditions in the present study were a voltage of 5 V, $S_2O_8^{2-}$ 0.4 mmol/g-VSS and Fe^{2+} 0.5 mmol/g-VSS, in which 96.7% or so of the total water was removed and the final dry-solids content of the filter-cake achieved more than 17.5 wt%.

From Fig. 2.26C, it can also be observed that the simultaneous combination of electrolysis (5 V) with $S_2O_8^{2-}/Fe(II)$ oxidation ($S_2O_8^{2-}$ 0.4 mmol/g-VSS and Fe^{2+} 0.5 mmol/g-VSS) resulted in the expulsion of an additional 6.0% water, which cannot be accessed by $S_2O_8^{2-}/Fe(II)$ oxidation alone. Based on the abovementioned results, it is therefore reasonable to note that the application of $S_2O_8^{2-}/Fe(II)$ oxidation to electrolysis during the sludge-dewatering indeed appears to be an effective alternative to reinforce the dewaterability and to increase the final dry-solids content. The pronounced improvement in sludge dewaterabilty could be explained by a synergetic effect of electro-osmosis and powerful $S_2O_8^{2-}/Fe(II)$ oxidation,

Table 2.8 Correlation Among the Sludge Physical Characteristics[a]: Pearson's Correlation Coefficients (*R*) (*p*-Values in Parenthesis). Values in Bold Refer to Variables for Which Correlation Is Significant

		DS	TSS	VSS	Slime EPS				LB-EPS				TB-EPS			
					PN	PS	Total EPS	PN/PS	PN	PS	Total EPS	PN/PS	PN	PS	Total EPS	PN/PS
SV		0.028 (0.895)	**0.797 (<0.001)**	**0.837 (<0.001)**	**-0.789 (<0.001)**	**-0.584 (0.003)**	**-0.774 (<0.001)**	-0.052 (0.809)	-0.282 (0.182)	-0.282 (0.182)	-0.285 (0.117)	0.166 (0.439)	0.351 (0.092)	0.430 (0.036)	0.378 (0.069)	-0.522 (0.009)
DS		-	0.028 (0.895)	-0.064 (0.767)	-0.106 (0.624)	0.288 (0.172)	0.110 (0.609)	**-0.456 (0.025)**	**-0.491 (0.015)**	**-0.403 (0.050)**	**-0.459 (0.024)**	-0.345 (0.098)	**-0.640 (<0.001)**	**-0.606 (0.002)**	**-0.631 (<0.001)**	-0.206 (0.334)
TSS		-	-	**0.923 (<0.001)**	**-0.780 (<0.001)**	**-0.485 (0.016)**	**-0.712 (<0.001)**	-0.031 (0.885)	**-0.452 (0.026)**	**-0.427 (0.037)**	**-0.447 (0.029)**	0.195 (0.361)	0.254 (0.231)	0.325 (0.121)	0.278 (0.189)	**-0.477 (0.018)**
VSS		-	-	-	**-0.738 (<0.001)**	**-0.705 (<0.001)**	**-0.817 (<0.001)**	0.281 (0.184)	-0.324 (0.123)	-0.367 (0.077)	-0.345 (0.098)	**0.434 (0.034)**	**0.493 (0.014)**	**0.540 (0.006)**	**0.510 (0.011)**	-0.313 (0.137)
Slime EPS	PN	-	-	-	-	**0.557 (0.005)**	**0.875 (<0.001)**	0.124 (0.564)	**0.549 (0.005)**	**0.512 (0.011)**	**0.539 (0.007)**	-0.157 (0.462)	-0.329 (0.116)	**-0.418 (0.042)-**	-0.359 (0.085)	**0.448 (0.028)**
	PS	-	-	-	-	-	**0.889 (<0.001)**	**-0.647 (<0.001)**	0.119 (0.581)	0.296 (0.160)	0.194 (0.364)	**-0.761 (<0.001)**	**-0.850 (<0.001)**	**-0.864 (<0.001)**	**-0.857 (<0.001)**	-0.090 (0.677)
	Total EPS	-	-	-	-	-	-	-0.308 (0.143)	0.372 (0.074)	**0.455 (0.026)**	**0.410 (0.047)**	**-0.530 (0.008)**	**-0.676 (<0.001)**	**-0.734 (<0.001)**	**-0.697 (<0.001)**	0.195 (0.362)
	PN/PS	-	-	-	-	-	-	-	0.100 (0.641)	-0.115 (0.592)	0.011 (0.959)	**0.844 (<0.001)**	**0.645 (<0.001)**	**0.571 (0.004)**	**0.623 (0.001)**	**0.430 (0.036)**
LB-EPS	PN	-	-	-	-	-	-	-	-	**0.959 (<0.001)**	**0.993 (<0.001)**	-0.199 (0.352)	0.272 (0.198)	0.243 (0.252)	0.263 (0.214)	0.270 (0.201)
	PS	-	-	-	-	-	-	-	-	-	**0.986 (<0.001)**	**-0.429 (0.037)**	0.108 (0.617)	0.088 (0.683)	0.102 (0.637)	0.164 (0.443)
	Total EPS	-	-	-	-	-	-	-	-	-	-	-0.297 (0.159)	0.206 (0.334)	0.181 (0.398)	0.198 (0.353)	0.229 (0.283)
	PN/PS	-	-	-	-	-	-	-	-	-	-	-	**0.659 (<0.001)**	**0.613 (0.001)**	**0.646 (<0.001))**	0.323 (0.123)
TB-EPS	PN	-	-	-	-	-	-	-	-	-	-	-	-	**0.988 (<0.001)**	**0.999 (<0.001)**	0.306 (0.145)
	PS	-	-	-	-	-	-	-	-	-	-	-	-	-	**0.994 (<0.001)**	0.191 (0.372)
	Total EPS	-	-	-	-	-	-	-	-	-	-	-	-	-	-	0.270 (0.202)
	PN/PS	-	-	-	-	-	-	-	-	-	-	-	-	-	-	-

[a]*DS*: Dry-solids content; *PN*: protein; *PS*: polysaccharide; Total *EPS*: *PN* + *PS*.

which is capable of removing water that cannot be accessed by dewatering by filtration alone. Electro-osmosis induced the substantial release of extracellular and intracellular substances that have a high affinity for water, most of which, however, could be degraded by powerful $SO_4^{-}\bullet$ generated from $S_2O_8^{2-}/Fe(II)$ oxidation. As a result, a large quantity of bound water trapped within the EPS and cells was released to the bulk liquid phase, which subsequently facilitated its removal. On the other hand, electro-osmosis may also have provided a driving force on the electrical double layer that forms adjacent to charged sludge particles, which may have caused the partial water in the double layer to move, and eventually increased the sludge dewaterability by decreasing its resistance to filtration.

2.5.3 Influence of Sludge Floc Properties on Dewaterability

Dewaterability is strongly affected by the physicochemical properties of the sludge. In this study, various influencing parameters including SV, TSS, VSS, and different EPS fractions were measured to identify their exact roles in the sludge-dewatering in order to develop a deep understanding of the principal mechanisms of the electro-dewatering process. To facilitate a statistical analysis, correlations among the different sludge properties have been verified by determining Pearson's correlation coefficients (R), which measure the strength of association between pairs of variables. The results of the statistical analysis in terms of Pearson's correlation coefficients (R) and p-values are reported in Table 2.8.

2.5.3.1 Settling Properties of Sludge After Pretreatment

Fig. 2.27 compares the settling behaviors of the sludge pretreated under the various electro-dewatering processes. Electrolysis appreciably modified the settling properties of sludge, and the settleability increased with the increasing applied voltages. As indicated in Fig. 2.27A, the best settleability for the sludge conditioned by electrolysis was obtained at 15 V, with an SV value below 40% after 30 min of settling, which is significant compared to the more than 94% for untreated sludge. In contrast, $S_2O_8^{2-}/Fe(II)$ oxidation alone did not exert a pronounced effect on the settling properties. Irrespective of applied $S_2O_8^{2-}/Fe(II)$ dosages, the conditioned sludge samples (Fig. 2.27B–D) revealed almost similar SV values (94%) as that of the untreated sample. By analogy with the results obtained by electrolysis or $S_2O_8^{2-}/Fe(II)$ oxidation alone, the utilization of electrolysis grouped with $S_2O_8^{2-}/Fe(II)$ oxidation tended to give rise to significantly better settling properties. For instance, at a fixed voltage of 25 V when $S_2O_8^{2-}$

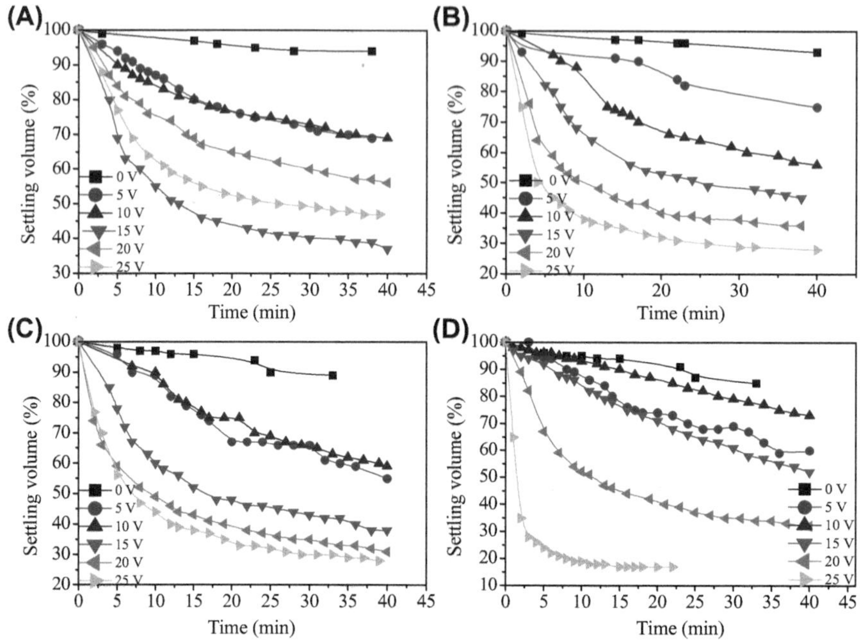

Figure 2.27 Settling curves of pretreated sludge: (A) electrolysis, (B) electrolysis and 0.2 mmol $S_2O_8^{2-}$/g-VSS + 0.25 mmol Fe^{2+}/g-VSS, (C) electrolysis and 0.4 mmol $S_2O_8^{2-}$/g-VSS + 0.5 mmol Fe^{2+}/g-VSS, and (D) electrolysis and 1.2 mmol $S_2O_8^{2-}$/g-VSS + 1.5 mmol Fe^{2+}/g-VSS.

and Fe(II) dosages were increased to 0.4 mmol/g-VSS and 0.5 mmol/g-VSS, respectively, the corresponding SV value decreased to around 30% after 30 min of settling, 18.4% lower than the sludge treated by electrolysis alone (Fig. 2.27A–C); and the further increase in $S_2O_8^{2-}$ and Fe(II) dosages to 1.2 mmol/g-VSS and 1.5 mmol/g-VSS, respectively, produced an additional 13% decrease in SV (Fig. 2.27D). Hence, it is reasonable to say that electrolysis integrated with $S_2O_8^{2-}$/Fe(II) oxidation is relatively more efficient and successful in promoting the compressibility and settleability of sludge.

Sludge settleability was strongly influenced by the composition and structure of the sludge surfaces. The statistical analysis of experimental data (Table 2.8) in this study indicated that the content of PN ($R = -0.789$, $p < .001$), PS ($R = -0.584$, $p = .003$) and total EPS ($R = -0.774$, $p < .001$) in slime EPS demonstrated a significant negative correlation with the SV, and the sludge samples with a higher slime EPS content had less difficulties with sedimentation. In contrast, no remarkable correlation existed between the SV and LB-EPS or TB-EPS content, revealing that slime EPS rather than LB-EPS or TB-EPS was crucial in controlling the settleability. This phenomenon can be partly explained by absorption and bridging mechanisms arising from biopolymers with a high molecular weight present in the slime EPS matrix. The biopolymers extended out from the sludge flocs' surface and could therefore act as a bridge to adsorb and link the smaller particles, which compressed the electrical double layers surrounding the flocs, reduced interparticle distances, and encouraged the formation of larger and compacter sludge flocs with better settling characteristics. In addition, the sedimentation was also strongly enhanced by iron ions from $S_2O_8{}^{2-}/Fe(II)$ oxidation since they were absorbed by vacant sites of the negatively charged sludge flocs, and provoked the agglomeration of the sludge flocs into stacks through charge neutralization and precipitation catching.

An interesting observation was that the correlation relating to the settleability and dewaterability (i.e., dry-solids content) ($R = 0.028$, $p = .895$) found was very poor. It should be noted that the absence of a correlation between the settleability and dewaterability should not be surprising since the dewaterability of sludge depends not only on the composition of solids and the content of slime EPS but also on the quantity and physical configuration of LB-EPS and TB-EPS, and the viscosity and particle size distribution, among others. The findings in this study are evidences that even when the settling characteristics of the sludge are good, filtration under pressure is not always easy.

2.5.3.2 Mass Composition of Various Pretreated Sludge

TSS and VSS changed substantially over the operating period in relation to the voltage and $S_2O_8{}^{2-}/Fe(II)$ dosages, as seen from Fig. 2.28. At low voltages (between 5 and 10 V), the removal of TSS and VSS was slightly low for all sludge samples conditioned by electrolysis. In contrast, higher voltages between 15 and 25 V contributed more to the removal of suspended solids, and the removal efficiencies under these conditions were at least 9.7% for TSS and 11.6% for VSS, respectively. This finding confirmed that the mineralization and evaporation effects occurred in this process. Furthermore, when in the presence of $S_2O_8{}^{2-}/Fe(II)$ oxidation, the effects tended to be significantly intensified (Fig. 2.28B–D). For example, at the applied voltage of 20 V, $S_2O_8{}^{2-}$ 0.4 mmol/g-VSS and Fe^{2+} 0.5 mmol/g-VSS, higher than 33.8% and 37.3% removal efficiencies of TSS and VSS were obtained, respectively. Based on the experimental results, it is thus apparent

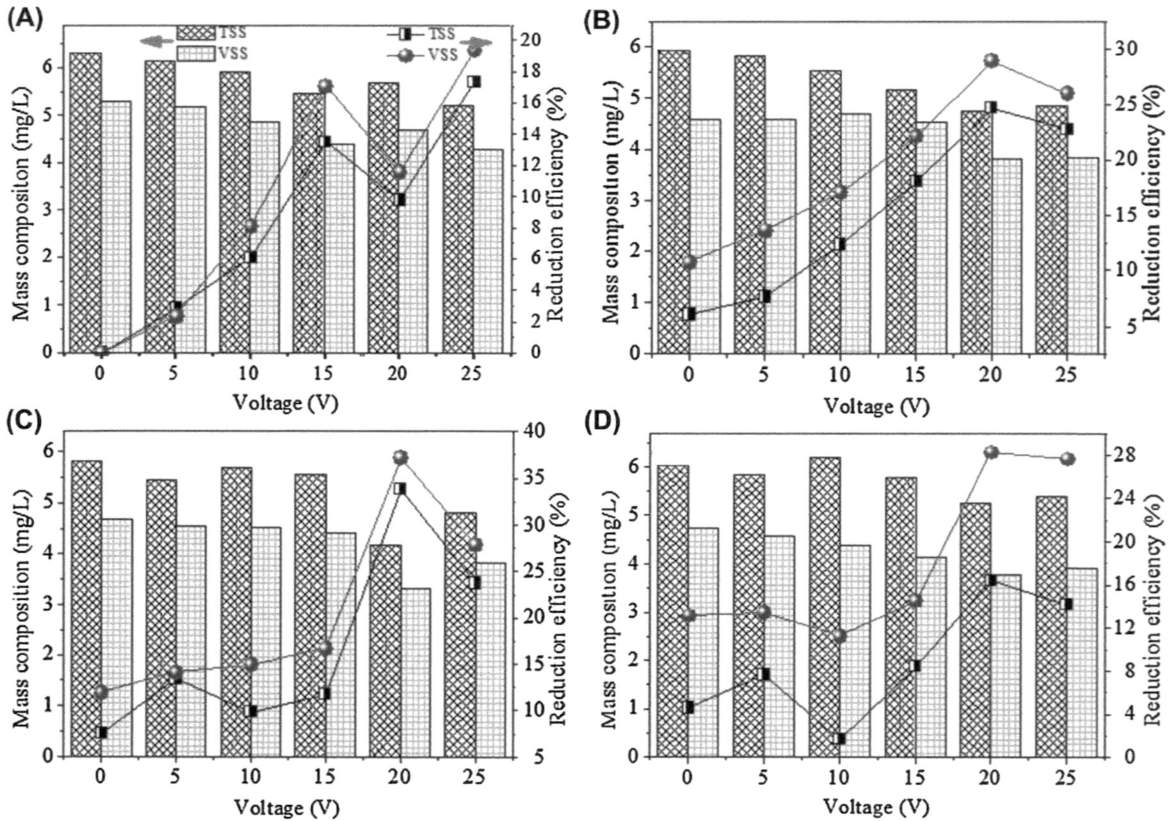

Figure 2.28 Variations of mass composition of the sludge during various electro-dewatering processes: (A) electrolysis, (B) electrolysis and 0.2 mmol $S_2O_8^{2-}$/g-VSS + 0.25 mmol Fe^{2+}/g-VSS, (C) electrolysis and 0.4 mmol $S_2O_8^{2-}$/g-VSS + 0.5 mmol Fe^{2+}/g-VSS, and (D) electrolysis and 1.2 mmol $S_2O_8^{2-}$/g-VSS + 1.5 mmol Fe^{2+}/g-VSS.

that $S_2O_8^{2-}/Fe(II)$ oxidation integrated with electrolysis disintegrated the sludge flocs and favored the mass transfer from the solids to liquid phase, which was considered to be particularly important to reduce the amount of sludge and the ultimate disposal costs. The clear positive correlation between VSS and total EPS content (i.e., PN + PS) in TB-EPS fraction ($R = 0.510$, $p = .011$; Table 2.8) further revealed that the removal of suspended solids during the combined process was achieved mainly through solubilizing TB-EPS.

Both TSS and VSS exhibited a relatively significant positive correlation with the SV ($R = 0.797$, $p < .001$; and $R = 0.837$, $p < .001$; respectively), indicating that the sludge with low values of solids concentration had better settling characteristics (i.e., low SV value). There was, however, no clear correlation between either TSS or VSS and dewaterability ($R = 0.028$, $p = .895$; and $R = -0.064$, $p < .767$; respectively), verifying that sludge dewaterability was independent of the solids' concentration.

2.5.3.3 Role of Different EPS Fractions on Sludge Dewaterability

Fig. 2.29A illustrates the effect of electrolysis on the different EPS fractions. The EPS content in TB-EPS fraction was reduced profoundly after the voltage was increased, whereas it was enhanced in LB-EPS, and in slime EPS in particular. As shown in the figure, the total EPS were originally estimated to be 23.9, 21.1 and 245.3 mg/L in slime EPS, LB-EPS, as well as the TB-EPS fractions for the untreated sludge, respectively. When the applied voltage was elevated to 20 V, the total EPS in TB-EPS fraction declined to 193.6 g/L, decreasing by around 21.1%; in contrast, they increased to 101.3 and 30.6 mg/L in slime EPS and LB-EPS fractions, increasing by roughly 323.8% and 45.0%, respectively. Moreover, a further increase of the voltage to 25 V produced another 11.9% reduction in TB-EPS, simultaneously another 195.4% and 15.7% increase in slime EPS and LB-EPS fractions, respectively. This can presumably be attributed to the fact that, when the sludge flocs were disrupted with electrolysis, a part of TB-EPS were solubilized and released into the liquid phase, thereby increasing the slime EPS and LB-EPS levels. The moderate opposite correlation between total EPS contents in slime EPS fraction and TB-EPS fraction ($R = -0.697$, $p < .001$) also partly verifies this hypothesis.

In comparison with electrolysis, the electrolysis coupled with $S_2O_8^{2-}/Fe(II)$ oxidation tended to have a more significant effect on the variations of EPS. For instance, at voltage 10 V, $S_2O_8^{2-}$ 0.2 mmol/g-VSS and Fe^{2+} 0.25 mmol/g-VSS (Fig. 2.29B), the total EPS in TB-EPS fraction

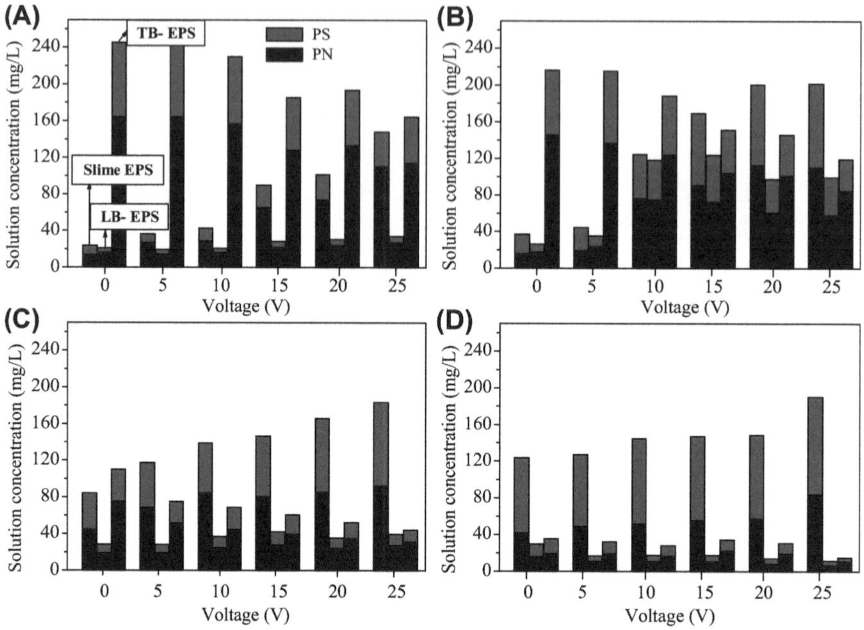

Figure 2.29 Variations of PN and PS concentrations in different EPS fractions: (A) electrolysis, (B) electrolysis and 0.2 mmol $S_2O_8^{2-}$/g-VSS + 0.25 mmol Fe^{2+}/g-VSS, (C) electrolysis and 0.4 mmol $S_2O_8^{2-}$/g-VSS + 0.5 mmol Fe^{2+}/g-VSS, and (D) electrolysis and 1.2 mmol $S_2O_8^{2-}$/g-VSS + 1.5 mmol Fe^{2+}/g-VSS.

dropped to 188.2 mg/L, drops of 18.7% compared with electrolysis alone, while it rapidly increased to 124.6 and 118.3 mg/L in slime EPS and LB-EPS, increases of 192.5% and 466.0%, respectively. The relatively significant increase in slime EPS and LB-EPS implies that the use of a combined process may decompose TB-EPS, and furthermore effectively disrupt the bacterial cells trapped in the sludge flocs, which contributed greatly to the release of intracellular substances and thus more pronounced increase in both slime EPS and LB-EPS. However, a high LB-EPS content is, in general, detrimental to the sludge-dewatering due to its stronger affinity for water through hydrophilic interactions and hydrogen bond forces. As a result, a large quantity of the released water could thus be reabsorbed by the increased EPS again, which influenced the sludge dewaterability, and as a consequence, resulted in an essential reduction of the dewatering rate (Fig. 2.26B). Nevertheless, the adverse impact on the dewatering could be basically controlled when electrolysis combined with a higher dosage of $S_2O_8^{2-}/Fe^{2+}$ oxidation was applied, as mentioned before (Fig. 2.26C and D). Both LB-EPS and TB-EPS displayed a relatively low EPS content, usually lower than 80.0 and 42.0 mg/L respectively in the presence of $S_2O_8^{2-}$ 0.4 mmol/g-VSS and Fe^{2+} 0.5 mmol/g-VSS (Fig. 2.29C); moreover, a further increase in $S_2O_8^{2-}$ and Fe^{2+} dosages could produce further reduction in EPS (Fig. 2.29D). These observations suggest that the combined process with higher dosages of $S_2O_8^{2-}/Fe^{2+}$ oxidation not only brings about the complete collapse of TB-EPS and the lysis of bacterial cells but also degrades most of the released TB-EPS and intracellular materials.

Different EPS fractions have different effects in the sludge-dewatering, depending on the EPS components and the types of EPS. The relationship between the dewaterability and different EPS fractions was thus established in an attempt to better understand their precise roles in the sludge-dewatering (Table 2.8). It was found that the dewaterability corresponded negatively to the individual polymers PN, PS and total EPS from LB-EPS ($R = -0.491$, $p = .015$; $R = -0.403$, $p = .050$; and $R = -0.459$, $p = .024$; respectively) and TB-EPS fractions ($R = -0.640$, $p < .001$; $R = -0.606$, $p = .002$; and $R = -0.631$, $p < .001$, respectively), while no recognizable correlation between slime EPS and dewaterability was found. This implies that sludge dewaterability is strongly governed by LB-EPS and TB-EPS rather than slime EPS. Slime EPS, which are evenly distributed in the liquid phase, are more easily removed together with free water during a filtration process and therefore its influence is definitely insignificant. In contrast, LB-EPS extending from TB-EPS have a highly porous and dispersible structure and can carry a large amount of bound water. A higher LB-EPS content thus brings more bound water into the sludge flocs and results in a worse performance in sludge-dewatering. TB-EPS, in comparison with LB-EPS, are often regarded as a less crucial factor among the influences on dewatering since they are located in the inner fraction of the flocs. In fact, the dewatering efficiency of sludge depends on not only LB-EPS but also the TB-EPS in the sludge flocs. This is because the TB-EPS covering the surface of the cell provides a protective barrier for bacteria inside the sludge flocs. The presence of bacterial cells, binding a large quantity of water of hydration, can result in high water content, and that sludge with a large number of cells will be consequently more difficult to dewater. Therefore, the destruction of TB-EPS observed in this study could induce the total destruction of sludge flocs, promote the release of EPS-bound water, and also make more bacterial cells available to harsh attack, which are conducive to the break-up of cell walls and the release of bound water trapped in cells into free water. This appears to be the core reason for the increased solid–liquid separation rate during electrolysis integrated with

$S_2O_8^{2-}/Fe^{2+}$ oxidation process. In addition, no apparent correlation was observed between PN/PS and sludge dewaterability (Table 2.8), further showing that the quantities of EPS are more important than the proportions of EPS components in controlling the dewatering.

2.5.4 UV—Vis Spectroscopic Properties of EPS

UV—vis spectroscopy has so far been mainly focused on the characterization of natural organic matter. Until now, little information on the UV—vis response of EPS derived from sludge is available in literature. Therefore, the first scientific attempt to use UV—vis spectroscopy for characterizing the EPS was done herein, in order to offer new insight into the contribution of EPS to sludge dewaterability. Typical UV—vis spectra of EPS extracted from the sludge flocs treated under various processes are depicted in Fig. 2.30. The two main absorbance bands at around 212—223 and 256—260 nm were identified in the ultraviolet region of 200—350 nm. Afterward, the UV—vis spectra of EPS exhibited a monotonically decreased absorbance as the wavelength increased at a low absorption of above 400 nm (visible region). The first absorbance band located at $\lambda_{abc} = 212$—223 nm can probably be assigned to the presence of carboxylic bonds (COOH). The second present at $\lambda_{abc} = 256$—260 nm may well belong to the $n \rightarrow \pi^\star$ and $\pi \rightarrow \pi^\star$ electron transition of C=O bonds and C=C double bonds (or chromophores).

Generally, the variation of UV—vis spectrum shape before and after pretreatment is a reflection of the transformation of organic components, configuration, and heterogeneity. It is apparent from Fig. 2.30, the bands in the TB-EPS fractions, which decreased and narrowed in the course of voltage and $S_2O_8^{2-}/Fe^{2+}$ dosage, thus highlight the decomposition and release of TB-EPS, as well as the modification of the functional groups during the combined process. In the case of LB-EPS, the absorbance band at $\lambda_{abc} = 256$—260 nm showed high relative intensities at low $S_2O_8^{2-}/Fe^{2+}$ dosages, whereas they nearly disappeared in the presence of $S_2O_8^{2-}$ 0.4 mmol/g-VSS and Fe^{2+} 0.5 mmol/g-VSS, which is consistent with the EPS results (Fig. 2.29). Increased dewaterability (Fig. 2.26C and D) accompanied by the reduced LB-EPS and TB-EPS indicated by absorbance bands at 212—223 and 256—260 nm was observed, revealing that the destruction of LB-EPS and TB-EPS was considerably favorable for the enhancement of sludge dewaterability, which is in good agreement with the results of the statistical analysis (Table 2.8). In contrast, the UV—vis spectra of slime EPS appeared to be more irregular and featureless, particularly when a high

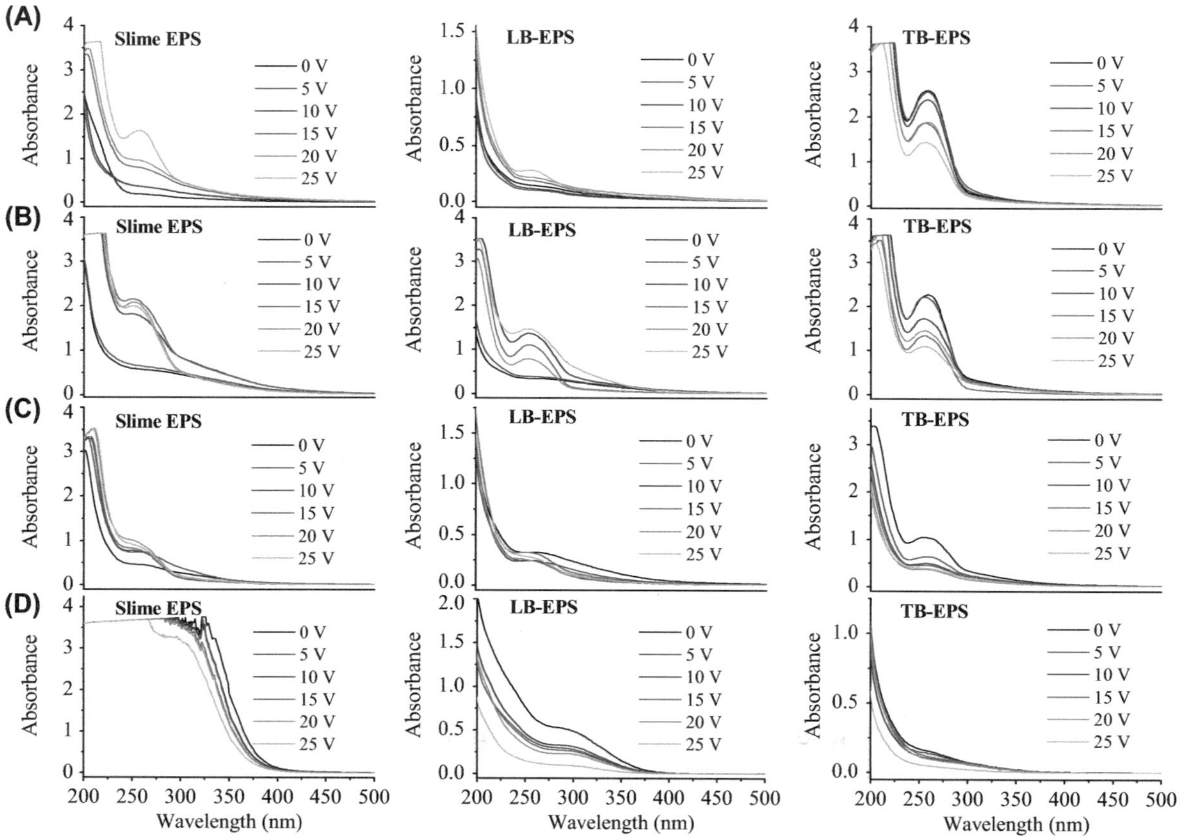

Figure 2.30 UV–vis absorbance spectra of different EPS fractions: (A) electrolysis, (B) electrolysis and 0.2 mmol $S_2O_8^{2-}$/g-VSS + 0.25 mmol Fe^{2+}/g-VSS, (C) electrolysis and 0.4 mmol $S_2O_8^{2-}$/g-VSS + 0.5 mmol Fe^{2+}/g-VSS, and (D) electrolysis and 1.2 mmol $S_2O_8^{2-}$/g-VSS + 1.5 mmol Fe^{2+}/g-VSS.

dosage of $S_2O_8^{2-}/Fe^{2+}$ was applied, showing no maxima or minima, which can possibly be linked to the interference of intermediates formed in the strong oxidation environment. These intermediates were distributed uniformly throughout the liquids, and were more easily extracted together with slime EPS, thus having a considerable impact on the UV−vis spectra of slime EPS. Nevertheless, this did not influence the application of the technique for determining the sludge dewaterability because of the negligible effect of slime EPS on the dewatering, as previously noted. With its high sensitivity and ability to rapidly determine LB-EPS and TB-EPS, and no need for chemical reagents, UV−vis spectroscopy is a useful and simple tool for characterizing sludge dewaterability.

2.5.5 Morphological (SEM-EDX) Analysis of Sludge Flocs

To more comprehensively understand the possible dewatering mechanisms, the morphology of sludge flocs after different pretreatments was characterized through SEM-EDX. As seen in Fig. 2.31, the rod-shaped bacterial cells associated with and entrapped inside the sludge matrix treated with the voltage of 5 V were identified. The flocs had a chaotic macro-porous structure. Their irregular morphology may well be related to the presence of EPS in the form of protein, polysaccharide, and lipids, etc., which was also demonstrated by the energy-dispersive X-ray (EDX) spectra analysis. The EDX results confirmed the presence of carbon and oxygen, which belonged to the EPS family, as described above, by the UV−vis spectra analysis. The close association between EPS and bacterial cells not only acts as an immobilization support but also provides some protective shielding, allowing cells to survive during the electrolysis process. When the voltage increased to 25 V, the morphology of sludge became relatively clear and regular, and large numbers of cells were exposed due to electrolysis impact, but still remained intact. This observation clarified our finding that electrolysis alone (5−25 V) could disrupt the partial TB-EPS surrounding the cells, but was still not able to release and crack the cell walls, which might be the main reason for its limited capability of improving sludge dewaterability.

Comparatively, in the presence of $S_2O_8^{2-}/Fe^{2+}$ oxidation ($S_2O_8^{2-}$ 0.4 mmol/g-VSS, Fe^{2+} 0.5 mmol/g-VSS), no sign of cells attaching to or entrapped by TB-EPS was detected after pretreatment, indicating that the combined process had already damaged the protective barrier (e.g., TB-EPS) of the cells, and finally broken the cells completely apart. Based on Fig. 2.31, it is clear that the enhanced dewaterability of sludge during the electrolysis integrated with $S_2O_8^{2-}/Fe^{2+}$ oxidation occurs in two steps

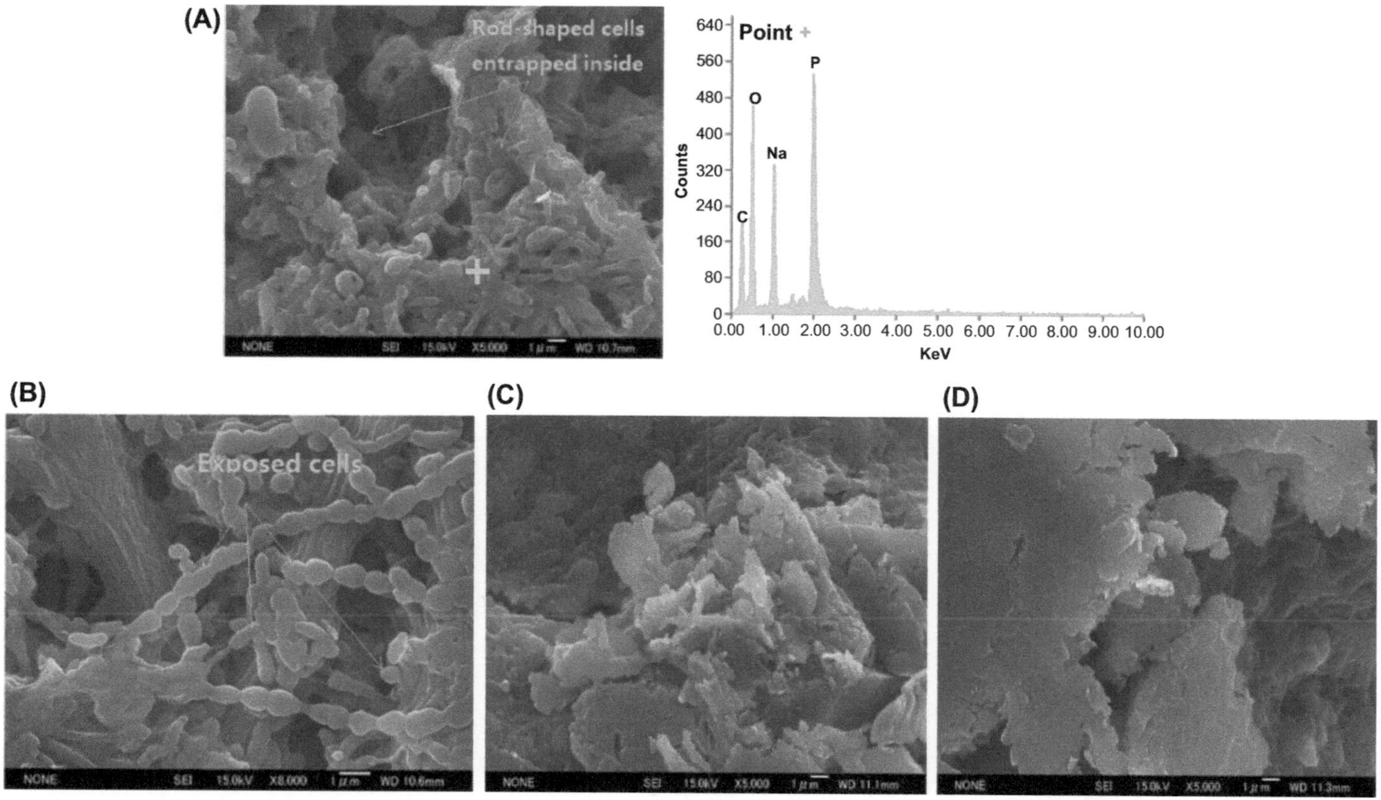

Figure 2.31 SEM-EDX spectra of pretreated sludge flocs: (A) electrolysis (5 V), (B) electrolysis (25 V), (C) electrolysis (5 V) and 0.4 mmol $S_2O_8^{2-}$/g-VSS + 0.5 mmol Fe^{2+}/g-VSS, and (D) electrolysis (25 V) and 0.4 mmol $S_2O_8^{2-}$/g-VSS + 0.5 mmol Fe^{2+}/g-VSS.

(Fig. 2.32): (1) the collapse of TB-EPS around the entrapped bacterial cells; and (2) the destruction of the exposed cells, resulting in the release of EPS-bound water and water of hydration inside cells into free water, subsequently improving sludge dewaterability considerably.

2.6 HYDROTHERMAL PRETREATMENT OF DEWATERED SLUDGE FOR DEWATERABILITY

Thermal treatment results in breakdown of the gel structure of sludge and the release of intracellular bound water. The main parameter for thermal treatment is temperature. This section will investigate impact and mechanism of hydrothermal pretreatment on two different sludge samples. Five temperatures were used: 60, 80, 120, 150, and 180°C.

2.6.1 Test Materials and Experimental Procedures

Dewatered sludge (DS) samples taken from two different WWTPs were used in this study (Table 2.9). A schematic diagram of the experimental setup is shown in Fig. 2.33. The high-temperature high-pressure reactor (GSH-1, Weihai, China) was made of stainless steel.

The pressure-dewatering device for evaluating dewaterability of samples was composed of a molding cylinder and a hydraulic press (Fig. 2.34). The wet solid-product sample from the reactor was sandwiched between two filter cloths and placed in the molding cylinder, and the solid product (SP) was then compressed with a hydraulic press. In this study, the pressure on SP was

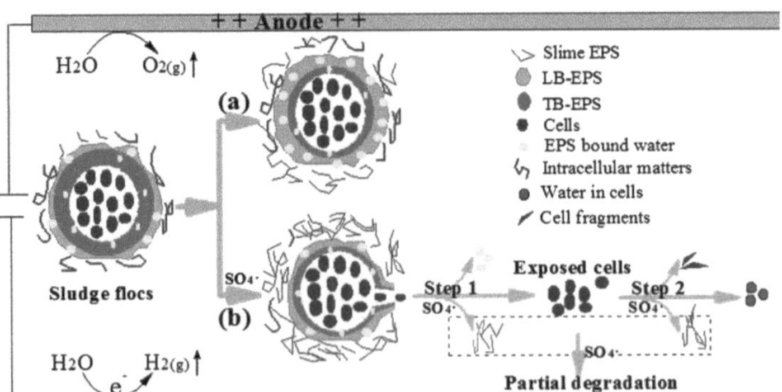

Figure 2.32 The possible dewatering mechanisms of sludge flocs: (A) electrolysis or integrated with low dosage of $S_2O_8^{2-}/Fe^{2+}$ oxidation (Fig. 2.26A and B), and (B) electrolysis integrated with high dosage of $S_2O_8^{2-}/Fe^{2+}$ oxidation (Fig. 2.26C and D).

Table 2.9 Dewatered Sludge (DS) Properties

	pH	VS (%/Dry Basis)	Water Content (%)
1#	7.4	33.2	82.2
2#	7.6	45.0	81.4

constant at 3.5 MPa. Semisolid product (SSP) represents the SP after hydrothermal pretreatment. SP represents the dewatered SP derived from the pressure-dewatering device. Separated liquid (SL) represents the liquid separated from SSP at the pressure-dewatering device. Separated liquid represents the liquid directly separated from no heating dewatered sewage sludge cake at the pressure-dewatering device.

2.6.2 Sludge Differential Thermogravimetry (DTG) Analysis

In order to make sure the temperature range of hydrothermal treatment, sludge thermogravimetry analysis was conducted (Fig. 2.35). A slight weight loss within the temperature range from 50 to 130°C is mainly caused by the dehydration reactions. In the range of 150 and 650°C, differential thermogravimetry peak is mainly caused by thermal decomposition and loss of volatiles. So, it can be seen that the sludge starts to decompose at about 150°C. In order to achieve hydrothermal pretreatment effect and avoid decomposition of organic matters, the upper limit temperature of hydrothermal was set as 180°C.

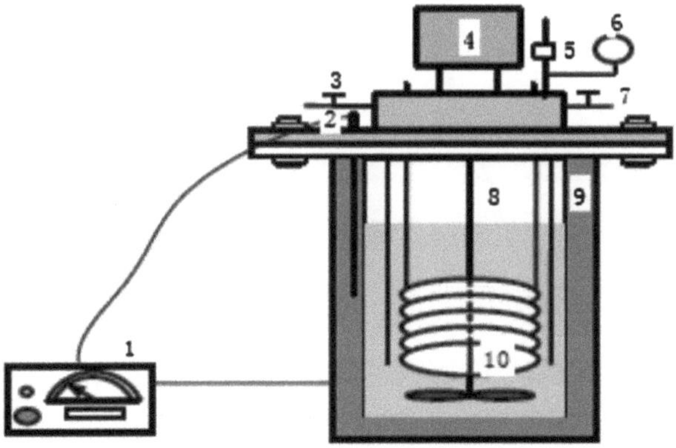

Figure 2.33 Schematic diagram of the experimental setup (1: controller; 2: thermoelectric couple; 3: sampling point; 4: electromagnetic dasher; 5: safety valve; 6: pressure meter; 7: blowhole; 8: stirrer; 9: electric jacket; and 10: condensing coil).

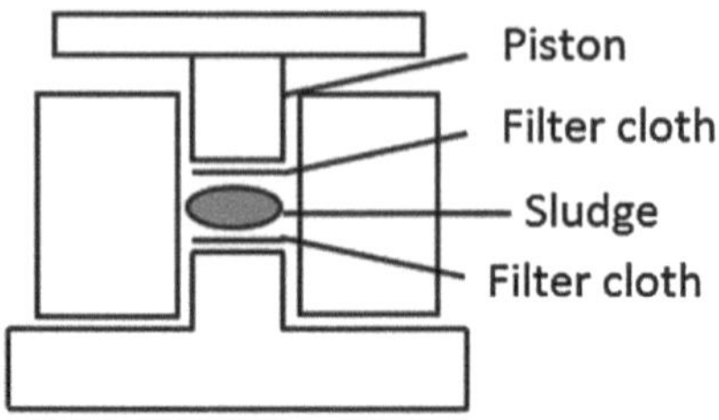

Figure 2.34 Molding cylinder of the pressure-dewatering device.

2.6.3 Effect of Hydrothermal Treatment Temperature and Holding Time on Dewaterability

When the reactor is heated to a predetermined temperature, then cut off the electric source. Effects of hydrothermal temperature on dewaterability of DS were listed in Fig. 2.36. The results indicated that the higher the hydrothermal temperature, the better dewaterability character was. For sludge sample 1#, the water content after high constant compression using pressure-dewatering device was 71.3%. The water contents of one# SP sample after 60, 80, 120, 150, and 180°C were 55.9%, 48.9%, 41.7%, 38.1%, and 26.7% respectively. For sludge sample2#, the water content after high constant compression using pressure-dewatering device was 73.3%. The water contents of 2# SP sample after 60, 80, 120, 150, and 180°C were 60.2%, 56.2%, 50.7%, 46.3%, and 42.7%, respectively. Sludge-dewaterability character of high VS content was better than that of low VS content. VS content was one of the important factors affecting the sludge dewaterability.

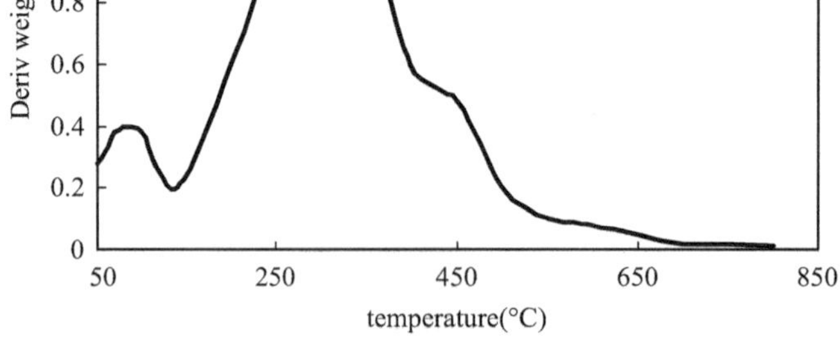

Figure 2.35 DTG curve of dewatered sludge.

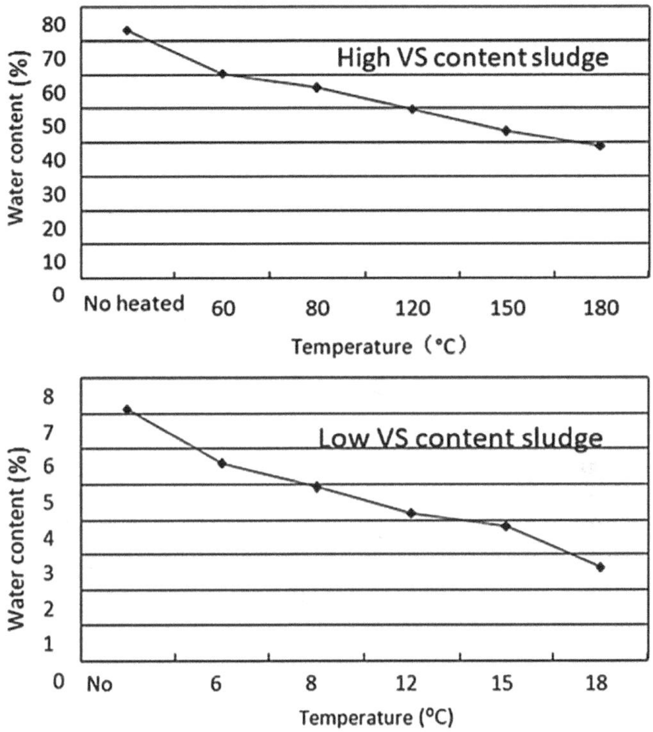

Figure 2.36 Impact of hydrothermal temperature on sludge dewaterability ($n = 3$, $p = .05$).

The holding time at predetermined temperature also has an important effect on the sludge dewaterability (Fig. 2.37). Taking sample 2# for example, the SP water contents were positively correlated with the hydrothermal holding time at predetermined temperatures in this study. Here, it is difficult to give the optimum hydrothermal treatment conditions, as this depends on the water content required by subsequent sludge disposal methods.

2.6.4 Mechanism of Hydrothermal Treatment for Dewaterability

2.6.4.1 Variation of Separated Liquid pH

Separated liquid (SL) pHs are listed in Table 2.10. The pH decreased as the temperature rose. This indicated that hydrothermal pretreatment destroyed zooglea, nucleic acid, humic acid, amino acid, etc., and released to the liquid phase. At the same time, carbon dioxide produced from decomposed organic matters also lead to pH decrease. Destruction of zooglea and decomposition of organic matters improved the sludge dewaterability.

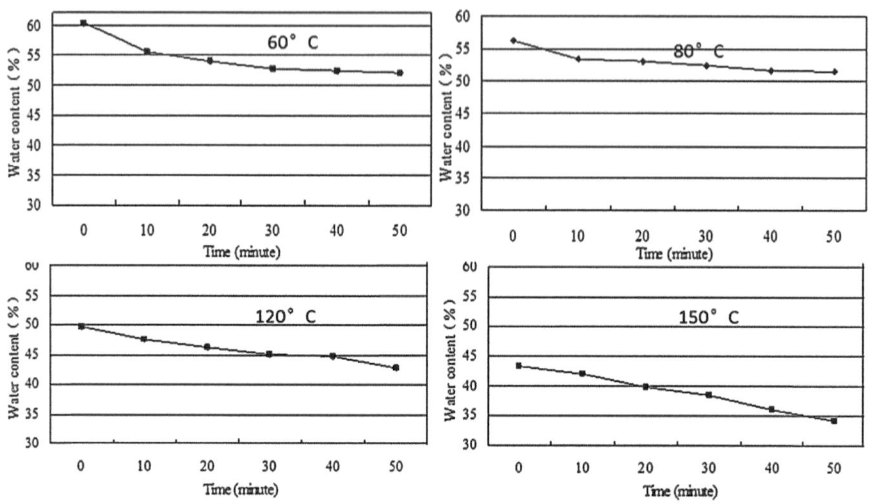

Figure 2.37 Impact of hydrothermal time on sludge dewaterability (n = 3, $p = .05$).

2.6.4.2 Variation of Extracellular Polymeric Substances (EPS) in SSP.

Proteins and polysaccharides are the main components of EPS, so sum of proteins and polysaccharides content represent EPS content in this study. Variations in proteins and polysaccharides content of SSP are listed in Fig. 2.38. Hydrothermal treatment process made organic matters in the sludge solid phase to dissolve and hydrolyze into liquid phase gradually. Proteins and polysaccharides content increased with the temperature rising. EPS content increased faster before 80°C than after 80°C—especially protein content showed decreased tendency after 80°C. This phenomenon indicated that some dissolved proteins and polysaccharides decomposed further into small molecule matters. EPS content of DS with high VS content were higher than that with low VS content. Proteins and polysaccharides are both hydrophilic materials. The existence of proteins and polysaccharides lead to an increase of sludge bound water. The destruction of cell membrane and thermal denaturalization of proteins improved sludge dewaterability. Statistical analysis indicated that sludge

Table 2.10 Impact of Hydrothermal Temperatures on pH (n = 3, $p = .05$) From Hydrothermal Pretreatment Sludge

pH	Separated Liquid	60°C	80°C	120°C	150°C	180°C
1#	7.4	6.55	6.34	6.36	5.79	5.92
2#	7.6	6.82	6.41	6.32	6.17	5.88

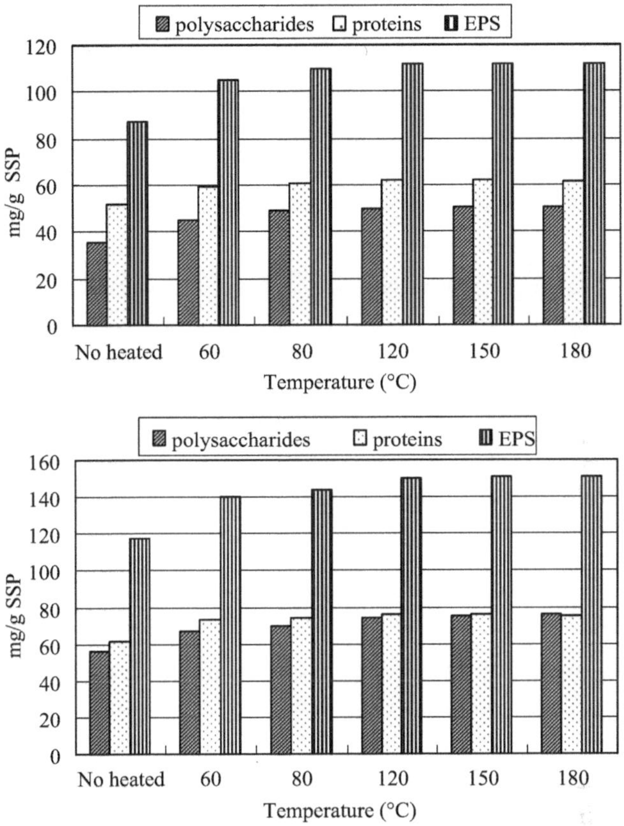

Figure 2.38 Impact of hydrothermal temperature on EPS of SSP (n = 3, $p = .05$) (up: low VS-content sludge; down: high VS-content sludge).

dewaterability-experienced hydrothermal processes were negatively correlated with EPS content. EPS and floc structure were the two main factors affecting sludge dewaterability.

2.6.4.3 Impact of Hydrothermal Pretreatment on Appearance of Sludge

Effect of hydrothermal pretreatment on morphology of sludge was investigated by SEM analysis (Fig. 2.39). It can be seen that raw DS has big flocculate and nonporous in appearance. After hydrothermal process, flocculate sludge disaggregated to small particles generally. The appearance was porous and the particle size decreased, which could be one of the important reasons for enhanced sludge dewaterability.

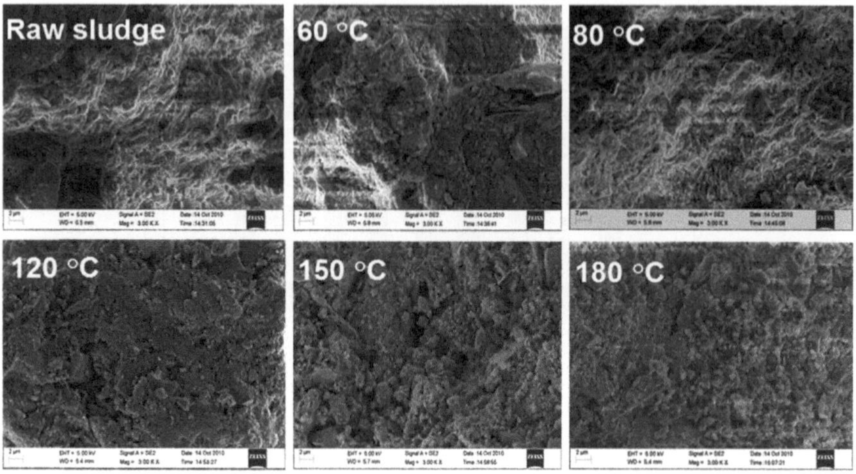

Figure 2.39 Impact of hydrothermal pretreatment on morphology of sludge (magnification: × 3000 times).

2.6.5 Resource Reuse of Separated Liquid and Solid Product

Fig. 2.40 lists the variations of SL COD from the sample 1# and 2#. With the increase of hydrothermal temperature, COD value of SL increased generally. For sample 1#, COD value increased from 11,964 in the SL without heating to 26,716 mg/L in 180°C hydrothermal sludge SL; correspondingly, for sample 2#, COD value increased from 13,215 to 29,716 mg/L. Some organic materials such as proteins, carbohydrate in solid phase transferred to liquid phase after hydrothermal pretreatment, which lead to increase of SL COD and decrease of C, H, N, and caloric value of SP 2 # (Table 2.11). Separated liquid can produce methane through anaerobic digestion for energy recovery or discharged back into the water treatment plant. SP can be used for incineration or landfill disposal.

2.7 FILTRATION IMPROVEMENT OF ZINC SLUDGE BY USING UNCONVENTIONAL ALKALIZATION SEQUENCE

Zinc is one of the heavy metals and should be removed from the wastes before these wastes can be disposed of safely. Conventionally, the zinc in solid wastes may be leached out by mineral acids, and then precipitated by the addition of hydroxide solution to the leach solutions and wastewaters. The sludge obtained is colloidal and very difficult to be filtered. Currently, the zinc sludge is usually dewatered by frame-plate press. The

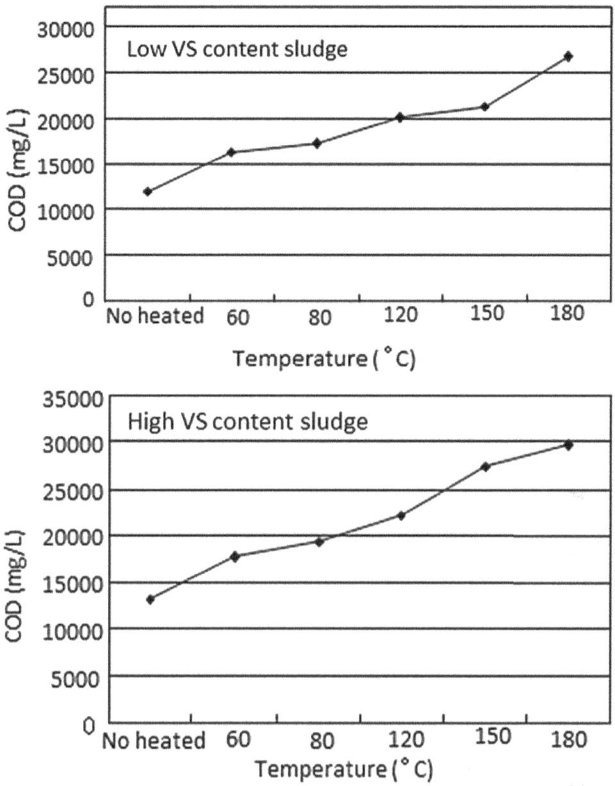

Figure 2.40 Separated liquid (SL) COD (n = 3, $p = .05$) subjected to hydrothermal treatment for low- and high VS sludges.

resultant cake using this process can be hardly rinsed, which is quite unfavorable to the recovery of relatively pure zinc hydroxide.

In order to improve the filtration efficiency of colloidal sludge including zinc, chromium, and other metals ions, filter aids such as diatomaceous silica and perlite may be mixed with the sludge to provide a great number of

Table 2.11 Impact of Hydrothermal Treatment on Element Content and Caloric Value of Solid Product Sample 2# (n = 3, $p = .05$)

Temperature (°C)	C (%)	H (%)	N (%)	Caloric Value (kJ/kg)
No heated	17.08	2.72	2.82	7.484×10^3
60	16.77	2.43	2.76	7.356×10^3
80	16.42	2.35	2.65	7.287×10^3
120	16.18	2.05	2.49	7.153×10^3
150	14.66	1.23	2.08	5.758×10^3
180	13.95	1.02	1.99	4.965×10^3

microscopic holes for the liquid to flow through. When the metals in the cakes could be redissolved into the aqueous solution, the recycling of the filter aids may be possible. Nevertheless, in the most cases, the cakes may be discarded or treated by other processes other than redissolution with acidic solution, which may result in uneconomic consequence and much less practical significance.

Obviously, the most attractive ways to improve the filtration efficiency of zinc hydroxide sludge may lie in the improvement of the filterability of the sludge without help of any additional filter aids. Taking advantage of the facts that zinc hydroxide can be dissolved in both acidic and alkaline solutions and the precipitates formed homogeneously are usually muddy and readily to filter, the zinc-containing solutions are added into excessive sodium hydroxide solution in this section. Consequently, it was found that the filterability of zinc sludge could be improved to an extent, depending on the molar ratios of OH^-/Zn used.

2.7.1 Parameters Affecting the Precipitation Efficiency of Zinc

The results are shown in Fig. 2.41. For the comparison of the effect of alkalization sequences on the precipitation of zinc, 20 mL of 18.9 g/L zinc solution was mixed with 10 mL of different concentrations of NaOH solution, and the results are shown in Fig. 2.41A. It can be seen that, for both two alkalization methods, zinc will be completely precipitated out from the solution, when the apparent concentration of NaOH (the total added NaOH divided by the final total volume) was lower than 1.35 mol/L (for alkalization method I, simplified as "zinc to NaOH method") or 1.15 mol/L (for alkalization method II, simplified as "NaOH to zinc method"), that is, the corresponding molar ratios of NaOH/Zn were in the ranges of 2—4 or 2—3, respectively. When the apparent concentration of NaOH increased further, part of the precipitates formed was dissolved and the precipitation efficiencies of zinc decreased. Such decrease was more rapid for the case of addition of NaOH solution to zinc solution than the reverse alkalization order. For the zinc to NaOH method, zinc will be precipitated completely when the apparent zinc concentration was higher than 14 g/L, that is, molar ratios of NaOH/Zn less than 4 (Fig. 2.41B), which can be further justified by the results shown in Fig. 2.41C in which the testing results using different NaOH and zinc concentrations are shown. Hence, a molar ratio of NaOH/Zn with less than 4 should be used when the alkalization method of zinc to NaOH solution is practiced in industry.

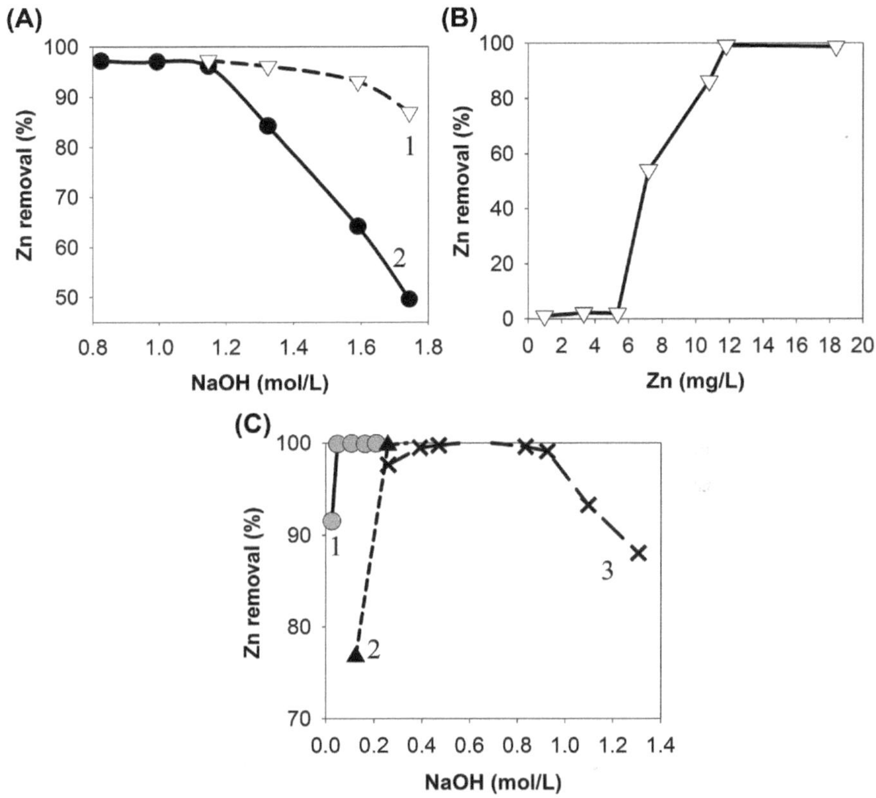

Figure 2.41 (A) Comparison of zinc precipitation by two alkalization sequences (Zn 18.9 g/L): sequences of alkalization: 1, addition of zinc solution to NaOH solution; and 2, addition of NaOH solution to zinc solution; (B) effect of the apparent concentration of zinc on the precipitation by the addition of zinc to NaOH solution (NaOH: 10 mL of 4 mol/L; Zn: 20 mL); (C) Effect of apparent concentration of NaOH on the precipitation at different apparent zinc concentrations by the addition of zinc to NaOH solution apparent zinc concentrations (1: 1.1 g/L; 2: 5.5 g/L; 3: 10.9 g/L).

2.7.2 Filtration Improvement of the Zinc Sludge

The results are shown in Figs. 2.42 and 2.43. The precipitates were formed under the conditions of NaOH/Zn molar ratio at 4. The filtration capacity of the precipitates formed by zinc-to-NaOH method increased as the vacuum pressure was increased (Fig. 2.42). The comparison of filtration capacity of the precipitates formed by zinc-to-NaOH and NaOH-to-zinc methods is shown in Fig. 2.43. It can be found that the former (line I in Fig. 2.43) is much higher than the latter (line II in Fig. 2.43) at the vacuum pressures used. In addition, it was found that the higher the vacuum pressure was

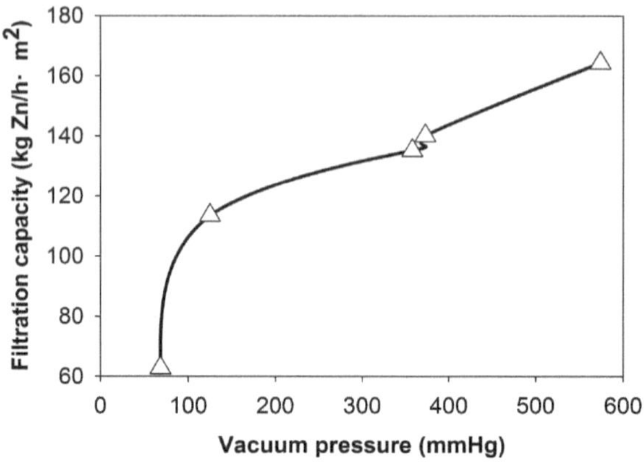

Figure 2.42 Effect of vacuum pressure on the filtration capacity of zinc precipitates formed by the addition of zinc solution to NaOH solution (apparent Zn 18.9 g/L).

used, the higher the filtration capacity would be obtained. When the NaOH/Zn molar ratios further decreases, the filtration capacity reduces considerably, as shown in Table 2.12. However, the filtration capacity of the precipitates formed by the zinc-to-NaOH method was always higher than that by the other method.

2.7.3 Comparisons of the Physical Properties of Precipitates

For the NaOH-to-zinc method, the precipitates formed settled down very slowly, and was voluminous. It is well-known that the precipitates formed in

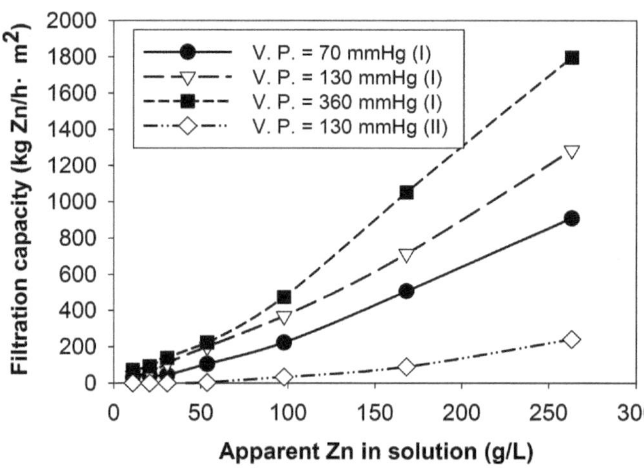

Figure 2.43 Relationship between the apparent zinc concentration or alkalization sequences and the filtration capacity alkalization sequence (I: addition of zinc solution to NaOH solution; II: addition of NaOH solution to zinc solution).

Table 2.12 The Filterability Comparison of the Precipitates

		Filtration Capacity ($Zn/h \cdot m^2$)	
No.	**NaOH/Zn Molar Ratios**	**Zinc to NaOH Solution Method**	**NaOH to Zinc Solution Method**
1	4.00	74.4	9.0
2	3.55	26.7	6.5
3	2.46	24.7	5.9

this way are colloidal, and very difficult to filter. In contrast, the precipitates formed homogeneously by the zinc-to-NaOH method were observed to be muddy, and could be filtered readily. In addition, the cake of colloidal precipitates will break when the filtration is finished or is dried. Nevertheless, the cake obtained by the zinc-to-NaOH method does not appear such phenomenon. The physical phenomena are summarized in Table 2.13.

2.7.4 Mechanisms of Filtration Improvement of Zinc Sludge by Alkalization Sequence

It is known that the properties of the precipitates formed homogeneously may be quite different from those formed heterogeneously. When zinc solution is added to NaOH solution, the alkalization is heterogeneous and the colloidal precipitates are obtained. However, when the mixing sequence of zinc and alkaline solutions is reversed, precipitates will not be formed until the molar ratios of NaOH/Zn are lower than 4. If the ratios are decreased a little slightly, precipitates will be formed rapidly. Hence, the precipitates are formed homogeneously in this way and are muddy.

Table 2.13 The Comparison of the Physical Properties of the Zinc Precipitates

	Precipitation Method Used	
Physical Phenomena	**Zinc-to-NaOH Solution Method**	**NaOH-to-Zinc Solution Method**
Settlement	Very quick	Slow
Filtrability	Very good	Very poor
Typical filtration capacity at vacuum pressure of 130 mm Hg (apparent Zn 98 g/L)	379.7 kg-$Zn/h \cdot m^2$	42.5 kg-$Zn/h \cdot m^2$
Rinsing	Very easy	Very difficult

The chemical analysis results of the two types of precipitates show that the precipitates consist of zinc hydroxide and sodium zincate with various ratios, depending on the molar ratios of NaOH/Zn and alkalization sequences. When NaOH solution is added into zinc solution and the molar ratios of NaOH/Zn are kept at around 2, only zinc hydroxide will be formed. Nevertheless, the precipitates usually contain a large amount of sodium salts, which may not be rinsed out readily, if the frame-plate filtration process is used for dewatering of the voluminous precipitates.

The precipitates formed by the zinc-to-NaOH solution method are usually pure sodium zincates, if the molar ratios of NaOH/Zn are lower than around 4. That is the reason why the precipitates formed in this way can be filtered readily, based on the following two alkalization reactions:

$$2NaOH + Zn(NO_3)_2 = Zn(OH)_2 + 2NaNO_3 \quad \text{(NaOH to zinc solution method)} \tag{2.20}$$

$$Zn(NO_3)_2 + 4NaOH = Na_2Zn(OH)_4 + 2NaNO_3 \text{(zinc to NaOH solution method)} \tag{2.21}$$

The actual consumption of NaOH for the latter method seems to be higher. It may be the disadvantage of using this alkalization method. As shown in Table 2.12, when even the molar ratios are kept at lower level when the alkalization is carried out, the filtration capacity of the resulting precipitates will be also improved by using the zinc to NaOH method. Obviously, such an alkalization should be strongly recommended in practical applications.

It can be concluded that the filtration capacity and filterability will be increased to some extent when the conventional alkalization method for the pouring of alkaline solution into the zinc containing solution is changed into the reverse alkalization sequence, that is, the addition of zinc solution to NaOH or other alkaline solutions.

2.8 ENHANCED DEWATERING OF WASTE-ACTIVATED SLUDGE BY COMPOSITE HYDROLYSIS ENZYMES

Hydrolysis enzymes such as protease, lipidase, and carbohydrase are gaining attention for use in improving WAS dewaterability by EPS hydrolysis.

The advantages of composite enzymes include nontoxicity, biodegradability, and low dosage. In this section, the feasibility of a composite hydrolysis enzyme, that is, *Enviro-Zyme 216* (Winston Company, Tulsa, USA) that contains protease and lipidase as well as one kind of anaerobic bacteria, *Aspergillus oryza*, in improving the dewaterability of sludge was verified.

2.8.1 Effect of Enzymatic Pretreatment on EPS Components

The composition variation of different EPS fractions in the presence of enzyme was investigated at the enzyme dosage of 1 mg/g-DS and temperature of 30°C. As shown in Fig. 2.44A, the PN concentrations in raw sludge

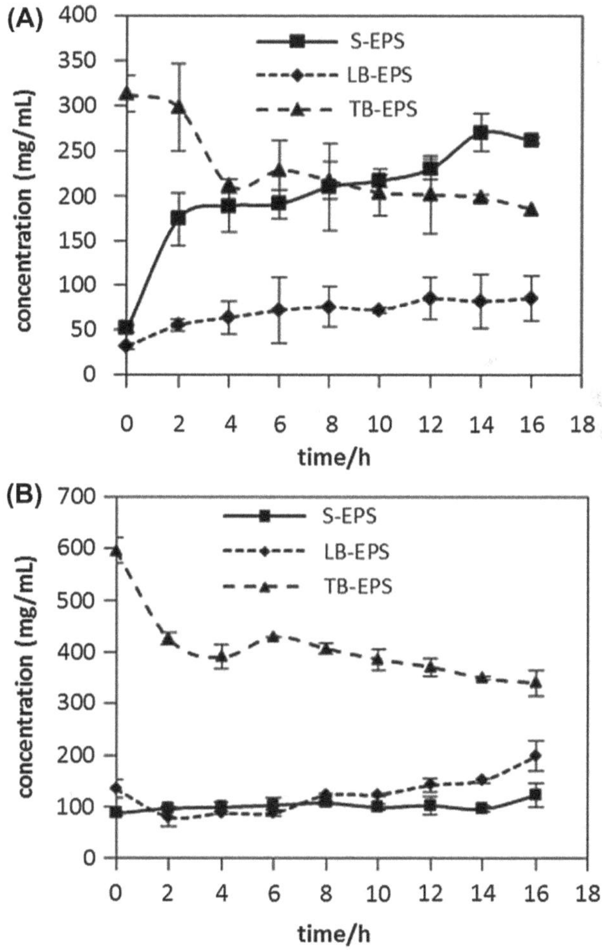

Figure 2.44 Composition variation of the different EPS fractions: (A) PN and (B) PS.

were estimated to be 51.84 ± 3.90, 32.13 ± 3.25 and 312.19 ± 20.14 mg/L of S-EPS, LB-EPS and TB-EPS for the untreated WAS, respectively. During enzymatic incubation, the PN concentration in TB-EPS decreased continuously from 312.19 ± 20.14 to 184.94 ± 3.25 mg/L; in contrast, it increased to 261.32 ± 3.90 and 86.10 ± 4.69 mg/L in S-EPS and LB-EPS, respectively. The hydrolysis of TB-EPS with the enzymatic proceeding would lead to the cleavage of linkages among macromolecules in TB-EPS, such as peptide bonds (—CO—NH—) and hydrogen bonds and partially released PN into the liquid phase. The released PN was isolated as the S-EPS and LB-EPS fractions, resulting in increasing contents of S-EPS and LB-EPS.

The concentration of PS in pretreated TB-EPS (Fig. 2.44B) showed a decreasing trend similar to that for PN of TB-EPS, and the concentration of PS in LB-EPS also decreased from 95.36 ± 0.45 to 81.10 ± 0.19 mg/L in the first 2 h. The decrease in PS in bound EPS can be attributed to the hydrolysis of macromolecular compounds. However, as a result of the great reduction of TB-EPS, the PS concentration of LB-EPS kept growing consistently in the following 14 h of incubation and PS in S-EPS also showed an obvious increase from 87.39 ± 3.88 to 123.33 ± 23.47 mg/L throughout the whole process, similar to the transformation of PN between bound EPS and S-EPS.

2.8.2 Effect of Enzymatic Pretreatment on Sludge Dewaterability

CST was measured to evaluate the dewaterability of WAS pretreated by the composite enzymes. As shown in Fig. 2.45, the CST decreased from 102.5 s to the lowest value, 53.5 s in the first 8 h of incubation, and then a slight deterioration of WAS dewaterability was observed from 8 to 10 h and CST stayed nearly constant from 10 to 16 h. An excessive solubilization of organic substances and break-down of the polymeric matrix may increase the viscosity of WAS and weaken the dewatering performance. Besides, due to the polymeric matrix breakdown, the release of supracolloidal particles resulted in a clogged filter medium, which limited the filtrate flow and increased the measured CST value.

2.8.3 Correlations Between Sludge Dewaterability and EPS Composition

The various components in EPS have different effects on the dewatering ability of microbial aggregates. An increasing carbohydrate fraction in EPS would enhance the aggregation of sludge flocs according to previously

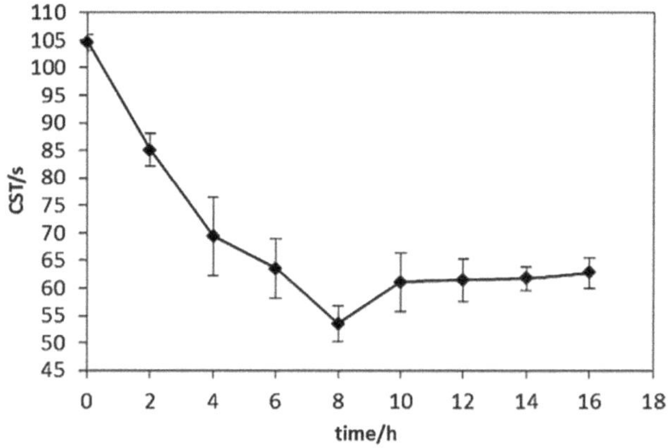

Figure 2.45 CST variation of the pretreated sludge.

reported results. On the contrary, the proteins had a high water-holding capacity due to the hydrogen bond induced by nitrogen-containing and oxygen-containing functional groups, entrapping water molecules in the gaps among chain-like macromolecules. The loosely packed sludge flocs consisting of hydrophobic proteins with a higher content of pore water and sticky matter would reduce the sludge compressibility and dewaterability. But if the macromolecules are hydrolyzed into small fragments, the hydrophobic substances will dissolve into liquid phase and have little effect on the hydrophilicity of WAS. Thus, the Pearson analysis was conducted, so that the relationship between the dewaterability and different EPS fractions could be established. The results are shown in Table 2.14. Values in bold refer to variables for which correlation is significant ($p < .05$).

Table 2.14 Correlation Between CST and EPS Fraction: Pearson's Correlation Coefficients (*Rp*) (*p* Values in Parenthesis)

	PN	PS	PN/PS	PN + PS
TB-EPS-CST				
Rp	0.9159	0.8545	0.3106	0.9094
p-value	5.231E−4	0.003350	0.4160	6.731E−4
LB-EPS-CST				
Rp	−0.9320	−0.2264	−0.5001	−0.5188
p-value	2.520E−4	0.5581	0.1704	0.1524
S-EPS-CST				
Rp	−0.8766	−0.6263	−0.8339	−0.8789
p-value	0.001920	0.07115	0.005210	0.001800

The statistical analysis in Table 2.14 indicates that the CST correlated positively with the individual polymers PN, PS from TB-EPS, and total TB-EPS (PN/PS). Also, CST correlated negatively with PN from S-EPS and LB-EPS, while no significant correlation was found between the dewaterability and PS, PN/PS from LB-EPS, or PS from S-EPS. These results revealed that WAS dewaterability is strongly governed by PN in EPS rather than PS. The CST correlated negatively with PN/PS from S-EPS, which implied that the hydrolysis of protein and transamination of amino acids are beneficial for the removal of hydrophilic function groups (i.e., $-NH_2$) and improve WAS dewaterability. Moreover, it is worth noting that PN located in different layers of the WAS floc structure could have positive or adverse effects on the dewaterability of WAS. This phenomenon may depend on the PN chemical structure change induced by the added enzyme. Therefore, investigating the migration and transformation of hydrophilic and hydrophobic components in different EPS fractions in the following sections will be valuable for further understanding the mechanism of enzymatic WAS dewatering.

Pearson correlation among different EPS fractions is shown in Table 2.15. It can be found that PN from S-EPS and LB-EPS both oppositely correlated with PN from TB-EPS ($Rp = -0.8673, p < .05$; $Rp = -0.9132, p < .05$). PS from TB-EPS and LB-EPS also showed a close correlation ($Rp = -0.8865$, $p < .05$). These results partly verified the hypothesis that when the WAS flocs were disrupted with hydrolysis, a part of TB-EPS were solubilized and released into the liquid phase, thereby increasing the S-EPS and LB-EPS levels.

2.8.4 Preliminary Study in the Enzymatic Degradation of EPS

A kinetic analysis was carried out to gain further insights into the kinetic variation of TB-EPS reduction and lay a theoretical foundation for

Table 2.15 Correlation Among Different EPS Fractions: Pearson's Correlation Coefficients (*Rp*) (*p* Values in Parenthesis)

	PN in S-EPS	PN in LB-EPS
PN in TB-EPS		
Rp	−0.8673	−0.9132
p-value	5.8100E−4	0.002460
	PS in S-EPS	**PS in LB-EPS**
PS in TB-EPS		
Rp	−0.6517	−0.8865
p-value	0.05720	0.003350

optimizing the enzymatic reaction conditions (temperature, pH, and enzyme dosage) from the perspective of the reaction kinetics. The kinetics of enzymatic reactions can be described by the *Michaelis–Menten* equation, that is, $V = V_{max} \times [S]/([S] + K_m)$, where V is the reaction rate, [S] is the substrate concentration, and K_m is the half-saturation constant. K_m and V_{max} are determined by the characteristics of the enzyme. Thus, applying an appropriate ratio of enzyme and substrate concentration is very important to improve the enzymatic reaction efficiency.

The dosage of composite enzymes was controlled at a fixed value, and then the effect of initial EPS concentration on PN and PS removal efficiency was investigated by kinetic analysis. The kinetic analysis can be used to evaluate the applicability of enzymes for the hydrolysis of substrates in WAS and lays a theoretical foundation for improving the enzymatic reaction efficiency. Three identical reactors were charged with initial MLSS of WAS of 3410, 6820, and 13,640 mg/L by diluting or concentrating the WAS and were maintained at 30°C. The dosage of enzyme was kept at 1 mg/g-DS. During the enzymatic reaction, the trends of the removal efficiency of PN and PS from TB-EPS with treatment time at different MLSS are illustrated in Fig. 2.46A and B, respectively. It can be found that at some specific time intervals within 16 h, reactors with different initial concentrations of TB-EPS attained similar removal efficiency. It seems that the removal efficiency was ultimately independent of the initial concentration of TB-EPS during the 16 h of incubation. According to recognized reports, the relation between the initial concentration of reactant and its conversion efficiency within the same reaction time and steady environmental conditions (temperature and pH) was as follows: negatively correlated during zero-order reaction, not correlated during first-order reaction and positively correlated during second-order reaction. Thus, it can be inferred that the enzymatic reaction matches the first-order kinetics. In addition, according to the *Michaelis–Menten* equation, only when [S] is far lower than K_m can the enzymatic reaction be approximated as a first-order reaction. Therefore, it can be determined that the introduced enzyme concentration of 1 mg/g-DS is adequate for the degradation of the corresponding substrate in WAS. Excessive enzyme is meaningless for the improvement of WAS dewaterability, and the environmental factors (temperature and pH) should be attached importance to the optimization of WAS enzymatic dewatering.

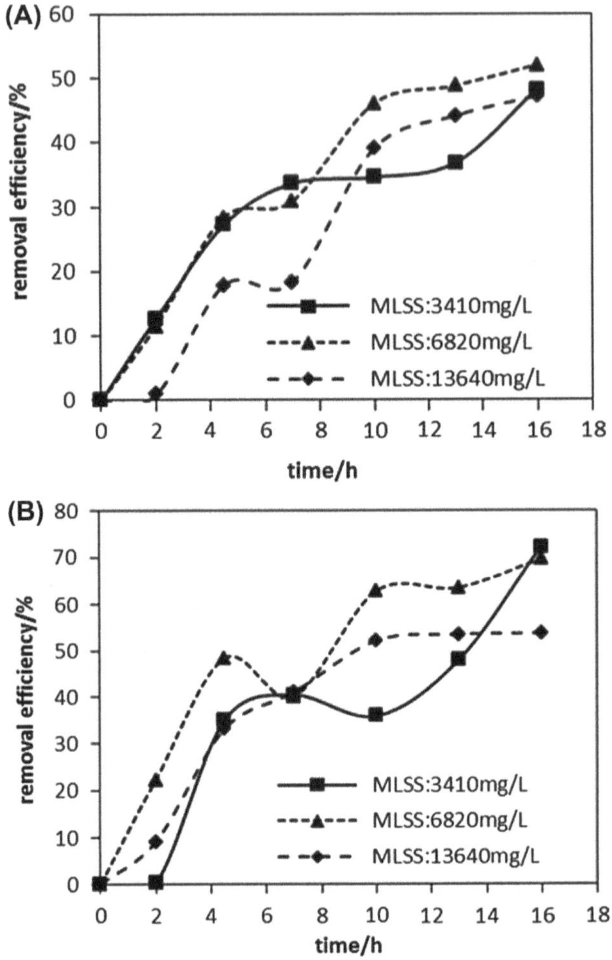

Figure 2.46 Removal efficiency of TB-EPS: (A) PN, and (B) PS.

2.8.5 Effect of Temperature on the Kinetic Constants

The composite enzyme used is secreted from Aspergillus oryzae; thus, the reaction temperature is chosen based on the preferential temperature (25–35°C) of the culture of Aspergillus oryzae and mega temperature should be avoided to limit its negative effect on enzyme activity.

Based on the above observation, the enzymatic removal of TB-EPS can be described by the following first-order kinetic equation:

$$V = -\frac{dX}{dt} = k_e \times X \tag{2.22}$$

$$\ln X = -k_e \times t + b \quad (2.23)$$

where V is the reaction rate, X is PN or PS of TB-EPS, k_e (h^{-1}) the enzymatic rate constant, and b the constant of integration. The effect of temperature on the rate constant was investigated under the condition as follows: the dosage of the enzyme was 1 mg/g-DS and pH was kept at the initial value. By plotting lnX versus t, the slope and the intercept can be obtained, which correspond, respectively, to the value of k_e and b in Eq. (2.22).

The values of k_e and b from 10 to 30°C are shown in Table 2.16. It is clear that the R-squared values at different temperatures were reassuringly in the range of 0.91−0.97, which is much higher than the R-squared value obtained from fitting the equations based on zero-order or second-order reactions (these data are not shown). This further demonstrated that the enzymatic release of TB-EPS in WAS obeyed first-order kinetics. Moreover, the rate constants k_e of PN and PS reduction both increased as temperature increased from 10 to 30°C, while similar kinetic constants were obtained at 25 and 30°C.

The Arrhenius equation attempts to establish the relationship between the rate constant and the reaction temperature as

$$k_e = A \times \exp\left(-\frac{E_a}{RT}\right) \quad (2.24)$$

where A is the pre-exponential factor, E_a (kJ/mol) the activation energy of the reaction, and T (K) the absolute temperature. When the enzymatic removal of TB-EPS in WAS proceeds at the initial pH and temperature in the range of 10−30°C, the relationship between $\ln k_e$ and $1/T$ is shown in Fig. 2.47. Obviously, the first-order hydrolysis rate constants obtained in this study followed an Arrhenius type of behavior; the activation energies E_a for PN and PS hydrolysis were 4.311 and 5.502 kJ/mol, respectively. Thus, the

Table 2.16 Kinetic Constants at Different Temperatures

	Protein			Polysaccharide		
Temperature (°C)	K_e	B	R^2	K_e	B	R^2
10	0.07630	6.139	0.9687	0.0892	5.398	0.9158
20	0.07900	6.228	0.9553	0.0978	5.471	0.9544
25	0.08500	6.326	0.9631	0.1028	5.462	0.9241
30	0.08520	6.297	0.9538	0.1031	5.404	0.9513

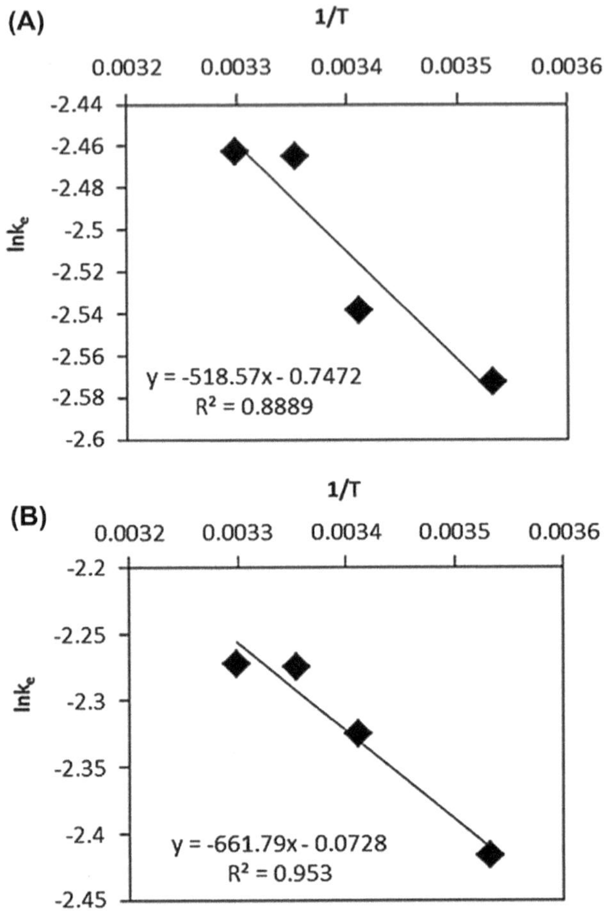

Figure 2.47 lnk_e as a function of 1/T: (A) PN in TB-EPS, and (B) PS in TB-EPS.

value of the first-order hydrolysis constant in the temperature range of 10–30°C, k_e, can be expressed by Eqs. (2.25) and (2.26):

$$k_e(\text{PN}) = 0.4737 \times \exp\left(-\frac{4.311 \times 10^3}{RT}\right) \tag{2.25}$$

$$k_e(\text{PS}) = 0.9298 \times \exp\left(-\frac{5.502 \times 10^3}{RT}\right) \tag{2.26}$$

From Table 2.16, it can be seen that the value of b remained almost constant in the temperature range of 10–30°C. The average values of b for PN and PS hydrolysis are 6.247 and 5.434, respectively. Accordingly, the kinetic models of PN and PS in TB-EPS can be simplified as follows (t is reaction time):

$$X(\text{PN in TB}-\text{EPS}) = 516.46 \times \exp\left(-0.474 \times \exp\left(\frac{4.311 \times 10^3}{RT}\right) \times t\right) \tag{2.27}$$

$$X(\text{PS in TB}-\text{EPS}) = 229.06 \times \exp\left(-0.9298 \times \exp\left(\frac{5.502 \times 10^3}{RT}\right) \times t\right) \tag{2.28}$$

2.8.6 Effect of pH on the Kinetic Constants

The values of k_e and b at different pH values are shown in Table 2.17. From the results, it can be seen that the highest rate constant of protease was 0.1035 at pH 6. Compared with the results collected from the experiment at the initial pH of WAS (Table 2.16), the slightly acidic condition was quite beneficial for improving the protease activity. In contrast, the rate constant of PS solubilization from TB-EPS to the liquid phase decreased with increasing pH. The pH can vary the rate constants by infiuencing the chemical structure (i.e., ionization state) of enzymes, which can limit the specific interaction between the enzymes and substrates. Moreover, an alkaline pH may increase the zeta potential of WAS and result in electrostatic repulsion among solid particles, which inhibits the adsorption or diffusion of carbohydrate enzyme to polysaccharides in the interior of the sludge floc. As a result, the rate constant of the enzymatic reaction decreased as the pH increased from 6 to 11.

2.9 PRACTICAL SIGNIFICANCE FOR MECHANICAL DEWATERING PROCESSES

Moisture in sludge determines the pollution control and resource recovery technology selections and parameters' optimization, including sanitary landfill, anaerobic digestion, and drying and incineration. The moistures in

Table 2.17 Kinetic Constants at Different pH Values

	Protein			Polysaccharide		
pH	K_e	B	R^2	K_e	B	R^2
3	0.08760	6.304	0.9661	0.1146	6.652	0.9620
6	0.1035	6.814	0.9768	0.1045	6.623	0.9584
9	0.08970	6.955	0.9619	0.09210	6.523	0.9540
11	0.08730	6.884	0.9745	0.08990	6.556	0.9672

sludge can be reduced to 80% using conventional equipment. To obtain lower moisture sludge, sewage sludge must be conditioned chemically, physically or biologically, before being dewatered by filters. Actually, addition of lime, soil, coal ash, refuse incineration bottom ash, construction wastes, etc., can improve the dewaterability of sludge. The problems for using these additives include the weight gains as over 10–30% (on dry basis of sludge) of the additives may be added so that an acceptable dewatering level can be obtained. Consequently, the heating values will be greatly reduced and make the DS incineration very difficult or even impossible. Lower dosage, higher efficiency, and acceptable cost should be the priority for the selection of chemical conditioning agents. Fenton pretreatment and Fe(II)-activated persulfate oxidation techniques developed in this work should be good alternatives for sludge conditioning, as the dosage and total costs of these agents are quite low and acceptable, with negligible weight gain and loss of heating values.

Mechanical dewatering of sludge is carried out after being conditioned. The principle methods include drying beds, centrifuges, belt filter presses, and chamber filter presses. Here, a new generation of chamber filter press named "elastic plate-filter press" was developed aiming to improve the sludge-dewatering efficiency.

2.9.1 Design and Working Principles of Elastic Plate-Filter Press

Fig. 2.48 shows diagrammatically the cross-section and working principles of the filter press, and Table 2.18 lists its models and operating parameters. The basic stages for each batch dewatering cycle include:

- *Filling*: after being conditioned for 5–10 min, the sludge is pumped into the empty chambers between consecutive plates, enveloped by filter cloths. The pumping of sludge offers an initial pumping pressure (0.8–1.5 MPa) in the spaces between plates for solid/liquid separation.
- *Filtration under maximum pressure*: during this stage, a hydraulic piston is used to provide mechanical pressure (Fig. 2.49). The sludge is squeezed and the filtrate passes through the filter cloths and exits through the filtrate ports of the filter plates. The applied pressure may reach roughly 5–7 MPa while it is only 0.6 MPa for the conventional plate-filter press, thereby leading to a 3–5-times rise in dewatering efficiency. Moreover, the filtration time of the new press is relatively short, just 30–50 min irrespective of the sludge types and characteristics, much lower than the conventional diaphragm plate-filter press (4 to 8 h).

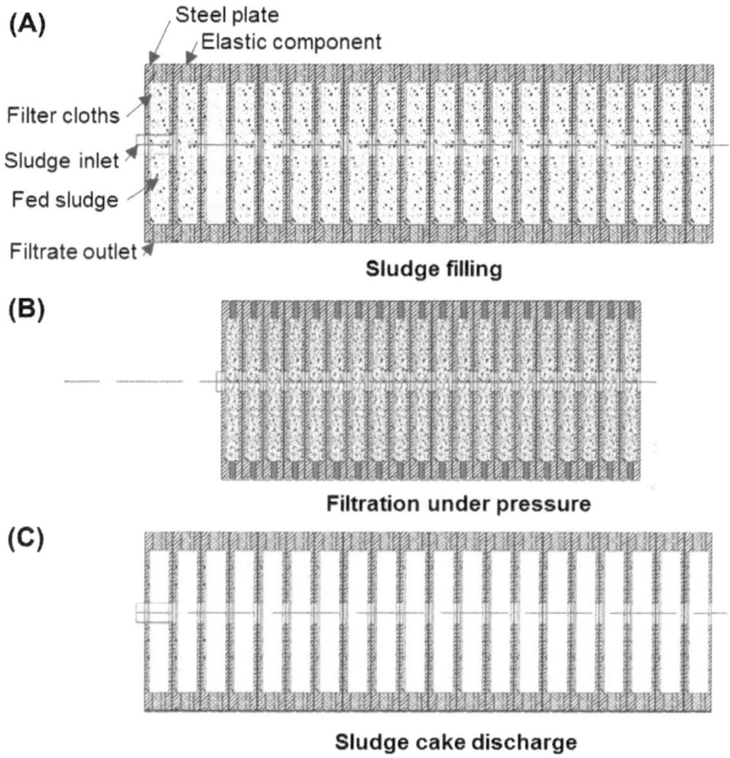

Figure 2.48 Design and principles of the elastic plate-filter press.

- *Filtrate collection and cake discharge*: the filtrate passing through the filter cloths is collected through drainage points and filtrate channels (Fig. 2.50), which is treated on site or conveyed into the nearby WWTPs. The sludge cake with moisture content of less than 40% can fall off automatically from the plates when the hydraulic piston moves back and the plates are separated from each other.

2.9.2 Commercial Applications of the Elastic Plate-Filter Presses in Sludge Mechanical Dewatering

The elastic plate-filter presses are proven and reliable processes that have been successfully appiled since 2010 in projects in many cities of China, like Shanghai, Nanjing, Wuxi, etc. For example, a new sludge-dewatering facility was bulit in 2012 in a WWTP, located in Zhangjiagang City, Jiangsu Province, China (Fig. 2.51). This sludge-dewatering scheme was composed of the following units: (1) screening for pretreatment

Table 2.18 Models and Operating Parameters of the Elastic Plate-Filter Presses

	Model			
Parameter	**TJYZ300/2000**	**TJYZ 200/1600**	**TJYZ 100/1250**	**TJYZ 40/1000**
Cross-sectional area of filtration (m^2)	300	200	100	40
Capacity (t/h) (cake water content 50%)	6–7	4–5	2.5–3	1.5–2
Size of plate (mm)	2000 × 2000 × 155	1600 × 1600 × 147	1250 × 1250 × 147	1000 × 1000 × 143
Cake thickness (mm)	35	35	35	35
Pressure system	Hydraulic	Hydraulic	Hydraulic	Hydraulic
Pumping pressure (MPa)	0.8–1.5	0.8–1.5	0.8–1.5	0.8–1.5
Filtration pressure (MPa)	<20	<20	<20	<20
Power of engine (kW)	Main 15/auxiliary 7.5	Main 11/auxiliary 5.5	Main 7.5/auxiliary 5.5	Main 7.5/auxiliary 5.5
Size of filter press (mm)	15,000 × 2800 × 4500	13,000 × 2100 × 3800	10,000 × 1800 × 3000	8000 × 1600 × 2600
Type of filter cloth (polypropylene fiber)	4218	4218	4218	4218
Weight of filter press (t)	62	55	34	28

Figure 2.49 An example of the elastic plate-filter press (A), sludge inset (B) and hydraulic piston (C).

(retention of coarse material like grit, sand, gravel, etc.), (2) conditioning and mixing tank (to enhance sludge dewaterability through chemical conditioning); (3) feeding pumps (to transport the conditioned sludge from the pretreatment tank to the filter press); and (4) elastic plate-filter press

Figure 2.50 Filtrate collection (left) and the produced dewatered sludge cake (right).

Figure 2.51 Sludge-dewatering project using elastic plate-filter presses in a local WWTP, located in Zhangjiagang City, Jiangsu Province: (A) sludge-dewatering units, and (B) its products.

that extracts water from the pretreated sludge. The dewatering units was capable of dewatering approximately 100 wet tons of sludge per day. Since the completion of this project, the WWTP produced dewatered cake at > 45% solids (odor strength < 3, and lower calorific value of 1500−2000 kcal/kg-dried sludge), which was much higher than its initial < 20% solids. The resulting by-products can be recycled as a fertilizer or as an alternative fuel or be landfilled if required.

The success of this project has invited the great interest of other municipalities in the surrounding area. The filter press has been adopted in the on-site dewatering of other special sludges such as petrochemical sludge, dyeing sludge (Fig. 2.52), etc. Since the utilization of this new dewatering units, the plants went from producing 20% solids to producing approximately 50%

(A)

(B)

Figure 2.52 Sludge on-site dewatering project, located in Jinshan District, Shanghai (a petrochemical sludge), and in Jiangsu Province (dyeing sludge), China.

solids, which dramatically reduced the hauling and disposal expense while reducing the carbon footprint and greenhouse gas emissions.

CHAPTER THREE

Sewage Sludge Solidification/ Stabilization and Drying/ Incineration Process

3.1 EFFECT OF CALCINED ALUMINUM SALTS ON THE ADVANCED DEWATERING AND SOLIDIFICATION/ STABILIZATION

The increasing number of waste-activated sludge devoted to waste immobilization reflects a genuine need for cost-effective solutions for restoring a safe and green environment. Solidification/stabilization (S/S) technologies are expected to increase in treating hazardous wastes, like heavy metal-contained wastes before a land disposal. A major factor in applying this process to wastes is that it improves the physical and chemical characteristics and reduces the mobility of contaminants. In this process, the identification of binders able to assume the fixation of contaminants is essential for the success of the technique. Binders or additive agents, such as Portland cement (PC), lime, MgO, fly ash mixture, or other additive, are usually used to form the solidified products.

MgO and cement mixtures had been successfully used for the solidification of sludge in Shanghai Laogang landfill site for years. The mixtures occupied around 4–8% of the sludge (80% moisture) and curing for one week was practiced to achieve moisture of below 55% required for a safe landfill without any sliding accident. Over 5 million tons of sludge had been solidified on site. Presence of calcined aluminum salts (AS), Fenton reagent, and Fe(II)-activated persulfate will speed up the moisture loss from sludge.

PC, or in combination with other materials, can be utilized to improve the dewatering performance and geotechnical properties of the sludge. As to the efficiency, however, some of these fixing agents are debatable since a number of the conventional S/S processes use over 20–30% binders of dewatered sludge (DS) to achieve the required moisture content and compressive strength for the landfill operation, leading to a larger volume increase, while others fail a satisfactory cost-benefit analysis.

Pollution Control and Resource Recovery: Sewage Sludge
ISBN: 978-0-12-811639-5
http://dx.doi.org/10.1016/B978-0-12-811639-5.00003-6

The aims of this section are to explore an effective chemical dewatering and S/S technique for the DS treatment and to assess the properties of the end products to decide whether they can be deposed of at a sanitary landfill site with a lower addition of binders and quicker solidification process.

3.1.1 Test Materials and Experimental Procedures

Dewatered sludge was obtained from the aeration basin at a full-scale waste water treatment plant (WWTP) in Shanghai, China (Table 3.1). The main elements present in dry sludge ash were silica (SiO_2), calcium oxide (CaO), and alumina (Al_2O_3).

$CaCO_3$, CaO, and $Al(OH)_3$ were applied as the starting materials, and CaF_2 was used as the mineralizer (Fig. 3.1). Meanwhile, $CaCl_2$, Na_2SO_4, and anhydrite $CaSO_4$ were used to investigate their influence on the strength development of solidified/stabilized sludge. The anhydrite $CaSO_4$ was prepared from $CaSO_4 \cdot 2H_2O$ at 500°C with a heating rate of 10–20°C/min and soaking time of 2 h. The solidifier for the advanced dewatering and S/S of DS (i.e., calcined AS) was prepared as described below. $CaCO_3$ and $Al(OH)_3$ were calcined at 950–1000°C for 2.5 h in an SX2-10-12 muffle furnace to produce reactive CaO and Al_2O_3, respectively, followed by the mixing of these two chemicals with a stoichiometric ratio of 12/7 after grinding to a size of 74 μm. Thereafter, the binary system was mixed with water at a liquid–solid ratio of 1/1 (mL/g), and hydrothermally heated to boiling for 1–1.5 h, dried at 65°C for 24 h. The product

Table 3.1 Physicochemical Properties of Dewatered Sludge Used in This Section

Property	Value								
Moisture content (%)	81.19 ± 0.24								
pH	6.70 ± 0.03								
Organic matter content (%)	45.67 ± 0.10								
Heavy Metal (g/kg Dry Basis)									
	Cu	Zn	Pb	Cd	Cr	Ni			
	0.20	1.00	0.04	No detected	0.10	0.03			
Chemical Composition (wt%[a])									
	SiO_2	MgO	Al_2O_3	Na_2O	K_2O	CaO	Fe_2O_3	TiO_2	Others
	46.40	3.97	10.30	1.76	3.19	7.85	5.66	0.79	20.08

[a]Dry dewatered sludge ash (ignition at 1100°C).

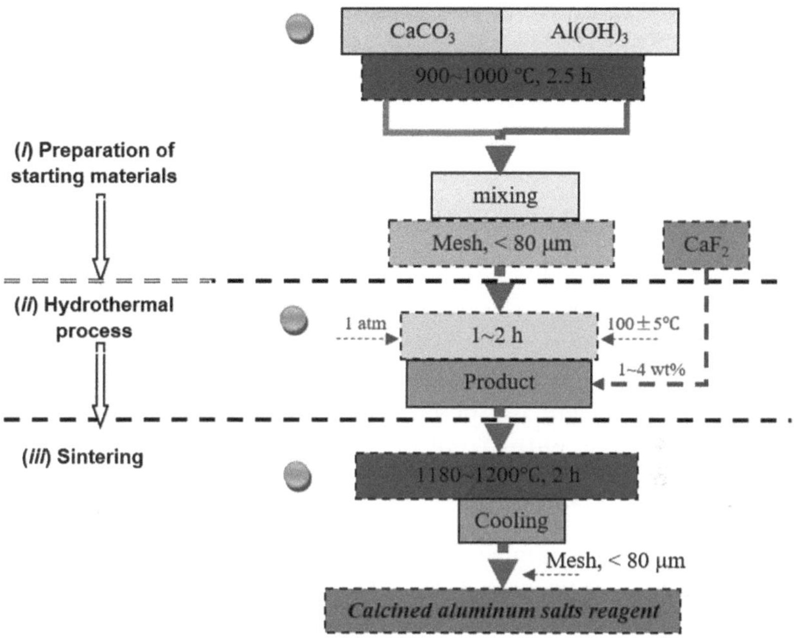

Figure 3.1 Synthesis of aluminate-based solidifying agent.

thus obtained was mixed 4 wt% CaF_2 and sintered in a muffle furnace at a rate of 10–20°C/min to 1180°C and soaked at the maximum temperature for 2 h. The calcined product, AS, finally obtained was ground into the sizes below 80 μm and employed as the cementitious media for the advanced dewatering and S/S of DS.

Table 3.2 Nomenclature and Components of the Pastes for Solidification/Stabilization of Sludge

Nomenclature	Mass Ratio of Binder/Sludge	Blended Binders	Additives
A_0	–	–	–
AS_0	5/100	100% AS	10% anhydrite $CaSO_4$
ASC_1	5/100	50% AS, 50% CaO	4.08% $CaCl_2$, 8.16% Na_2SO_4
ASC_2	5/100	50% AS, 50% CaO	5.22% $CaCl_2$, 10.44% Na_2SO_4
ASC_3	5/100	50% AS, 50% CaO	5% anhydrite $CaSO_4$
ASC_4	5/100	50% AS, 50% CaO	10% anhydrite $CaSO_4$

For the preparation of the solidified samples the DS, blended binders, and a little amount of additives were mixed and cured in molds with a size of 39.1 mm in diameter and 80.0 mm in height for 24 h under air-tight condition at 25 ± 0.5°C. A reference sample, without any binder added, was cast for the comparison purpose (Table 3.2).

3.1.2 Characterization of Calcined Aluminum Salts

The predominant crystalline constituents of the calcined AS are shown in Fig. 3.2. It can be found that the crystalline materials, such as mayentie ($12CaO \cdot 7Al_2O_3$) and $11CaO \cdot 7Al_2O_3 \cdot CaF_2$, which were desirable as they possess hydraulic properties, are present, while $Ca(OH)_2$ and CaO, as minor compounds, were also observed. However, the phases of ordinary Portland and sulfoaluminate cement such as C_3A and C_2A were not detected. Hence, when reactive CaO and Al_2O_3 with a stoichiometric ratio of 12/7 were calcined at 1180°C, a large quantity of $12CaO \cdot 7Al_2O_3$ could be induced due to the occurrence as shown in Eq. (3.1).

$$12CaO + 7Al_2O_3 \xrightarrow[2\ hr]{1180^\circ C} 12CaO \cdot 7Al_2O_3 \tag{3.1}$$

3.1.3 Effect of Various Binders on Moisture Content

The moisture contents were determined as a function of the curing time (Fig. 3.3). The results signified that the time required for the solidified

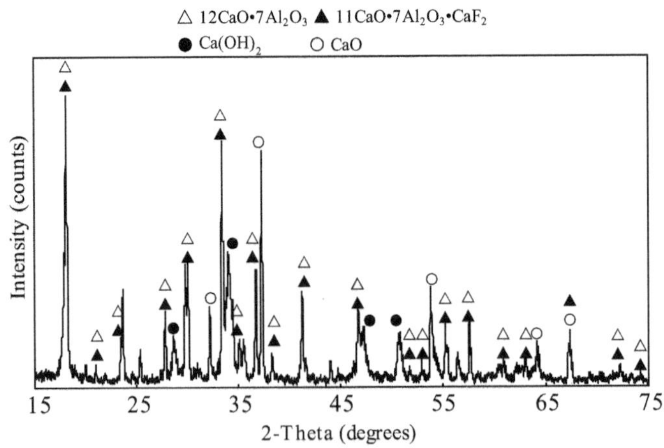

Figure 3.2 X-ray diffraction spectra of the crystalline phases in aluminum salt with 4 wt% CaF_2.

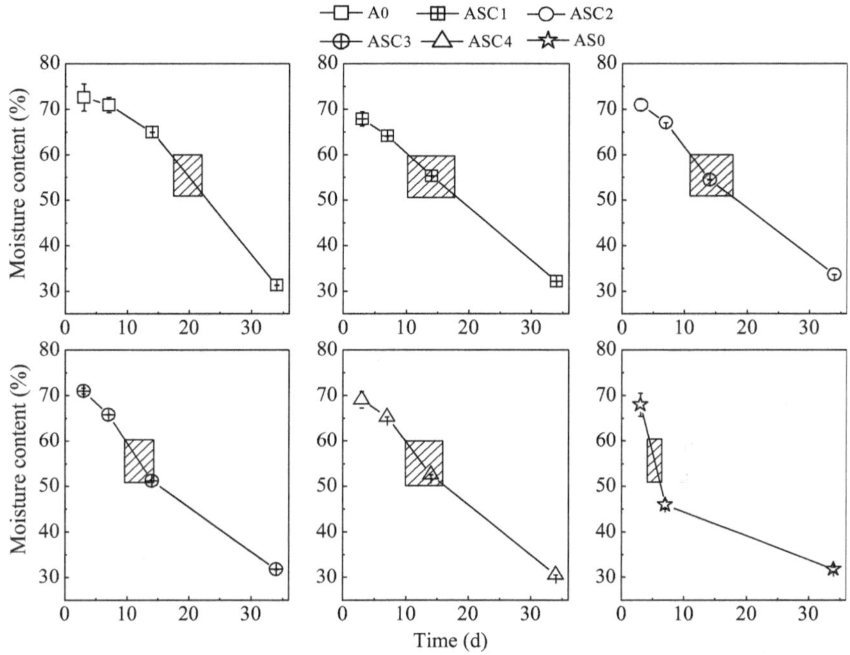

Figure 3.3 Moisture contents of solidified/stabilized sludge during the curing period.

samples to reach a moisture content of 50–60% varied. For the paste AS_0 constructed with 5% AS and 10% $CaSO_4$ of AS by weight, moisture content could be decreased to 50–60% after a 5–7 days of curing, and corresponding time was increased to 10–17 days for mortars ASC_1 and ASC_2, and 10–14 days for mortars ASC_3 and ASC_4, respectively. The presence of high content of free-CaO in the S/S products might lead to the increase in the curing time and delay the depletion of evaporable water.

3.1.4 Unconfined Compressive Strength of Solidified/Stabilized Sludge

Unconfined compressive strength (UCS) value is an indication of the rate at which hydration reactions are occurring. These tests were performed to investigate the effect of various binders on the strength of the resultant solidified sludge (Table 3.3). It can be noticed that the mortar AS_0 presented an UCS of 51.32 ± 2.9 kPa after 5–7 days of curing, meeting the minimum strength requirement for a sanitary landfill operation. This indicated that in this case the reactive pozzolanic minerals hydrated rapidly and yielded enough cementitious products binding the solidified sludge particles into a

Table 3.3 Variation of the Unconfined Compressive Strength of Different Solidified Mortars According to Lapsed Curing Time

Curing Time (Day)	UCS (kPa)					
	A_0	ASC_1	ASC_2	ASC_3	ASC_4	AS_0
3	0.24 ± 0.02	0.62 ± 0.02	0.54 ± 0.06	2.31 ± 0.03	1.92 ± 0.18	2.96 ± 0.13
7	0.67 ± 0.09	0.89 ± 0.15	0.65 ± 0.07	2.81 ± 0.06	2.28 ± 0.25	51.32 ± 2.9
14	2.14 ± 0.31	4.09 ± 0.57	2.90 ± 0.85	15.52 ± 0.24	12.33 ± 0.58	111.39 ± 7.4
28	7.14 ± 4.17	18.87 ± 4.82	12.82 ± 1.06	62.97 ± 0.99	51.95 ± 8.70	146.42 ± 12.73

stiff matrix, and subsequently induced a satisfying strength. In contrast, the specimens solidified using the mixture of AS and CaO obtained the strength of below 2.81 ± 0.06 kPa at the same curing time. Hence, the substitution of superfluous CaO could not compensate for the AS loss in binders, contrarily delayed, and even hindered hydration process, subsequently reducing the consumption of the evaporable water and severely weakening the compressive strength in the solidified pastes.

Further analysis was carried out to investigate the efficiency of anhydrite $CaSO_4$ and equivalent mass of $CaSO_4$ ($CaCl_2 + Na_2SO_4 \rightarrow CaSO_4 +$ $2NaCl$) in enhancing the strength improvement of matrixes. Results displayed that the mortar ASC_3 fabricated with the addition of 5% anhydrite $CaSO_4$ achieved the strength of 62.97 ± 0.99 kPa after 28 days of curing, but the corresponding value of 18.87 ± 4.82 kPa for the sample ASC_1 with the incorporation of an equivalent mass of $CaSO_4$. Moreover, surface cracks were also observed in the latter. These results suggested that the presence of anhydrite $CaSO_4$ in AS-CaO system improved the rate of hydration reactions, accelerated the generation of the hydration products such as gismondine ($CaAl_2Si_2O_8 \cdot 4H_2O$) and $CaCO_3$ (Fig. 3.4) and also manifested that it played a much more significant role in the enhancement of the compressive strength.

3.1.5 Development of Crystalline Phases in Solidified Specimens

The solidified specimens ASC_1, ASC_3, and AS_0, with a higher compressive strength were further studied using XRD. From Fig. 3.4, it can be seen that the pozzolanic gismondine ($CaAl_2Si_2O_8 \cdot 4H_2O$ or C—A—Si—H), a hydrated compound containing silicon dioxide and $CaCO_3$, was identified. These results might be explained by an acceleration of unbound quicklime (Fig. 3.2) in the calcined AS that initiated the faster dissolution and

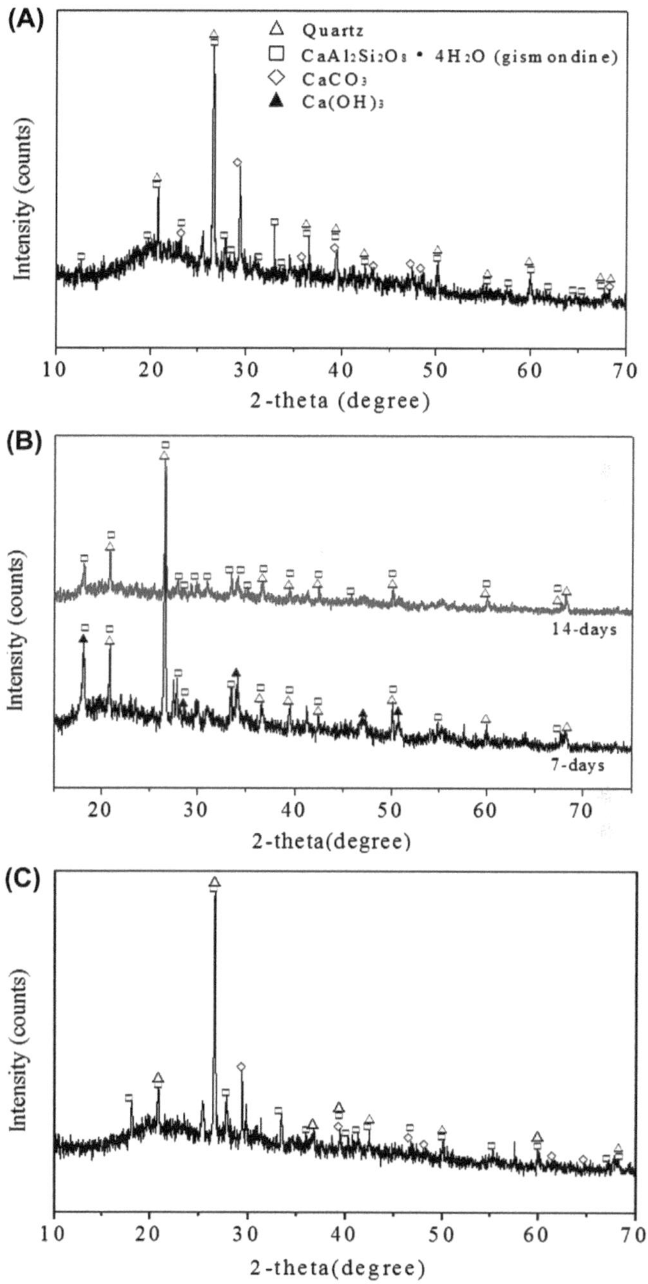

Figure 3.4 X-ray diffraction spectra of the crystalline phases in specimens AS_0 after 14 days of curing (A), ASC_1 after 7 and 14 days (B), and ASC_3 after 14 days (C).

incorporation of silica from high siliceous sludge into gismondine structure, and as expected, normal crystalline phases such as ettringite, monosulphate hydrates, and katoite ($Ca_3Al_2(SiO_4)_{3-x}(OH)_{4x}$) were absent at this age. This could be related to the complex chemical composition of the DS sample (Table 3.1) and gismondine's intercepting massive C—S—H gel.

The peak intensity is an indication of a given hydrate content. The type of binders resulted in considerable differences in the XRD spectra. For gismondine, the highest intensity of peak could be recorded in the mortar AS_0, followed by ASC_3 and ASC_1. As to $CaCO_3$, the highest peak intensity appeared in the mortar AS_0 but not in the samples ASC_1 and ASC_3. This verified that a high content of AS with the aid of an appropriate dosage of $CaSO_4$ could offset the interference of organic substances, favor the hydration reaction, induce the formation of a large quantity of hydrates, and subsequently produce a hard and resistant solidified paste.

Furthermore, the formation process of hydration crystals was also further analyzed to explore the conversion mechanism among various hydrates. Taking the mortar ASC_1 for example, it was evident that the formation of $CaAl_2Si_2O_8 \cdot 4H_2O$ was quick, its presence was identified after 7 days of curing, and $Ca(OH)_2$ crystals were also detected at this stage. As time extended, $Ca(OH)_2$ reduced gradually and nearly disappeared on day 14; meanwhile, a slight increase could be observed in the peak intensities of $CaAl_2Si_2O_8 \cdot 4H_2O$ and $CaCO_3$. These observations could be interpreted in principle that the thermodynamically stable hydrates, that is, $CaAl_2Si_2O_8 \cdot 4H_2O$ and $Ca(OH)_2$, could be precipitated at the earlier stage, partially ascribed to the presence of quicklime and subsequent acceleration of pozzolanic reaction, and that the transformation of the portlandite was responsible for the formation of gismondine and calcite, owing to atmospheric carbonation.

3.1.6 Scanning Electron Microscopy Observations

During the scanning electron microscopy (SEM) examination, special attention was paid to the sludge-binder pastes to illustrate the details of the mortar microstructure and confirm the XRD analysis results. The raw DS A_0 was also monitored for a comparison purpose. Fig. 3.5A is a typical image of the raw DS after 28 days of curing time. It can be noted that the sludge particle still retaining its spherical shape was easily distinguishable, while it was possible to verify numerous voids visible, indicating the presence of large pores and providing the basic evidence why the reference still exhibited a relatively low compressive strength (Table 3.3) at this stage.

Figure 3.5 Scanning electron microscope of spherical sludge particles and visible voids in reference sample A_0 (A), specimens ASC_1 (B), ASC_3 (C), and AS_0 (D) cured under laboratory conditions for 28 days.

In Fig. 3.5B, C and D, the SEM photos of the solidified pastes ASC_1, ASC_3, and AS_0 are displayed. Honeycomb-like hydrated products (viz., gismondine crystals) were revealed with the aid of microanalysis, which coincided with the data derived from the XRD analysis (Fig. 3.4A, B and C). Such crystals, which were not observed in the reference before solidification, were found to be widely dispersed and finely incorporated in the solidified pastes, binding together sludge and additives into a solid matrix as physically or chemically as possible to fill the pores of the mortars. The filling effect of the hydration products decreased the pores, altered the morphological characteristics of the formed hydrates and thus efficiently hardened the mortars. This observation was valid for the superior strengths of the solidified specimens, especially for the specimen AS_0. While higher UCS value was obtained for the AS_0 sample, SEM analysis revealed that intact AS particles were still present, even at the end of the curing time. That was in part attributed to the presence of a little amount of portlandite caused by quicklime

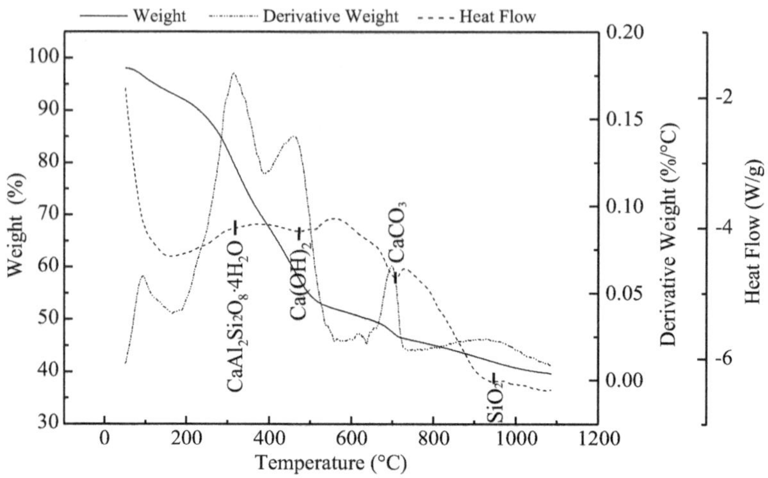

Figure 3.6 Thermogravimetric analysis/differential scanning calorimetry curves of the mortar AS_0 at 28 days.

introduction. The excess portlandite acted inhibitory for the action of AS particles, which demonstrated the truth of the negative effect of excessive CaO on reaction rate of AS particles and the strength of the ASC_1 and ASC_3 pastes.

3.1.7 Thermal Analysis

Thermogravimetry-differential scanning calorimetry (TG-DSC) was done for the specimen AS_0 aged 28 days to validate the results obtained from XRD and SEM analysis. The thermogram is depicted in Fig. 3.6. At around 100 °C, water was moved from hydrated products. When temperature went up to 300°C, a little endothermic peak appeared, corresponding to the dehydration of C—A—Si—H. Furthermore, the endothermic peaks at 450°C and 720°C were attributable to the dehydroxylation of few $Ca(OH)_2$ and the decarbonation of calcium carbonate, respectively. In addition, because of the SiO_2 conversion from anhydrous crystalline to β-SiO_2 and cristobalite, a small endothermic peak was found at 950°C.

3.1.8 Effect of Binders on Leachability of Solidified/Stabilized Sludge

The final solidified/stabilized products were evaluated by the leaching test, aiming to understand the potential toxic contaminants leaching hazard. The toxic contaminant leaching hazard (TCLP) and horizontal vibration (HJ

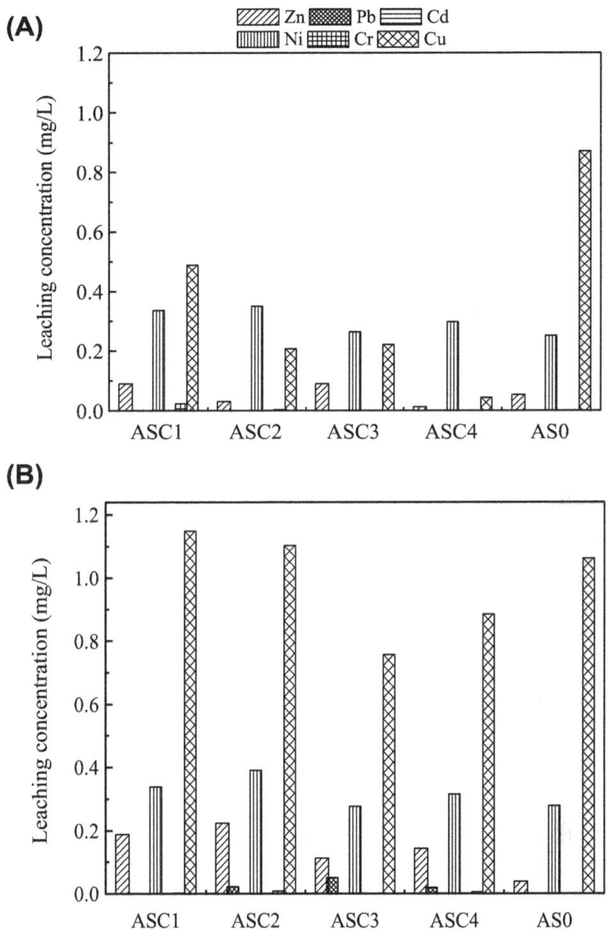

Figure 3.7 Heavy metals' concentrations in eluates after the application of the toxicity characteristic leaching procedure (A), and the HJ 557—2009 test (B) for the pretreated mortars after 28 days of curing time.

557—2009) tests were applied to evaluate the extraction potential of certain metallic elements from the S/S products. The corresponding results are illustrated in Fig. 3.7A and B, respectively. It can be found that the leachates from mortars ASC_1 and ASC_2 presented higher content of metals elements, followed by samples ASC_3, ASC_4, and AS_0. In general, high concentrations of Ni were observed in the all mortars, exceeding 0.3 mg/L. Meanwhile, the China HJ 557—2009 test resulted in a higher release of Cu and Pb, probably thanks to a high pH (8.83—9.39) maintained at the interface, than the US

TCLP test (pH 4.93 of the extraction liquor). These results confirmed that the amphoteric character of Pb at higher pH led to an increase in solubility and consequently a much larger release. In addition, the Cd and Cr concentrations were found to be much lower than regulatory limits (Cd: 1 mg/L; Cr: 15 mg/L) in China (GB 5085.3−2007) as well.

The retention of elemental metals in the solidified sludge appeared to be a combination of more than one process. The leachability was substantially dependent on their chemical speciation, the association to the respective organic structure, or physical retention in a porous structure and especially pH value. Thus, as curing process was proceeding, the pH change of eluates would inevitably affect the solubility of the trace metal species. Additionally, some of them were also considered to be ionic adsorption to the hydrated products in the hydrated paste, such as ionic incorporation into the crystalline network of some compounds of the hydrated binders, which, therefore, induced a reduction in availability of heavy metals.

As far as the heavy metal retention was concerned, it could be concluded that the heavy metals' concentrations did not exceed the toxicity characteristic criteria in China (Cu: 100 mg/L; Pb: 5 mg/L; Cd: 1 mg/L; Ni mg/L; Cr: 15 mg/L), suggesting that there were no notable hazardous effects that could be caused to potential environmental and human health.

3.1.9 Economic Analysis

The results provided the valid evidence that AS solidifier obtained in this study could successfully dewater and solidify/stabilize DS after a short time of curing. Nonetheless, regarding the potential of such an original approach to modify the DS, its treatment cost would be a key consideration. The research work has been applied at Shanghai Laogang Sludge Landfill. Five percent of the calcined AS was firstly mixed with the DS with 80% moisture in a closed workshop, after that the mixture was transferred by sealed vehicles to a special curing site for the next 5−7 days of curing to obtain the solidified sludge with about 50% moisture and more than 50 kPa strength, which enhanced the slope stability of landfill and thus avoided the occurrence of secondary disasters. The corresponding distribution of costs is given in Table 3.4.

The data in Table 3.4 displayed that the deposal costs were approximately 71 CNY/ton sludge (USD $11), much lower than those for drying-incineration (300−500 CNY/ton) and anaerobic digestion, etc. Also, the sludge volume shrank considerably after mixing with AS additives,

Table 3.4 State-of-the-art Cost Distribution for Solidification/Stabilization of Dewatered Sludge Using the Calcined Aluminum Salt

Items	Costs (CNY/Ton Sludge With 80% Moisture)
Blended AS solidifier	50
Wages and benefits for workers	10
Equipment depreciation and maintenance	11
Total	71

in comparison with the volume increase after mixing with ordinary PC, lime, and bottom ash. It should be mentioned that the raw DS with 80% moisture might keep its water content for weeks or even months, without addition of any modified chemicals.

3.2 ALUMINATE $12CAO \cdot 7AL_2O_3$-ASSISTED PORTLAND CEMENT-BASED SOLIDIFICATION/ STABILIZATION

Mayenite ($12CaO \cdot 7Al_2O_3$), one among aluminates, can be detected in some high-performance cement. It owns the outstanding pozzolanic and rapid hardening activity, which makes it sufficiently possible to be employed as the accelerator of cement to enhance its resistance to organics. Whereas very few attentions have been paid to the possibility, thanks to the exiguity of mayenite, as thus to date, there have been no comprehensive studies in the pertinent literature on how the PC—mayenite systems function.

The specific objective of the present section was to investigate the potentials of mayenite as the modifier for PC, and also the PC—mayenite systems as a binder for sludge S/S. This involved varying the mix design (aluminate/ PC ratios and the binder content), evaluating the mechanical and microstructural properties as well as leachability based on pH, also exploring the mechanisms of sludge S/S.

3.2.1 Characterization of Binders

The binders used for S/S of the DS were PC type CEM II 32.5 and mayenite. The mineralogical analysis of PC and synthesized aluminate were done by XRD, and the results are presented in Fig. 3.8. The diffractogram of PC showed the presence of quartz, tricalcium silicate (C_3S), and tricalcium aluminate (C_3A). In reference to the aluminate, mayenite $12CaO \cdot 7Al_2O_3$

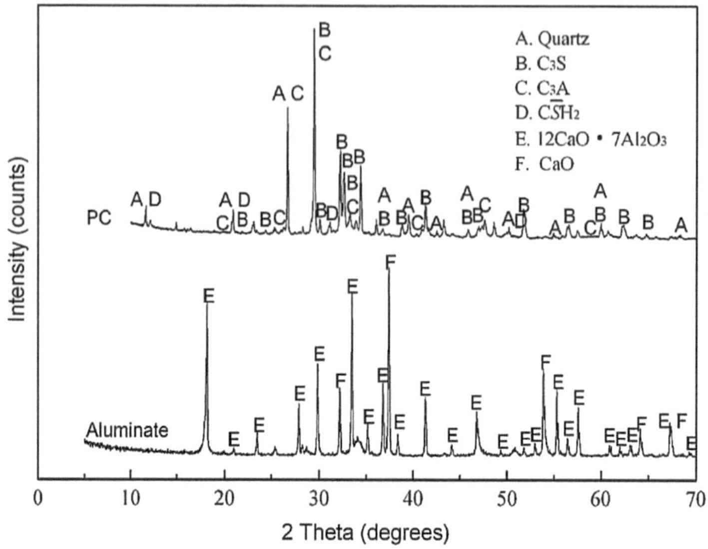

Figure 3.8 X-ray diffractograms of the Portland cement (PC) and aluminate.

was identified as the primary phase, with the main ray at the 18.82 and 33.48 degrees of 2θ.

3.2.2 Confined Compressive Strength

Compressive strength of the pastes solidified by aluminate/PC-blended binders is given in Fig. 3.9. It can be seen that the mixture ratio of aluminate/PC played an important role in strength development of solidified sludge and the presence of aluminate resulted in a strength increase at all ages and more importantly during the early ages. At a given addition of solidifiers (e.g., 10 wt% of sludge), the highest strength of 157.2 kPa could be obtained for paste $AC_{4/6-10}$ after 28 days of hydration. Also the corresponding values were 62.3, 92.8, and 108.8 kPa, respectively, for pastes $AC_{2/8-10}$, $AC_{3/7-10}$, and $AC_{5/5-10}$. The reference $AC_{0/10-10}$, with pure PC, however, exhibited only a strength value of around 25.1 kPa at the same age. Based on the aforementioned observations, it was possibly argued that the aluminate effectively improved the performance of PC, accelerated dissolution and transformation of the silicon and aluminum into the compacted matrices, and hence reduced the interconnectivity of the pores in matrixes, leading to a denser microstructure and higher compressive strength.

Furthermore, as the addition of solidifiers increased to 20 wt% pastes, $AC_{4/6\text{-}20}$ and $AC_{5/5\text{-}20}$ showed outstanding early strength. As shown in

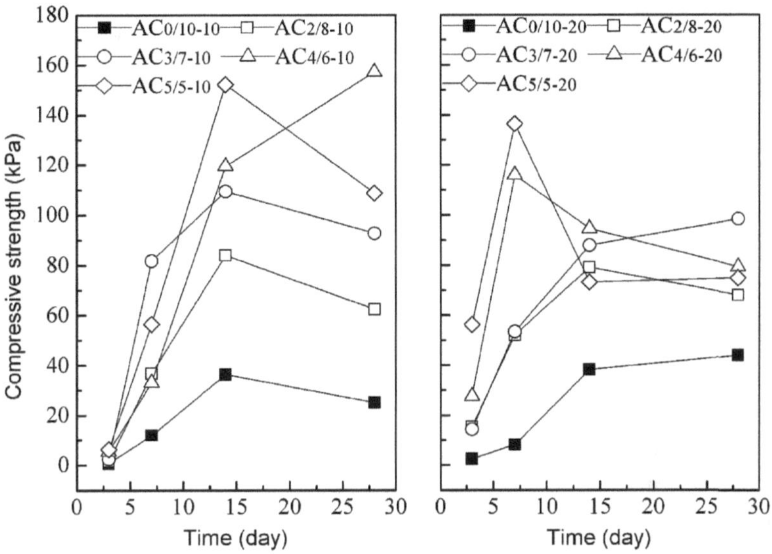

Figure 3.9 Effect of the ratios of aluminate/Portland cement (PC) on the strengths of solidified/stabilized pastes (for the sample ACx/y-z, x/y represents the ratio of aluminate/PC, and z is the addition dose of the aluminate-PC binder).

Fig. 3.9, at 7 days, the activating effect of aluminate tended to reach the maximum, where the two samples exhibited the compressive strengths of 115.9 and 136.4 kPa respectively. Afterward, a rapid decrease was detected with the final strengths of approximately 79.3 and 74.9 kPa, respectively, at 28 days of age, which were noticeably higher as compared to reference $AC_{0/10-20}$ where the strength was 43.9 kPa, and yet were, respectively, around 24.1% and 31.4% lower than specimen $AC_{3/7-20}$. The abrupt reduction observed in strength was probably contributable to the fact that the sulfate present in PC was depleted, and therefore insufficient to complete with the remaining of aluminate. In addition, the conversion of main hydration products in the solid matrix was responsible likewise.

3.2.3 Crystalline Phase and Chemical Composition Analysis

Many factors influence the strength development of aluminate—PC/sludge systems. It is apparent that the aluminate/PC ratio is critical to the hydration mechanisms. To determine the influence of aluminate/PC ratio on hydration, XRD analysis was therefore carried out. Fig. 3.10A displays the X-ray

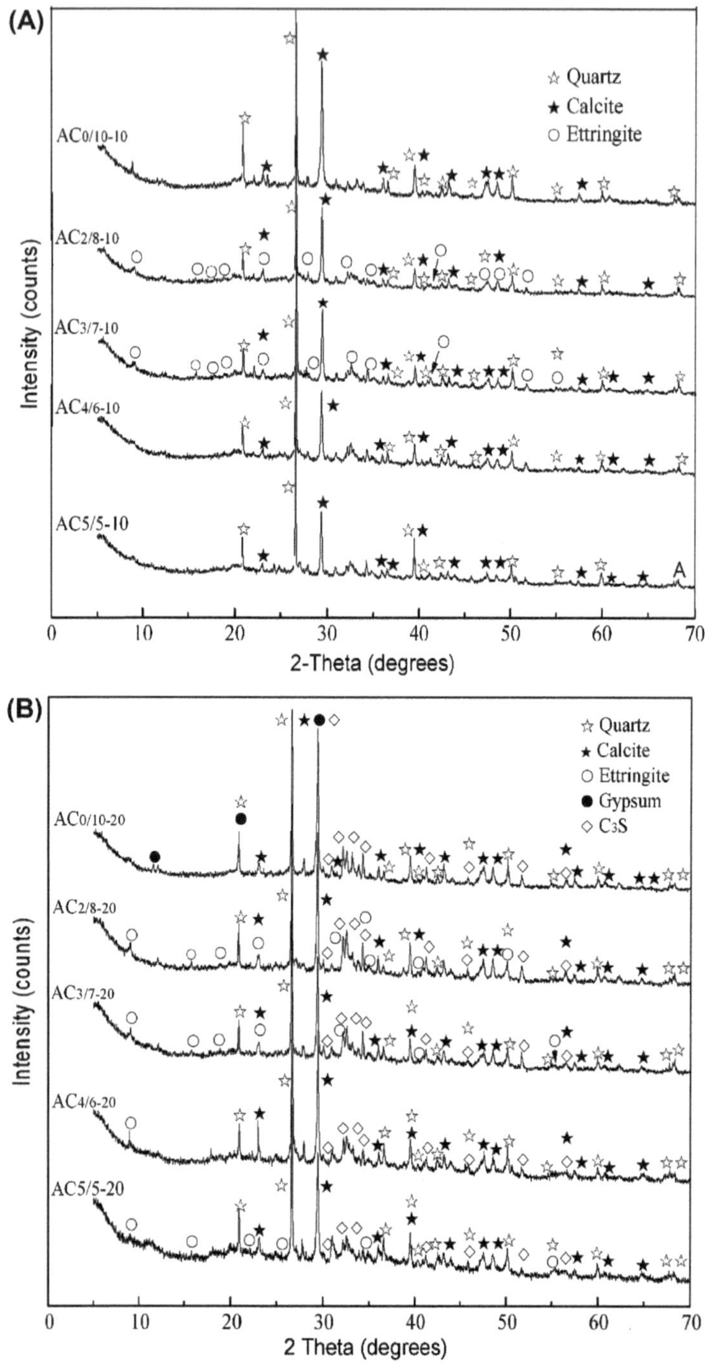

Figure 3.10 X-ray diffractograms of the various solidified/stabilized pastes.

diffractograms of S/S pastes with the addition of 10 wt% solidifier, revealing the strong presence of ettringite ($C_6A\overline{S}_3H_{32}$) in $AC_{2/8-10}$ and $AC_{3/7-10}$ samples. As expected, the hydrate was not detected in reference $AC_{0/10-10}$ probably thanks to its high organic content (e.g., 37.13 wt% of dried sludge), whereas a very interesting finding was that the similar organic contents were measured in $AC_{2/8-10}$ and $AC_{3/7-10}$ samples (around 39.04 and 40.66 wt%, respectively). It became very clear that the presence of aluminate $12CaO{\cdot}7Al_2O_3$ eliminated the interference from organic compounds, and enhanced the formation of ettringite through Eqs. (3.2) and (3.3). The crystalline compound was able to fill the pores of the pastes making a contribution to the setting and conspicuous strength development.

$$C_3A + 3C\overline{S}H_2 + 26H \rightarrow C_6A\overline{S}_3H_{32} \quad (3.2)$$

$$C_{12}A_7 + 3C\overline{S}H_2 + 53H \rightarrow C_6A\overline{S}_3H_{32} + 3AH_3 + 3C_3AH_6 \quad (3.3)$$

with: $C = CaO$; $\overline{C} = CO_2$; $A = Al_2O_3$; $\overline{S} = SO_3$; $H = H_2O$.

Regarding $AC_{4/6-10}$ and $AC_{5/5-10}$ pastes, more appreciable strength was developed at 28 days, however, XRD failed to identify ettringite. The formation and stability of ettringite was closely related to the ratio of sulfate and aluminates in binders as well, implying that the sulfate from PC was the limiting reactant. Moreover, the conversion of ettringite to monosulphoaluminate hydrate $C_4A\overline{S}H_{12}$, C_4AH_{13}, and C_2AH_8 might take place due to the presence of superfluous $12CaO{\cdot}7Al_2O_3$ (Eq. 3.4) in aforementioned pastes. Additionally, since the samples have been cured at ambient temperature, the carbonation of samples might have made ettringite turn to calcite (Eq. 3.5), leading to disappearance of ettringite. This could be partially confirmed by the XRD analysis that higher peak intensities of calcite could be detected in $AC_{4/6-10}$ and $AC_{5/5-10}$ pastes in comparison with those in $AC_{2/8-10}$ and $AC_{3/7-10}$, as shown in Fig. 3.10A. However, XRD failed to identify calcium-bearing hydrates too (i.e., $C_4A\overline{S}H_{12}$, C–S–H, C_4AH_{13}, and C_2AH_8) probably due to their small amount and low crystallinity.

$$C_{12}A_7 + C_6A\overline{S}_3H_{32} + 34H \rightarrow 3C_4A\overline{S}H_{12} + 1/2C_4AH_{13} + 2C_2AH_8 + 5/2AH_3(\text{amorphous}) \quad (3.4)$$

$$3C_6A\overline{S}_3H_{32} + \overline{C} \rightarrow 3C\overline{S}H_2 + 3C\overline{C} + A{\cdot}XH + (26 - X)H \quad (3.5)$$

The main hydration products with the same aluminate/PC ratio and different additions of blended binder in aluminate/PC-sludge systems seemed much more similar, as shown in Fig. 3.10A and B. Only difference was that C_3S peaks appeared in the pastes with the 20 wt% addition of binder. The observation that C_3S remained inactive until the termination of curing revealed that C_3S hydration had been delayed or inhibited owing to the poison of sludge, and that an incompatibility between the aluminate and PC hydration products took place when addition of binder was higher than 20%. From these findings, it can be assumed that both the consumption of ettringite and delay of C_3S hydration might be especially responsible for the considerable decrease in the strengths of $AC_{4/6\text{-}20}$ and $AC_{5/5\text{-}20}$ pastes.

3.2.4 Thermal Analysis

Fig. 3.11 compares derivative thermogravimetric analysis/differential scanning calorimetry data of S/S products with different aluminate/PC ratios after 28 days of curing. As in the XRD results, the thermal analysis reflected the differences in phase composition of hydrated pastes.

The weight loss occurred at about 120°C, exhibiting a weak peak in the DTG curves, which might be attributed to the evaporation of free and

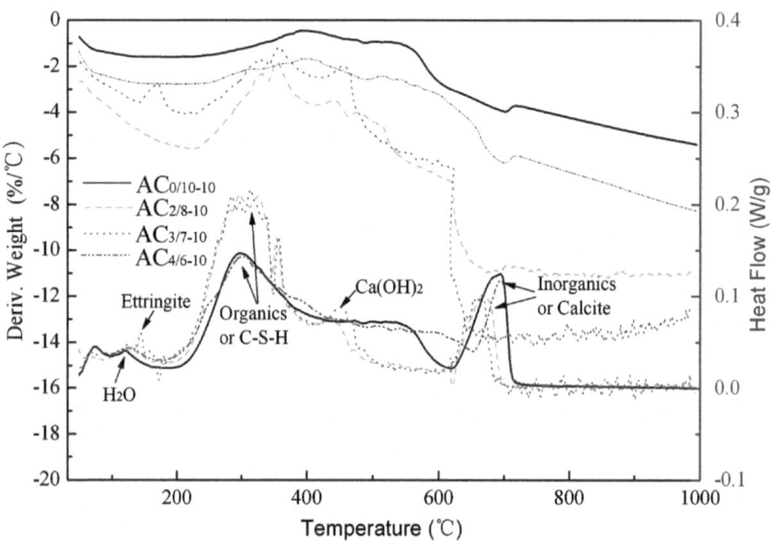

Figure 3.11 Differential thermogravimetry/differential scanning calorimetry curves of the various solidified/stabilized pastes at 28 days.

bound water of sludge. When temperature was up to 160°C, a weight loss was detected in $AC_{3/7\text{-}10}$, but not in other samples, indicating that considerable ettringite has formed in the specimen, corresponding exactly with the results from XRD analysis. The peaks at 250–350°C referred to the pyrolysis of organics from sludge and dehydration of C–S–H with much lower level of crystallization. Also, several other peaks appeared for the solidified pastes except $AC_{0/10\text{-}10}$ at around 450°C, which was mainly because of the water loss of portlandite in traces formed by hydration of free lime from the synthesized aluminate. Depending on the aluminate/PC ratio, the peaks with various intensities were visible at 700°C or so in the DTG curves, relating to CO_2-loss of calcium carbonate and pyrolysis of some inorganic matters within sludge.

3.2.5 Scanning Electron Microscopy Analysis

The pastes with the formulation aluminate/PC = 0:10, 3:7 and 4:6 after 28 days of hydration respectively were subjected to SEM analysis. The results are displayed in Fig. 3.12. In the case of low-aluminate binder (Fig. 3.12A), the hydrate phases covering the sludge particles exhibited the much lower crystallization and no enough of the gel-like hydration products has formed to fill the pore space and to embed sludge particles, suggesting that the presence of organic materials introduced from sludge retarded the hydration of PC. As a result of this, the matrix appeared extremely loose, and thus it was possible to verify numerous voids visible. The high porosity and low crystallinity of hydration products provided the basic evidence for the low strengths of samples $A_{0/10\text{-}10}$ (Fig. 3.9) at this stage.

Fig. 3.12C shows a special SEM photo of the paste $AC_{4/6\text{-}10}$. SEM observations confirmed the presence of a large amount of vitreous and prismatic-shaped ettringite with high crystallinity, homogenously distributed in the paste. This finding corresponded well with those observed from XRD and DTG/DSC analysis, as shown in Figs. 3.9 and 3.10. When added, aluminate $12CaO \cdot 7Al_2O$ rapidly reacted with $C\overline{S}H_2$ to form ettringite gels as shown in Eq. (3.2), which covered the sludge particles, reduced, and even offset the harmful organics interference providing a safe environment for PC hydration. As a result of this, the PC hydration and ettringite generation were considerably accelerated. These hydration phases filled the pores in the mortar and interconnected the sludge particles as physically or chemically as possible. The filling effect decreased the porosity, altered the morphological characteristics, and subsequently made the mortar

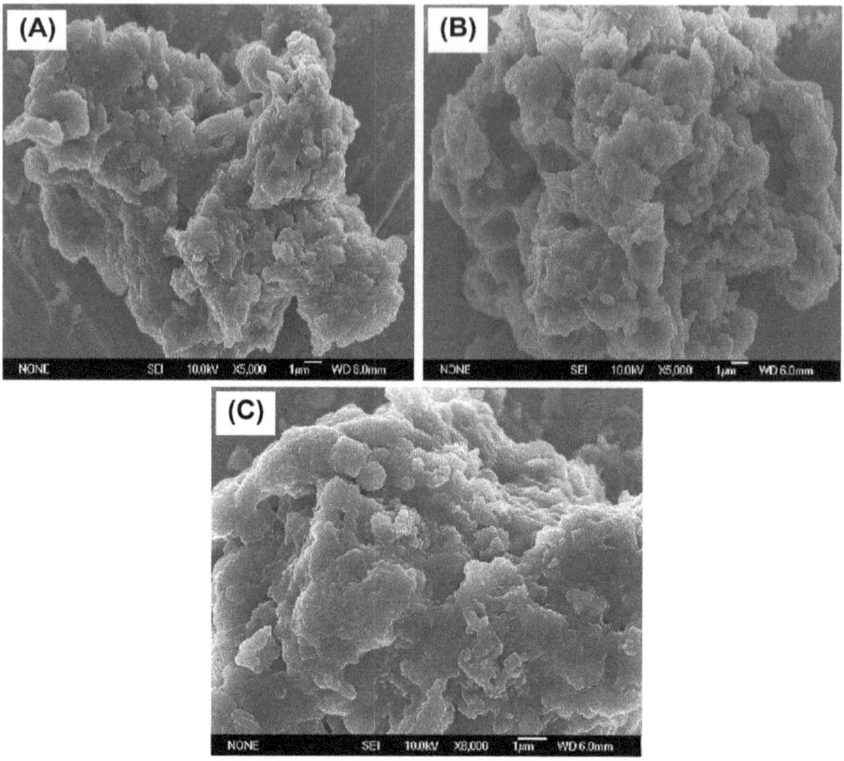

Figure 3.12 Scanning electron microscopy images of the samples $AC_{0/10\text{-}10}$ (A), $AC_{3/7\text{-}10}$ (B), and $AC_{4/6\text{-}10}$ (C) cured under room temperature for 28 days.

denser and stronger in compression. The low porosity of the gel-like matrix and the tight connection between the sludge particles might be the reason for the significantly high compressive strength of mortar $AC_{4/6\text{-}10}$.

3.2.6 Acid Neutralization Capacity

The acid neutralization capacity (ANC) can be used to evaluate the ability of a certain material to neutralize acid and the likely performance of contaminant immobilization in an acidic environment, since precipitation/dissolution of metals is directly affected. The amount of acid added in each solution was converted into equivalents of acid per dry weight of the material so as to obtain ANC. Fig. 3.13 shows ANC curves obtained from the 10% binder matrixes cured for 28 days. It is obvious that the starting pH values were extremely low (pH 8.8–7.8) in part due to the absence of hydrates $Ca(OH)_2$, with a steep initial gradient to pH 7, followed by a plateau at

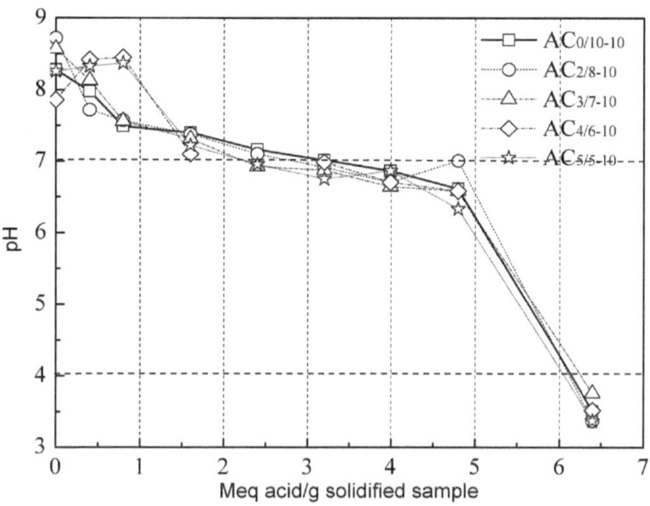

Figure 3.13 Acid neutralization capacity curves of the 10 wt% binder samples.

pH 7−6. The acid neutralization capacities of pH 7 (referred to as ANC_{pH7}) and pH 4 (ANC_{pH4}) were considered to evaluate their acid resistance. As depicted in Fig. 3.13, though the similar ANC curves were obtained for all the solidified pastes, the capability of resisting to acid attack varied slightly, dependent on aluminate/ANC ratio. The ANC_{pH4} values (6.1−6.3 meq/g) were relatively similar for all the solidified samples; however, a small reduction in ANC_{pH7} values was observed. The corresponding ANC_{pH7} values were around 2.7, 2.6, 2.2, and 2.1 meq/g for samples $AC_{2/8\text{-}10}$, $AC_{3/7\text{-}10}$, $AC_{4/6\text{-}10}$, and $AC_{5/5\text{-}10}$, respectively, which were somewhat lower than that of reference $AC_{0/10\text{-}10}$ (3.2 meq/g). The buffering capability of calcite, low Ca/Si ratio C−S−H, and silica gel were, due to their low dissolution characteristics, mostly responsible for the ANC of solidified mortars, which was observed to occur between pH 7−5. The introduction of aluminate in the present study was observed to induce the generation of ettringite ($C_6A\overline{S}_3H_{32}$) and yet result in low calcite, C−S−H and silica gel contents as previously stated. This presumably aroused the small decrement of ANC values and lowered the resistance to long-term acid attack.

3.2.7 Leaching Behavior of Heavy Metals in Solidified Sludge Exposed to Nitric Acid

To evaluate the influence of pH on leaching behavior of toxic heavy metals, we plotted the leachate concentration at key points in the ANC curves versus pH (Fig. 3.14). These data clearly depicted that leaching levels showed to be strongly pH-dependent, and the presence of aluminate caused slightly higher solubility of Pb, Cd, Ni, and Cr in a wide range of pH 3–6. For instance, 20% replacement of cement with aluminate increased leachable metal concentrations for metals such as Pb and Cd up to about 300–400% at pH around 3.5 compared with the ANC only-sludge mixture, while the dissolution/retention of heavy metals in solidified sludge appeared to be a relatively complex process. It depended on their chemical speciation, association to the respective organic complexes, adsorption on specific binding sites of mineral surface, and physical retention in a porous structure etc.

In respect to Pb, the increased solubility in the S/S pastes with the addition of aluminate, as seen in Fig. 3.14A, was most likely due to the competition between aluminate and Pb for the active components from ANC, which influenced the ionic incorporation/complexation into the crystalline network of hydrated products. That was a possible explanation that the blended additives were less effective for lead fixing than ANC (Fig. 3.14A). Furthermore, a

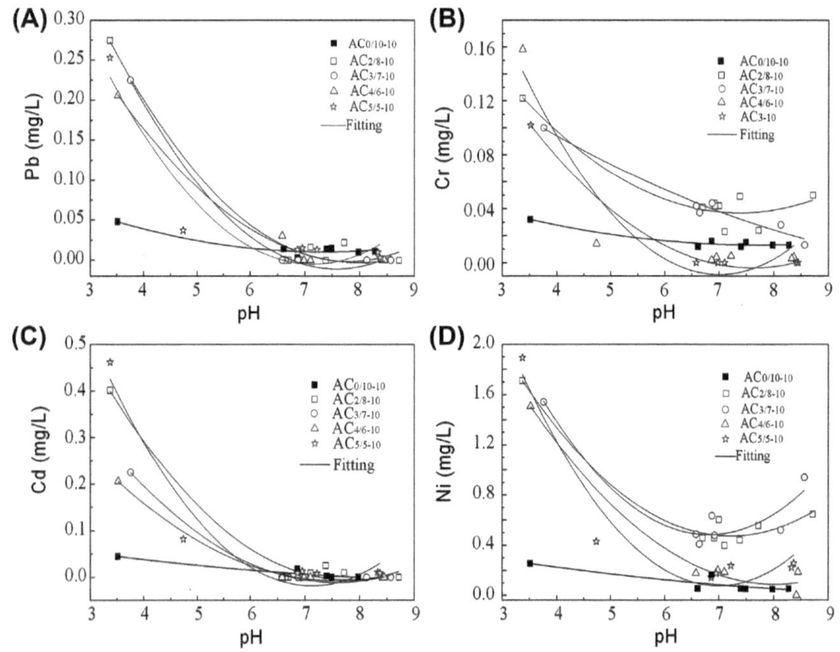

Figure 3.14 Leached amount (mg/L) of Pb (A), Cr (B), Cd (C), and Ni (D) as a function of pH in solidified sludge exposed to nitric acid.

similar leaching behavior could be observed for Cr, Cd, and Ni (Fig. 3.14B, C and D). The results also showed that all mortars had lower leaching of heavy metals, clearly below the legal limits in China.

For the retention of Cr and Cd, XRD analysis (Fig. 3.15) indicated the presence of $Fe_2(CrO_4)_3(H_2O)_3$ and $Al_{13}(OH)_{11}(CrO_4)_4 \cdot 36H_2O$, accordingly supporting the coprecipitation mechanism. Meanwhile, the encapsulation into hydrates phases (e.g., $Ca_6Al_2Cr_3O_{18} \cdot 32H_2O$ and $Ca_4Al_2CrO_{10} \cdot 12H_2O$) partially corresponded to the leaching control of Cr. The configuration of Cd in S/S pastes, however, possibly because of the low crystallization, was not detected. In addition, the complexation with organics and silicates from the decalcification of C—S—H might also affect the leachability of Cr and Cd. Referring to Ni, surface adsorption reaction, however, was generally considered as the most concernful controlling factor.

Nevertheless, leachate levels of approximately 2.0 mg/L or below were observed, although the leaching of these elements increased with decreasing pH. The relatively low extract amount for acidic pH made them not critical ones when solidified sludge was disposed in a sanitary landfill, in spite of the low legal limits (Pb: 5 mg/L; Cr: 15 mg/L; Cd: 1 mg/L; Ni 5 mg/L) in China (GB 5085.3—2007). Evidently, the addition of aluminate as a modifier for ANC during the S/S process of sludge was not markedly detrimental to the leaching control of heavy metals.

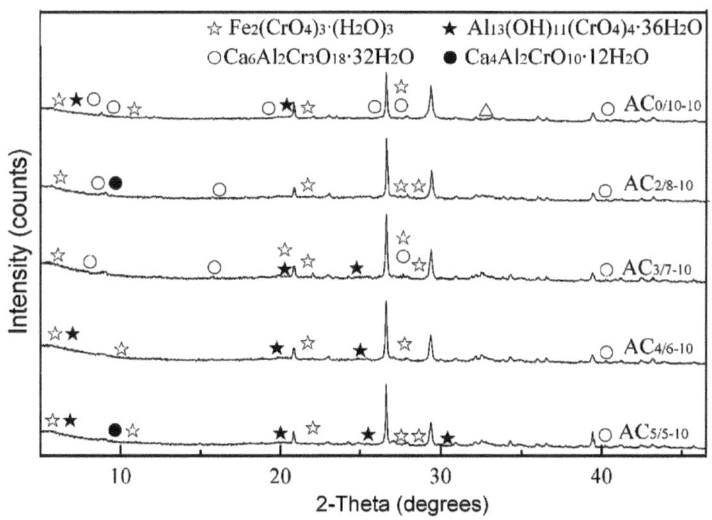

Figure 3.15 X-ray diffractograms of the S/S pastes between $2\theta = 5.2$ and 46.8 degrees.

3.3 HYBRID CEMENT-ASSISTED DEWATERING AND SOLIDIFICATION/STABILIZATION OF SEWAGE SLUDGE WITH HIGH ORGANIC CONTENT

A main barrier for the effective solidification of sludge is its high organic content; high content of organics (e.g., phenols) can retard the hydration of cement paste and inhibit the formation of cement hydrates, accordingly leading to a slow strength development.

Mayenite, a newly synthesized binder as described in Section 3.2, showed a promising ability on dewatering and solidification of sludge when combined with PC. In this section, its potential as the modifier for sulfoaluminate cement in sludge solidification treatment was further assessed. Sulfoaluminate and ferroaluminate cements contain high content of aluminates and ferrates, respectively, and they can rapidly develop strength with ideal mechanical properties while producing low greenhouse gases. More importantly, the cements have the good compatibility to other materials, and thus can be an alternative to PC for waste stabilization. Here, seven kinds of binders consisting of different oxides [MgO, mayenite ($12CaO \cdot 7Al_2O_3$)] and cements (sulfoaluminate, ferroaluminate, and PC) were investigated and a novel S/S technology using CASC ($12CaO \cdot 7Al_2O_3$/sulfoaluminate cement) binder was developed to dewater and stabilize sewage sludge with high organic content.

3.3.1 Test Material and Experimental Procedures

The sewage sludge was sampled at sludge dewatering units of two sewage treatment plants in Guiyang City, China. The sludge *R* was taken from sequencing batch reactorprocess of the *Small River Sewage Plant* with a daily capacity of 80,000 tons; the sludge *V* was from reversed A^2O process of the *New Village Sewage Plant* with a daily treatment capacity of 250,000 tons. The main characteristics of the samples are listed in Table 3.5.

Totally, seven kinds of hybrid binders were prepared by mixing MgO and mayenite ($12CaO \cdot 7Al_2O_3$) with ordinary Portland, sulfoaluminate, or ferroaluminate cement according to Table 3.6. The binders with mass ratios from 3% to 15% were added into sludge samples, and one sample without adding any binder was also prepared as a control.

3.3.2 Variation in Water Content of Solidified Sludge Over Curing Time

After 7 days of curing, water contents of the two sludge samples that had been solidified by different hybrid binders with different rates of

Table 3.5 Basic Properties of the Dewatered Sludge

Item	Sludge R	Sludge V
Water content (%)	73.64	72.49
pH	6.20	6.60
VS (%)	62.50	72.02
Total organic carbon (TOC, %)	27.10	32.41
Heavy Metals (g/kg Dry Sludge)		
Zn	0.74	6.76
Cu	0.15	0.14
Cr	0.11	0.46
Pb	0.04	0.03
As	0.03	0.03
Ni	0.03	0.03
Cd	Not detected	0.01

binder/sludge were measured. In general, the water contents of solidified sludge decreased with the increase of binder addition rates. During dewatering process, the water of sludge was removed by two pathways; one part of water was evaporated with the heat produced by hydration reaction, and the other part of water was incorporated with hydration products in the cement matrix of sludge.

Fig. 3.16 shows that the binders of MSC3/2, CAPC2/3, and CASC2/3 exhibited a high sludge-water removal efficiency. These binders contained high content of aluminum compounds such as synthesized mayenite ($12CaO\cdot7Al_2O_3$) and/or sulfoaluminate cement, indicating that aluminum compounds played a key role on dewatering of sludge even with high

Table 3.6 Nomenclature and Components of the Binders

Nomenclature	Components	Mass ratio (m: m)
MPC3/2	MgO/Portland cement	3:2
MSC3/2	MgO/sulfoaluminate cement	3:2
MFC3/2	MgO/ferroaluminate cement	3:2
CAPC2/3	Mayenite ($12CaO\cdot7Al_2O_3$)/ Portland cement	2:3
CASC2/3	Mayenite ($12CaO\cdot7Al_2O_3$)/ sulfoaluminate cement	2:3
CAFC2/3	Mayenite ($12CaO\cdot7Al_2O_3$)/ ferroaluminate cement	2:3
CASC3/2	Mayenite ($12CaO\cdot7Al_2O_3$)/ sulfoaluminate cement	3:2

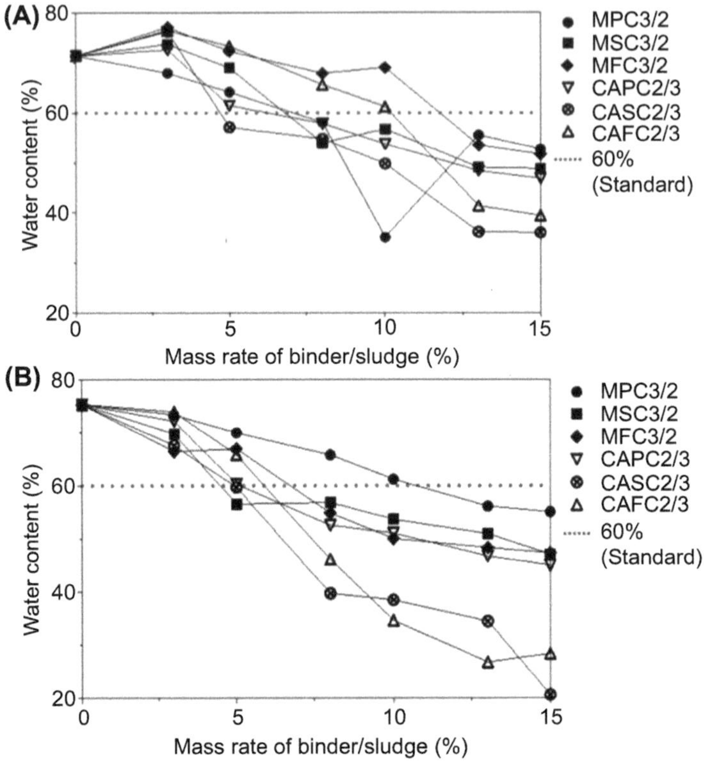

Figure 3.16 Effect of mass ratio of binder/sludge on the water content of solidified sludge R (A) and V (B) after 7 days of curing time.

organic content. It has been observed that mayenite could quickly deplete the interstitial water to form hydrated compounds through moderate pozzolanic activity, and it was confirmed that main components of sulfoaluminate cement—yeelimite ($Ca_3Al_6O_{12}CaSO_4$) and mayenite—are responsible for accelerating hydration process. However, the CAFC2/3 binder, which comprised mayenite, showed a poor efficiency in dewatering for sludge R. The water content was still above 60% even the binder rate was higher than 10%, but decreased abruptly when the addition rate of CAFC2/3 was raised to 13%. The CASC2/3 binder, of which the ratio of mayenite and sulfoaluminate cement was 2/3, showed the best performance in both sludge samples. For sample R, the water content was reduced to lower than 60% with only 5% of addition ratio. As the ratio was increased to 15%, the water content was ultimately decreased to 36%. For sample V, although the water content was slightly higher than 60% at addition ratio 5%, it decreased rapidly to 40% with the addition rate of 8%.

3.3.3 Variation in Strength of Solidified Sludge Over Curing Time

To evaluate the solidification effect, UCS tests were conducted to investigate the optimal type of binder and its addition ratio. Generally, the lower the water content of sludge, the higher the compressive strength. The results showed that both MSC3/2 and CASC2/3 were more efficient in strength development for sludge R while CASC2/3 and CAFC2/3 delivered higher UCS values in sludge V. The UCS increment was consistent with water depletion, which can be jointly attributable to the effect of binders on sludge dewatering and solidification.

In Figs. 3.17 and 3.18, the UCS values of all solidified mortars were higher than 50 kPa with only 8% of binder addition after 7 days of curing, except for MFC3/2 and CAFC2/3-solidified sludge R, and this strength met the UCS requirement of sanitary landfill operation. These results suggested that cement components could hydrate rapidly and yield enough reactive minerals that incorporate the sludge particles into the stiff matrix. The solidification process consequently improved structural integrity and induced ample strength. Sulfoaluminate and ferroaluminate cements, as quick hardening cements, are favored to form hard cement matrix, and therefore present a good efficiency in strength development. However, it should be noticed that the strength development of mortar is closely related to water content of sludge. Insufficient dewatering of sludge led to poor strength performance of binders, as the case of CAFC2/3-solidified sludge

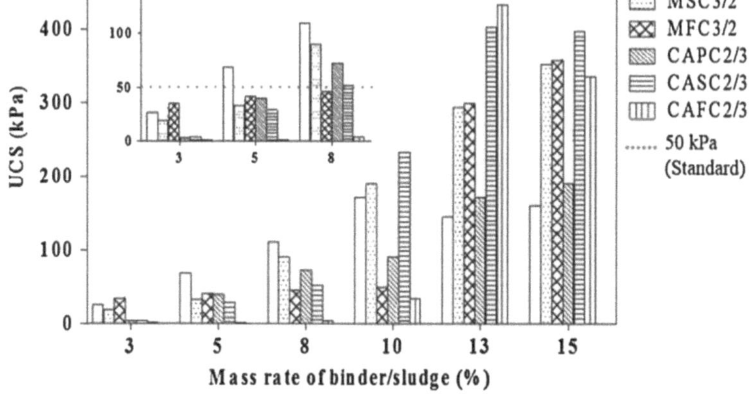

Figure 3.17 Effect of mass ratio of binder/sludge on the unconfined compressive strength of solidified sludge R after 7 days of curing time.

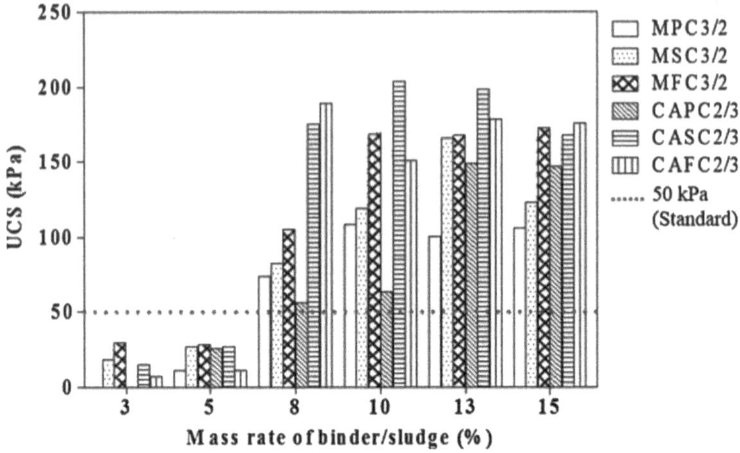

Figure 3.18 Effect of mass ratio of binder/sludge on the unconfined compressive strength of solidified sludge V after 7 days of curing time.

R at 8% addition. Besides, low UCS value of the normal PC-added mortar might be mainly due to the interference of organic substances in the sludge during the hydration process. The organics may be responsible for the retardation of normal cement hydration, and thus affect the strength development of mortars.

The UCS test showed that CASC2/3 binder had the optimal strength; the sludge R exhibited the highest UCS value of 402.90 kPa with the addition ratio of 13%. For sludge V, the UCS value reached 203.59 kPa with CASC2/3 addition ratio of 10%.

3.3.4 Selection of Binder for Dewatering and Solidification of Sewage Sludge With High Organic Content

The results indicated that mortars solidified by CAPC2/3 and CASC2/3 at 8% of binder/sludge addition after 7 days of curing can meet the requirements for sanitary landfill operation. Therefore, based on the performance of binders on water depletion and strength development, CAPC2/3 and CASC2/3 were selected for further study on sewage sludge with high organic content. In addition, concerning the effectiveness of mayenite, the mayenite/sulfoaluminate ratio of CASC2/3 was modified from 2/3 to 3/2, and defined as CASC3/2.

Fig. 3.19 shows the variation in sludge dewatering and strength development of CAPC2/3, CASC2/3, and CASC3/2. The water contents of CASC2/3-solidified sludge R and V were 55% and 50%, respectively,

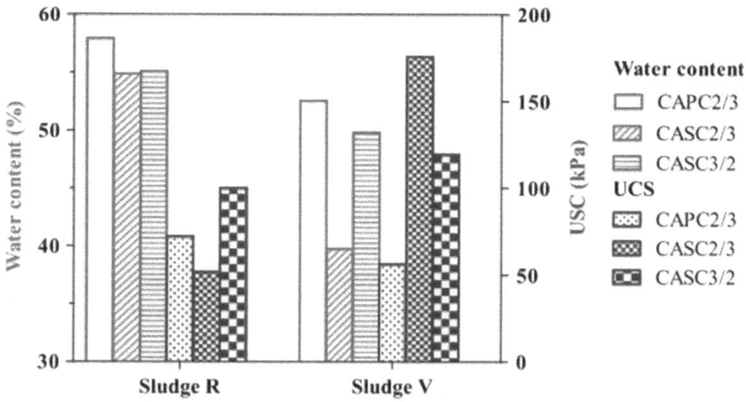

Figure 3.19 Selection of binder according to water content and unconfined compressive strength tests.

which were better in dewatering efficiencies than those of CASC3/2. However, in the UCS test, high percentage of aluminum compounds in CASC3/2 effectively strengthened the mortar of sludge R with the UCS value above 100 kPa. Nevertheless, considering the combined effects of strength development and dewatering, CASC3/2 is still the recommended binder for dewatering and solidification of high organic content sludge.

3.3.5 Leaching Behaviors of Heavy Metals

The stability of heavy metals in the solidified sludge is a crucial factor in determining the efficiency of S/S technology. A horizontal vibration leaching test (HJ 557—2009) was carried out to determine the leachability of heavy metals in the original, CAPC2/3, and CASC3/2-solidified sludge. The results showed that heavy metal concentrations in the extracts of solidified sludge were far below that of original sludge as well as regulatory limits in China (GB 5085.3—2007), indicating that heavy metals were well confined and contained in cement matrix of solidified sludge. In the well-solidified sludge, heavy metals were physically entrapped by encapsulation or chemically immobilized in the cement hydrates. As presented in Fig. 3.20, the S/S process increases the stability of sludge, evidently in the case of CASC3/2-solidified sludge and confirms that CASC3/2 is an excellent binder in stabilization of sludge with high organic content.

Besides, the pH of solidified sludge reached 10 after S/S while the original pH values of sludge R and V were 6.2 and 6.4. The alkalinity of sludge increased the binding of metals on solid surfaces and even possibly induced

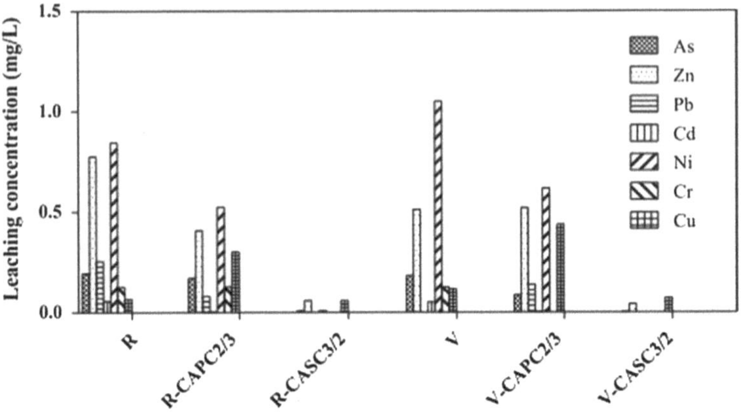

Figure 3.20 Heavy metals' concentrations in extracts of sludge samples.

the metal precipitation, because most metals, such as As, Zn, Ni, Cr, and Cu, have minimal solubility within pH range of 9–11. From the aforementioned results, considering the stability of heavy metals of extracts, it could be concluded that the CASC3/2 binder is the excellent option for S/S of sewage sludge with high organic content.

3.3.6 Characterizations of Hydrated Phases

SEM images and XRD patterns of original sludge and CASC-solidified sludge were analyzed to investigate the mechanisms of sludge S/S process. The XRD patterns of CASC showed that mayenite and $Ca(OH)_2$ were the effective components, besides cement components in the sludge.

Comparing the SEM images between original and solidified sludge samples, it could be observed that solidification favored the consolidation of sludge particles, resulting in reduction of particle size. In Fig. 3.21, the construction of original sludge was incompact and disoriented, whereas a homogeneous network of sludge particles became visible after the S/S process. The formation of smaller and more condensed clots enhanced the containment and therefore promoted the ability of encapsulation. In addition, needle-like or long tree-like crystals were found at SEM images of CASC-solidified sludge. The crystals were well connected and aggregated through hydration process, forming a macroencapsulation of sludge into cement matrix.

Later, XRD analysis further revealed that the heavy metals were chemically fixed into insoluble cement components. It is noted that trace elements

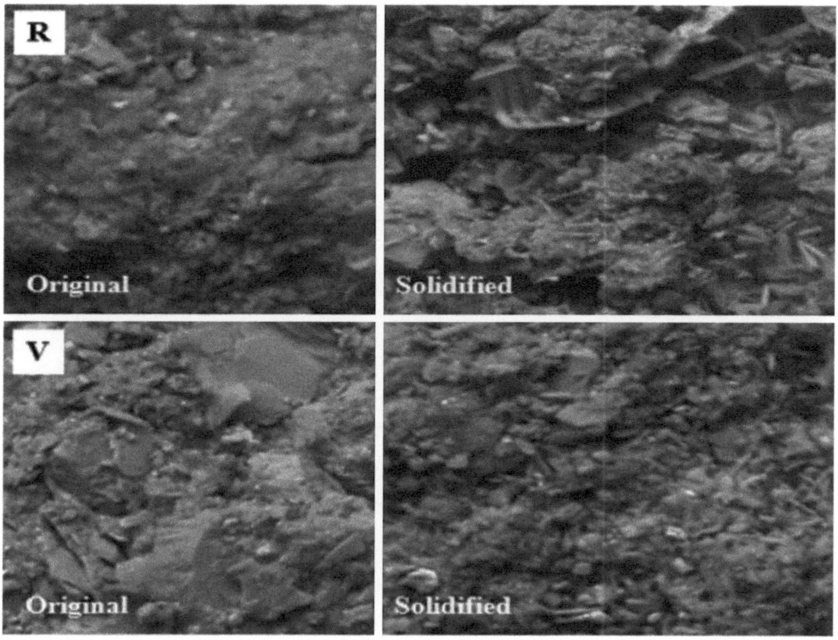

Figure 3.21 Comparative scanning electron microscopy images between original and CASC-solidified sludge samples (×10,000).

can be integrated in calcium silicate hydrates (CSHs) and Ca—Al hydroxy-sulfate. From Fig. 3.22, it could be confirmed that CSHs, such as illite, were synthesized through cement hydration reaction. In addition, chromian was found in solidified sludge R, which might have contributed to the negligible concentration of metals in extract of the solidified R in Fig. 3.20. The formation of hydrates with morphology change in solidified sludge could jointly explain the advanced solidification and stabilization capability of CASC binder.

3.3.7 Economic Analysis

The original approach of applying S/S technology to treat with sewage sludge is the economic feasibility in a developing country. Therefore, the treatment cost is a key factor of practical application. The results showed that the binder of CASC3/2 successfully dewatered and solidified sewage sludge with high organic content at only 8% addition rate after 7 days of curing.

Table 3.7 presented a simple economic analysis on condition that the CASC3/2 binder was actually applied in dewatering and solidification of

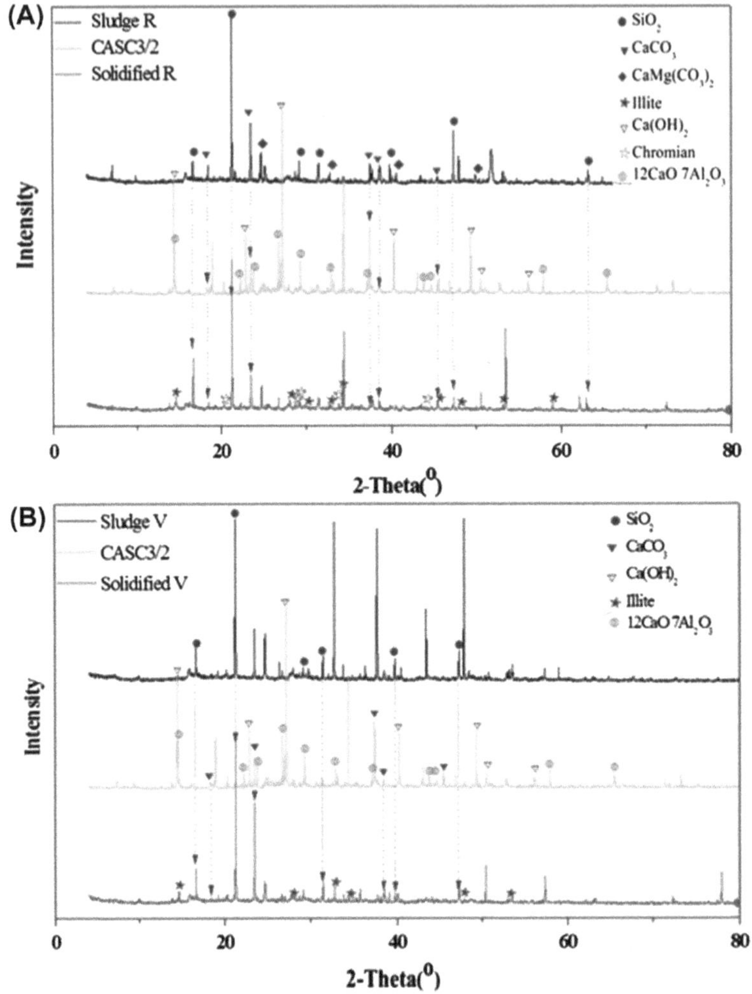

Figure 3.22 Comparative X-ray diffraction patterns between original, CASC binder and solidified sludge R (A) and V (B).

sludge V and sludge R. The CASC3/2 was firstly mixed with the DS at binder/sludge rate of 8% in a closed workshop, and the mixture was transported by special vehicles to the curing site. As the unit value of CASC3/2 is expected to be around 880 CNY/ton (143 USD/ton), which was based on the unit value of each component, the disposal cost was about 90 CNY/ton sludge (15 USD/ton sludge), lower than operation costs of anaerobic digestion drying (100–260 CNY/ton sludge) and drying-incineration (300–500 CNY/ton sludge).

Table 3.7 Cost Analysis for Solidification/Stabilization of Dewatered Sludge Using the CASC3/2 (1USD = 6–7 CNY)

Content	Unit Price (CNY/Ton)	Costs (CNY/Ton Sludge)
Mayenite ($12CaO \cdot 7Al_2O_3$)	1000	
Sulfoaluminate cement	600	
CASC3/2	880	70
Wages and benefits for workers		10
Equipment depreciation and maintenance		10
Total		90

3.4 STABILIZATION/SOLIDIFICATION USING MAGNESIUM OXYCHLORIDES CEMENT

Magnesium oxychlorides cement (MOC), also known as Sorel cement, is a type of nonhydraulic and rapid hardening cement. It is manufactured by blending powdered magnesium oxide (MgO) with magnesium chloride ($MgCl_2$) in certain mixing ratios. In the presence of water, it reacts mutually to form $MgO-MgCl_2-H_2O$ ternary system and becomes a kind of high-strength material. Two main reaction phases, $3Mg(OH)_2 \cdot MgCl_2 \cdot 8H_2O$ (phase 3) and $5Mg(OH)_2 \cdot MgCl_2 \cdot 8H_2O$ (phase 5), particularly exist as reinforced components in the ternary system at ambient temperature. Shapes of phase 3 and phase 5 may be observed as needle-like or long tree-like crystal at SEM images. The crystals are aggregated and connected with each other at crystallization point even like net structure on a condition that hydration would be reacted completely. Compared with PC or other binders, MOC may have a higher solidified effect and short solidified time. When it is mixed with sewage sludge, powdered MgO and $MgCl_2$ may react rapidly on water contained in the sludge and the crystal structure allows the solidified products to be of high mechanical strength.

In this section, MOC is investigated as a new type binder in the S/S process of sewage sludge, including the influences of the molar ratios of MgO/$MgCl_2$ and MOC/sludge on the S/S process, strength development, XRD, SEM, and leaching test.

3.4.1 Test Material and Experimental Procedures

Dewatered sludge (DS) was taken from a full-scale WWTP in Shanghai, China (Table 3.8). MgO with a purity of 98%, bulk density of 0.35 g/

Table 3.8 Characteristics of Dewatered Sludge Used for Stabilization/Solidification Using Magnesium Oxychlorides Cement

Parameters	Data
Moisture content (%)	85.4
Total organic carbon (%)	23
pH	6.65
Coefficient of permeability (cm/s)	1.2×10^{-8}
Unconfined compressive strength (kg/cm^2)	0.085
Heavy Metals (mg/kg Dry Basis)	
Cu	3500
Zn	5984
Pb	112
Cd	8.57
Cr	670
Ni	34.8
As	224
Chemical Compositions (%)[a]	
Silica (SiO_2)	24.52
Lime (CaO)	16.33
Alumina (Al_2O_3)	35.65
Magnesia (MgO)	2.84
Iron oxide (Fe_2O_3)	16.19
Loss on ignition[b]	9.67

[a]Dry sludge ash.
[b]Loss on ignition at 1100°C.

cm^3, and $MgCl_2$ of hexahydrate crystal, with a high hygroscopic adsorption and deliquescence effect even in moist air, were used as binders for the sludge. Components of MOC were MgO and $MgCl_2$ in different molar ratios ranging from 1:1 to 5:1, with a mass ratio of MOC/sludge between 100: 1 and 100:20 (Table 3.9).

Table 3.9 The Study Matrix of Sludge and Magnesium Oxychlorides Cement With Different Ratios

Molar Ratio	Mass Ratio of Magnesium Oxychlorides Cement/Sludge				
of $MgO/MgCl_2$	DS1 = 1:100	DS2 = 3:100	DS3 = 5:100	DS4 = 10:100	DS5 = 20:100
M1 = 1:1	√	√	√	√	√
M2 = 2:1	√	√	√	√	√
M3 = 3:1	√	√	√	√	√
M4 = 4:1	√	√	√	√	√
M5 = 5:1	√	√	√	√	√

√- Components and ratio of mixture.

$MgCl_2$ was added in the surface of sludge with centrifugal swinging for 5 min until it absorbed water from sludge and deliquescence. Then the centrifugal swing machine was stopped and MgO was added. MOC, including $MgCl_2$ and MgO, dispersed in sludge with a mixer at higher speed (800 r/min) for 3 min. The mixtures were solidified in steel molds with a size of $40 \times 40 \times 160$ mm for 24 h, and the resultant products were extruded and cured in airtight condition of 20°C for 10 days.

3.4.2 Unconfined Compressive Strength

As shown in Fig. 3.23, with increase of the proportion of MOC, the sludge strength increased concomitantly. At a fixed molar ratio of $MgO/MgCl_2$ [e.g., 3: 1 (M3)], the corresponding UCSs of DS1, DS2, DS3, DS4, and DS5 series (mass ratios of MOC: sludge was 1: 100, 3:100, 5: 100, 10: 100, 20: 100) were found to be 1.36, 8.96, 20.17, 70.67, 85.14 kg/cm^2, respectively, meaning the strength of sludge increased significantly with the addition of MOC. The same trend also occurred in other molar ratios of $MgO/MgCl_2$. It is also found that the minimum UCS of 0.75 kg/cm^2 was M1-DS1, and the other solidified products were higher than 5.99 kg/cm^2, meeting the environmental protection agency's limit of a solidified material with strength of 50 psi (pounds per square inch) = 3.44 kg/cm^2 (US EPA OWSER Directive, No. 9437.00-2A).

Unconfined compressive strengths of DS3 series were observed in 1-day, 3-day, 7-day, 10-day and compared in Fig. 3.24. It is found that molar ratio of $MgO:MgCl_2$ can affect the strength of solidified sludge. Because of the

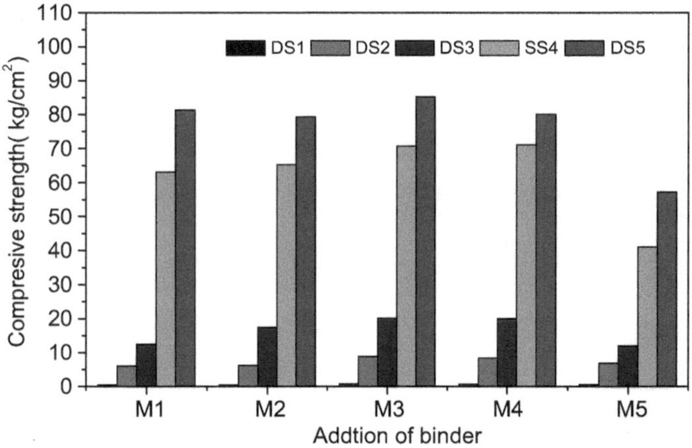

Figure 3.23 Compressive strengths of different mixtures after air curing for 10 days.

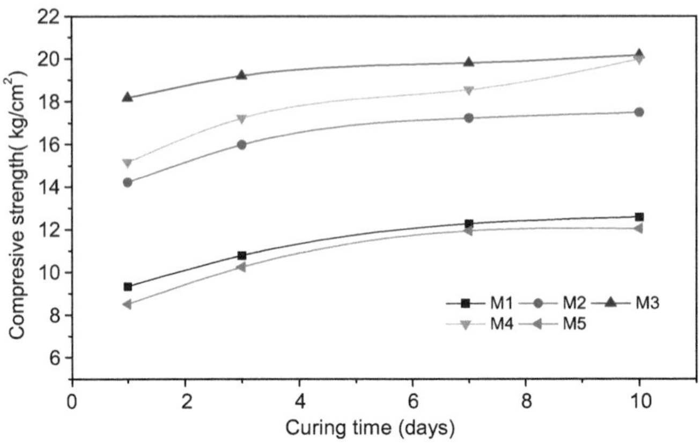

Figure 3.24 Strength developments of the samples of magnesium oxychlorides cement/dewatered sludge DS2.

higher strength of M3 ($MgO:MgCl_2 = 3:1$) than others, the optimal molar ratio of $MgO:MgCl_2$ in DS3 series can be obtained. It is also found that the strength development curves were ascending but the rising trend was flat, which may be contributed to the reinforced reaction of mixture occurred and finished in steel molds in the initial period. As a result, the strength later in the curing time increased slowly.

3.4.3 Setting Time

The hydration crystal process of MOC is the vital step for the fabrication of a dense and chemically stable material in the MOC—sludge system. Table 3.10 showed that the initial setting times could be determined by the ratio of

Table 3.10 Setting Time of the Hydration Crystal Process of Magnesium Oxychlorides Cement

Samples	Initial Setting Time (h)	Final Setting Time (h)	Moisture (%)
M1-DS2	2.1 ± 1	17.6 ± 0	64.90
M3-DS2	2.4 ± 0	17.0 ± 0	64.38
M3-DS3	1.0 ± 0.5	11.6 ± 0	65.53
M5-DS3	1.2 ± 0	12.7 ± 0	62.75
M3-DS4	1.1 ± 0	11.7 ± 0	55.20
M4-DS4	1.5 ± 0	12.3 ± 0	53.17
M3-DS5	0.8 ± 0	13.4 ± 1	42.33
M5-DS5	0.9 ± 0	11.8 ± 0	45.19

MOC/sludge and the initial and the final setting times were shortened with raising the proportion of MOC, especially in DS5 series, with the initial setting times of less than 1 h and the final time of 11.8 h. The longest final setting time was 17.6 h for M1-DS1, while the shortest one was 11.6 h under the case of M3-DS3. The variation of water contents in the solidified products ascribed to the amount of MOC. With increasing amount of the binder, the water contents in the products were lower than unsolidified sludge, with the minimal value of 42.33% for M3-DS5. These results confirmed that the hydration reaction of MOC can occur under the condition of less addition of MOC, meaning that the dilution of MOC in sludge had no adverse influences on the solidification of MOC, and that the water contents of sludge apparently decreased in the solidified process.

3.4.4 Crystalline Phase and Chemical Composition Analysis

XRD analysis (Fig. 3.25) was conducted for samples with a fixed molar ratio of M3 (MgO:$MgCl_2$ = 3:1), which was the optimal molar ratio of MOC and represented a good set of solidified products. In XRD diffractogram, several typical hydration products were identified, such as phase 3 ($3Mg(OH)_2 \cdot MgCl_2 \cdot 8H_2O$) and phase 5 ($5Mg(OH)_2 \cdot MgCl_2 \cdot 8H_2O$) and $Mg(OH)_2$. For the reason of the dilution effect of MOC in sludge, the gelling reaction was an incomplete reaction and unreacted product of $Mg(OH)_2$ existed. Some of $Mg(OH)_2$ reacted with SiO_2, Al_2O_3, or MgO in sludge and pozzolanic effect occurred in the solidified process. As a result, a stable

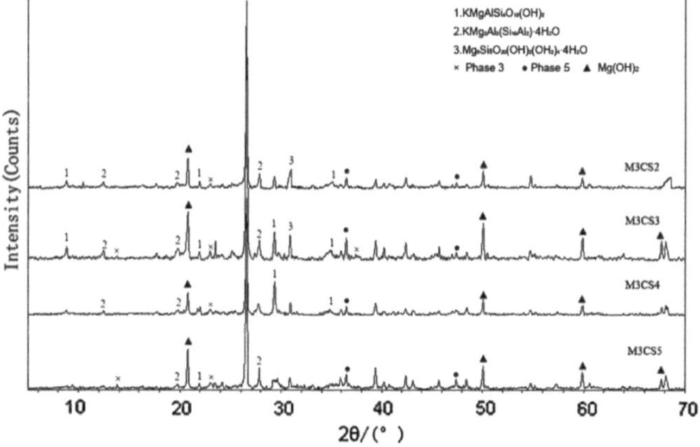

Figure 3.25 X-ray diffraction of solidified products with fixed molar ratio of M3 (MgO:$MgCl_2$ = 3:1) at the optimal molar ratio of magnesium oxychlorides cement and sludge.

crystalline phase of Mg—Si—Al was generally formed, such as Musicovite Magnesian 1M ($KMgAlSi_4O_{10}(OH)_2$), Osumilite-(Mg) ($KMg_2Al_3(Si_{10}Al_2)$ $4H_2O$), Magnesium Aluminum Silicate Hydrate ($5MgOAl_2O_3$ $3SiO_24H_2O$). It can be seen that phase 3 and phase 5 were the main reinforced compounds in sludge.

3.4.5 Scanning Electron Microscopy

SEM observations show that the construction of unsolidified sludge was disordered and incompact (Fig. 3.26A). Typical MOC—sludge product was characterized by the needle-like crystals and honeycomb-like hydrated microstructures on the surface and interior (Fig. 3.26B). In particular, the hydrated compounds, besides the ternary system $MgO—MgCl_2—H_2O$ in MOC, often contained Mg—Si—Al flaky hydration in the micropore space of sludge. The crystals were aggregated and connected each other, which distributed in sludge and formed a homogeneous network. Under the coaction of $MgO—MgCl_2—H_2O$ and Mg—Si—Al gel, the high mechanical strength of sludge can be obtained.

3.4.6 Heavy Metal Leaching Behaviors of Solidified/Stabilized Sludge

The heavy metal leachability of DS2, DS3, DS4, and DS5 series were evaluated. Cu, Zn, Cd, Cr, and As were determined. Fig. 3.27 shows the variation of final leaching concentrations with successive extractions, which are lower than the detection limit (Table 3.11) in China and the US EPA. In all cases, the different percent retention of heavy metal molecules made the extraction concentration decreased with the addition of MOC. This fact

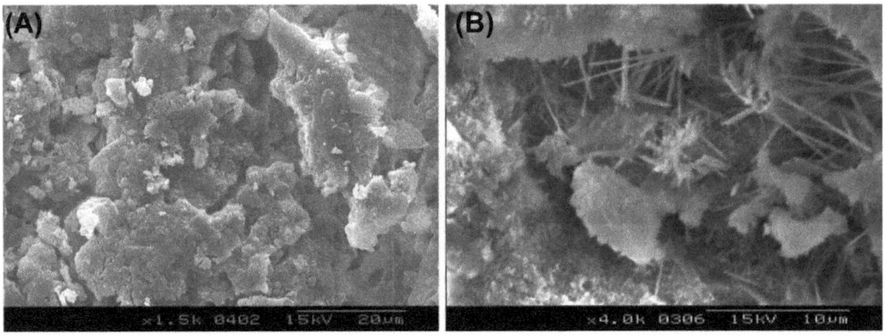

Figure 3.26 SEM photographs of magnesium oxychlorides cement—sludge product: (A) unsolidified sludge, (B) solidified sludge.

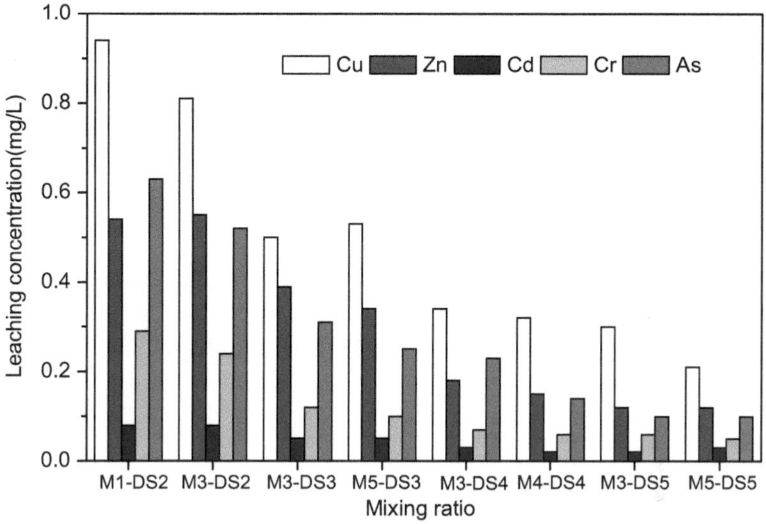

Figure 3.27 Heavy metal leachability of DS2, DS3, DS4, and DS5 series for 10-day-curing magnesium oxychlorides cement/sludge products.

was corroborated when using a greater amount of MOC (e.g., MOC: sludge = 20:100 and MgO:$MgCl_2$ = 5:1). The obtained leaching concentrations of Cu, Zn, Cd, Cr, and As were found to be 0.21, 0.12, 0.03, 0.05, 0.1 mg/L, with 22.3%, 22.2%, 37.5%, 17.2%, 15.9% of leaching concentrations of M1-DS2, respectively. As far as the heavy metal retention was concerned, it can be concluded that greater the amount of MOC added, the greater the percentage of heavy metal retention in the mixed ratio range of MOC: sludge of 3:100—20:100 obtained.

3.4.7 Main Mechanisms of Solidification/Stabilization

The moisture content of sludge was as high as 85.4%, and quite difficult to be dewatered further by common dewater machines. Under the action of $MgCl_2$, partial desorption water was released from sludge and became

Table 3.11 Toxicity Characteristic Criteria in China and the United States Environmental Protection Agency (mg/L)

Criteria	Cu	Zn	Cd	Cr	As
IEPT	50	50	0.3	10	1.5
EPA TCLP	—	—	1.0	5.0	5.0

magnesium chloride solution. Then magnesium oxide dissolves in magnesium chloride solution and comes into being a gel paste, and the hydration reaction has occurred at the same time. In the course of reaction, part of desorption water evaporated with the hydration heat, and the others enter the hydration products, thus decreasing the water contents in the products. These hydrations are generated in the ternary gelatinization system of $MgO{-}MgCl_2{-}H_2O$ as below:

$$5MgO + Mg^{2+} + 2Cl^- + 13H_2O \rightarrow 5Mg(OH)_2 \cdot MgCl_2 \cdot 8H_2O \quad (3.6)$$

$$3MgO + Mg^{2+} + 2Cl^- + 11H_2O \rightarrow 3Mg(OH)_2 \cdot MgCl_2 \cdot 8H_2O \quad (3.7)$$

When MOC is dispersed in sludge, the main components, SiO_2, Al_2O_3, or MgO contained in sludge can therefore form Mg−Si−Al ($MgO \cdot SiO_2 \cdot Al_2O_3 \cdot nH_2O$) hydration with unreacted $Mg(OH)_2$. After a period of curing time, the hydration of phase 3, phase 5, and other hydration components in sludge develop a crystallizing network, which has been observed as by the microstructural study. The crystallizing network not only enhances the strength of sludge but also prevents precipitation of heavy metal ions from sludge.

3.5 VAPORIZATION AND DEPRESSION CONTROL OF HEAVY METALS IN SLUDGE SUBJECT TO INCINERATION

Drying-incineration is becoming more popular throughout the world. However, during drying-incineration, heavy metals are not destroyed by high-temperature thermal treatment and the air pollution control devices may not always trap the submicrometer metal particulate and gaseous compounds. More volatile metals such as Hg, Pb, and Cd are significantly present in the stack emission, despite the scrubber stages, but in line with the volatility of the heavy metals. Heavy metal emissions from waste incineration plants have become a great environmental concern because of their toxicity to both human health and environment.

Previous studies have been conducted to seek ways to convert the heavy metals present in the waste into chlorides, sulfates, and other compounds or to transfer as much as possible of the heavy metals in fuel into flue gas, from which they can be captured by condensation in the filter-ash or the scrubber water and then recovered as metal-rich residues. A lot of effort has been put into this innovative concept, and the separation and recovery technology of metals from combustion residues has also been studied. Besides, a contrast option for reducing the emission of heavy metals is to weaken vaporization (i.e., by immobilization), making the flue gas much easier to clean as well. A profound knowledge of the behavior of heavy metals in the combustion process is the most important factor for selecting waste disposal alternatives and flue gas-cleaning systems. Therefore, vaporization and depression control mechanism of heavy metals during sludge incineration process will be introduced in this section.

3.5.1 Chloride Effect on Vaporization and Depression Control of Heavy Metals in Sludge

3.5.1.1 Vaporization of Heavy Metals

Polyvinyl chloride (PVC), $FeCl_3$, $CaCl_2$, and NaCl were added as chlorine donors to the combustion experiments to study the vaporization potential of heavy metals in sludge. This included an organic chlorine compound and inorganic chlorides with different lattice energies. The background Cl concentration in sludge was neglected because of its low content, and then the $FeCl_3$, $CaCl_2$, NaCl, and PVC were spiked according to different molar mass ratios between Cl and dried sludge (Fig. 3.28).

As shown in Fig. 3.28A, the vaporization percentage of Zn, Pb, Cr, and Cu increases with increasing amounts of doped $CaCl_2$. The biggest increase of vaporization of Zn (13.0%) was found at a Cl/sludge ratio of 0.54 mol/kg followed by Pb (6.5%), Cr (6.2%), and Cu (4.5%). Heavy metals with refractory volatility, such as Cu and Cr, were partitioned mostly in the bottom ash, although an enhancement of emissions was noted. In the case of Cu, about 64% were distributed in the bottom ash.

The increases of vaporization percentage of heavy metals caused by the addition of NaCl were less than that caused by spiking with $CaCl_2$ (Fig. 3.28B). The volatilities of zinc (Zn) and lead (Pb) appear to increase slightly from about 44.5% to 49.7% for Zn and from 48.9% to 53.8% for Pb when the ratio of Cl/sludge was increased from 0 to 0.72 mol/kg. In the case of the refractory volatility of heavy metals, the vaporization percentage of copper (Cu) increased from 32.8% to 38.4%, whereas that of chromium (Cr) was almost unchanged (about 14%).

Increased chlorine content in the form of $FeCl_3$ promoted an increase of vaporization percentage with a consequent decrease in the bottom ash (Fig. 3.28C). The vaporization percentages of Pb and Cu increased significantly, from 48.9% to 56.2% for Pb and from 32.8% to 41.1% for Cu when the ratio of Cl/sludge increased from 0 to 0.72 mol/kg. Moreover, it was

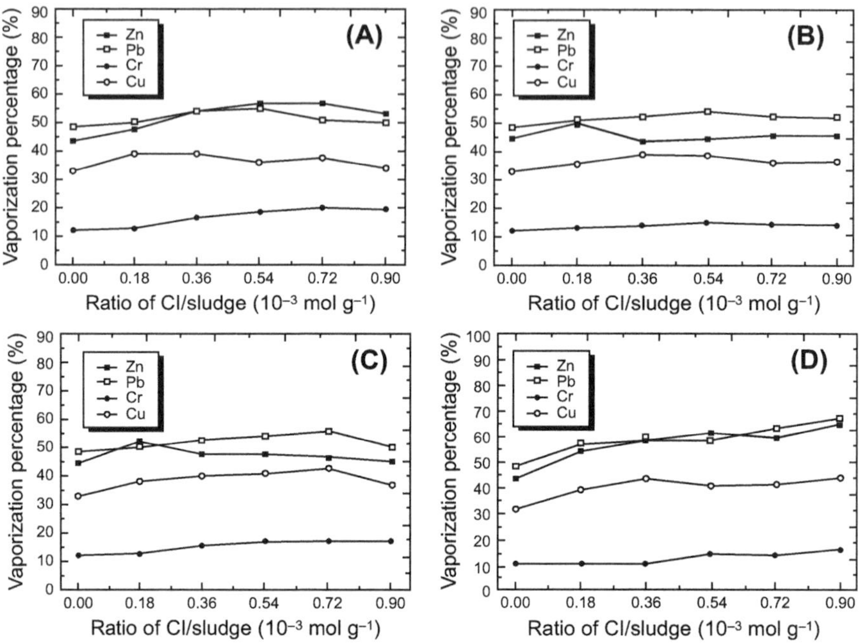

Figure 3.28 Effects of $CaCl_2$, NaCl, $FeCl_3$, and polyvinyl chloride addition on vaporization of heavy metals in sludge.

interesting to find that the vaporization of Pb, Zn, and Cu decreased slightly when the ratio of Cl/sludge exceeded 0.54 or 0.72 mol/kg.

The effect of the addition of an organic chlorine compound of PVC on the behavior of heavy metals during thermal treatment of sludge was also investigated. The results clearly show the accelerating and increasing effect of PVC addition on the vaporization of Zn, Pb, Cu, and Cr (Fig. 3.28D). The increases in the vaporization percentage of Zn (11.2%), Pb (18.4%), and Cu (12.0%) were larger than those obtained by adding $CaCl_2$, NaCl, and $FeCl_3$, whereas the vaporization percentage of Cr increased slightly in comparison with the addition of inorganic chlorides.

3.5.1.2 Combustion Temperature Effects on the Vaporization of Heavy Metals in Sludge

Combustion temperature is a significant factor for heavy metal vaporization. Fig. 3.29 shows the relationship between the combustion temperatures and vaporization percentages of Zn, Pb, Cu as well as Cr in the incineration process of sludge. The vaporization percentage with PVC addition increased

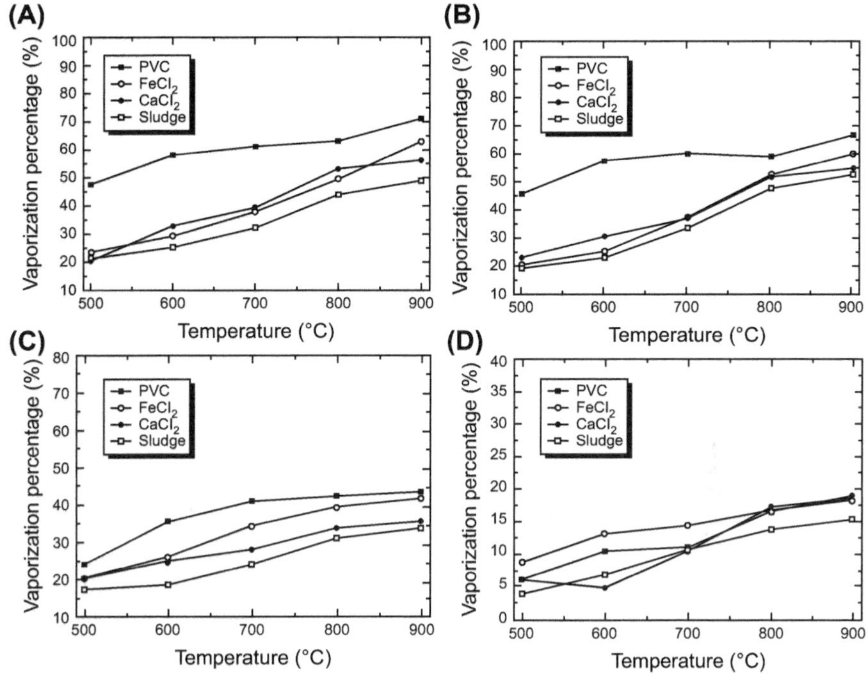

Figure 3.29 Combustion temperature effect on vaporization of Zn, Pb, Cu, and Cr in sludge (Cl/sludge = 0.54 mol/kg).

from 47% to 71% for Zn, from 45% to 65% for Pb, and from 24% to 43% for Cu when the combustion temperature was increased from 500°C to 900°C, whereas those with $FeCl_3$ addition increased considerably from 23% to 64% for Zn, from 20% to 59% for Pb, and from 20% to 41% for Cu. Moreover, the vaporization percentage for $CaCl_2$ addition increased significantly from 20% to 56% for Zn, from 22% to 53% for Pb, and from 20% to 34% for Cu when the combustion temperature was increased from 500°C to 900°C. In the case of the less volatile element Cr, its vaporization percentage increased similarly from 6% to 17% with PVC, $FeCl_3$, and $CaCl_2$ additions.

Because of the deposition on the furnace walls or low concentration of heavy metals, the average metals' mass-balance closures were in the range of 75–90%. Although the balance for all the heavy metals was not complete, its closure was considered to be of acceptable quality to allow the derivation of valid results.

During thermal treatment of sludge doped with $CaCl_2$, NaCl, $FeCl_3$, and PVC, the vaporization percentages of heavy metals increased when the ratio of Cl/sludge was raised from 0 to 0.54 mol/kg. Zn was the most volatile metal followed by Pb. In the case of the so-called refractory-volatility heavy metals, the affinity of Cu to chlorine was greater than that of Cr to chlorine; hence, the vaporization percentage of Cu showed a larger increase with increasing chlorine content as shown in Fig. 3.28. The results revealed that the presence of chlorine promoted the formation of volatile metal chlorides in the sludge during combustion. Therefore, the physical–chemical properties and the volatility of heavy metals and/or their compounds can change during the combustion process, and this will eventually govern their partitioning behavior in the incineration residues.

Moreover, the dominant mechanism involved in Cu emission could include the direct role of the entrainment process; and thus the vaporization percentage of Cu was subject to minor fluctuations, although the enhancement of emissions was noted as a whole with an increasing Cl/sludge ratio. Similarly, the more refractory-like Cr was partitioned up to 85% to the bottom ash, and the vaporization percentage increased slightly when the chlorine content was increased because of the effects of entrainment and the formation of volatile CrO_2Cl_2.

Comparing Fig. 3.28, it becomes evident that the accelerating and increasing effects of the addition of chlorides on the vaporization of heavy metals during combustion are dependent on the release capacity of chlorine radicals. Although NaCl and $CaCl_2$ belong to ionic compounds, the alkaline metals have a greater affinity to chlorine than the alkali-earth metal. Hence, the increased rate of vaporization percentage of heavy metals when NaCl was added was smaller than those spiked with $CaCl_2$. With regard to the $FeCl_3$ addition, since the energy of the molecular bond (Fe—Cl) was lower than for the ionic bond (Na—Cl, Ca—Cl), the release capacity of chlorine ions was higher than for other inorganic chlorides. Therefore, the tendency of metal chloride formation and volatilization percentage of heavy metals was increased, especially for Cu and Pb. Due to having the lowest energy of molecular bond, as expected, the free chlorine radicals provided by PVC were more available than those contributed by inorganic chloride. Therefore, the increasing vaporization percentage of heavy metals with increasing Cl concentrations originating from the presence and decomposition of PVC was more significant than that originating from other inorganic chlorides. It can be seen that the largest increase of vaporization percentage was 20.7% for Zn, 18.3% for Pb, 12% for Cu, and 4.9% for Cr.

Moreover, it was interesting to find that for a Cl/sludge range between 0.54 and 0.9 mol/kg, the vaporization of Pb, Zn, and Cu decreased slightly with increasing Cl/sludge ratio. This might be due to the increased moisture content in the sludge, caused by the addition of NaCl, $CaCl_2$, and $FeCl_3$ (deliquescence effect). The presence of moisture in the waste may facilitate the oxidization of metallic Pb and Zn and consequently the volatilities of these metals will decrease as the moisture content increases.

$$Cl + H_2O \leftrightarrow HCl + OH \tag{3.8}$$

$$Pb + Cl \leftrightarrow PbCl \tag{3.9}$$

$$PbCl + Cl \leftrightarrow PbCl_2 \tag{3.10}$$

$$PbCl + OH \leftrightarrow PbO + HCl \tag{3.11}$$

Unlike NaCl, $FeCl_3$, and $CaCl_2$, PVC has no water-absorption capacity and so there was no increase in moisture by spiking PVC. Therefore, the volatilization of Pb and Zn increased with raising the Cl content originating from PVC. This demonstrates that moisture in the sludge can obstruct and reduce the volatility of heavy metals during combustion.

As expected, increasing the incineration temperature enhanced the volatilization of heavy metals in each case. PVC releases Cl radicals when subjected to thermal decomposition at 100–250°C. Due to the low bonding energy, the availability of hydrogen radicals, heavy metals, and Cl radicals can enhance the formation of HCl and heavy metal chlorides, which explains the accelerating and increasing effect of PVC addition on the volatilization of Zn, Pb, and Cu, which was more remarkable than those of other tested chlorides at low temperatures. When the combustion temperature was more than 700°C, $CaCl_2$ and $FeCl_3$ also played an increasingly important role in the vaporization of heavy metals. In the case of the refractory-volatility heavy metal Cr, the effects of various chlorides on the vaporization was relatively small. Probably the so-called entrainment process played a leading role in the Cr emissions, but with increasing temperature the volatility also increased.

The experimental results show that the volatilization of heavy metals (Pb, Zn, Cu, and Cr) during incineration of sludge increased with chlorine addition (PVC, $FeCl_3$, $CaCl_2$, and NaCl) and the observed enhancement effects of chlorides on vaporization depended on the bonding energy of the chlorine compounds. Due to the relatively low molecular bond energy, PVC released more free Cl-radicals for the volatilization of heavy metals at low temperatures in comparison with other chlorine compounds. As sludge contains mainly high concentrations of medium- and refractory-volatility heavy metals, a complete mobilization of Pb, Zn, Cu, and Cr in the combustion process cannot be expected despite the high Cl/sludge ratio. Moreover, due to the so-called deliquescence effect of chlorides, the volatilization of Zn and Pb again decreased with increasing moisture caused by the effect of high levels of chloride addition on the ratio of Cl/sludge.

3.5.2 Phosphoric Acid Effect on Vaporization and Depression Control of Heavy Metals in Sludge

Phosphatization was carried out by adding phosphoric acid (85% H_3PO_4) to sludge containing sufficient water (50%) to exhibit the rheology of a thixotropic slurry, thereby achieving the best compromise between minimum water input and optimal contact between the reactant and the solids, and this mixing process was followed by the formation of foam. After mixing sludge with H_3PO_4, the phosphatized sludge was put in a drying bed to avoid moisture interference.

3.5.2.1 Phosphoric Acid Effect on Thermal Behavior of Sludge

TG analysis provides a prior knowledge of the initial and final temperatures for the thermal degradation of sludge. The results of TG/DTG of sludge are shown in Fig. 3.30A. It can be seen that the DTG curve of sludge changes

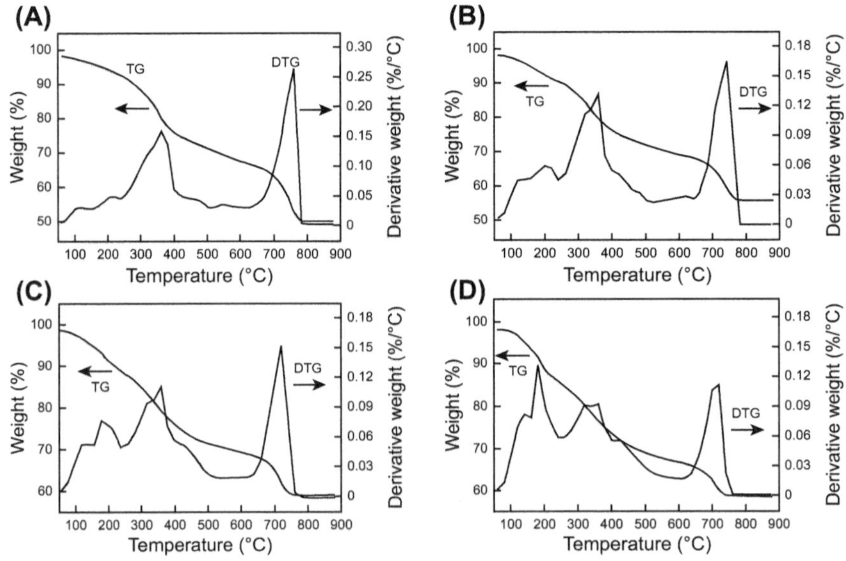

Figure 3.30 Thermogravimetry and differential scanning calorimetry curves of sludge at the heating rate of 10°C/min. (A) Sludge; (B) Sludge A; (C) Sludge B; (D) Sludge C.

slightly relatively from 50 to 200°C and has 5.8% weight loss in TG analysis. Two obvious peaks are observed in the DTG curve from 200 to 400°C with 18.5% weight loss and from 660 to 780°C with 15.4% weight loss, respectively. After being finally held from 780 to 900°C, the rate of mass loss was approximately zero.

Fig. 3.30B shows the TG and DTG measurements of sludge pretreated with 85% H_3PO_4, which accounts for 10% of dry basis of sludge (Sludge A). There was a small peak in the temperature range 120–220°C and the corresponding weight loss of the sludge was 5.3%. Two other obvious peaks were also observed in the DTG curve from 240 to 400°C with 14.0% weight loss and from 660°C to 780°C with 11.5% weight loss, respectively.

Fig. 3.30C shows the TG and DTG measurements of sludge pretreated with 85% H_3PO_4, in which H_3PO_4 accounted for 20% of the dry basis of sludge (Sludge B). For Sludge B, the maximum weight loss was observed in the temperature range 140–440°C. In this temperature range, two different peaks were observed in the DTG curve. The first peak occurred in the temperature range 120–240°C and the corresponding weight loss was 6.8%, while the second occurred in the temperature range 240–440°C and the corresponding weight loss was 13.9%. Another main decomposition peak was observed between 660 and 780°C with a weight loss of 8.9% in TG analysis.

Fig. 3.30D shows the TG and DTG measurements of sludge pretreated with 85% H_3PO_4, in which H_3PO_4 accounted for 30% of the dry basis of sludge (Sludge C). The main weight loss of Sludge C generally occurred over the temperature of 120–440°C. In this temperature, two distinct peaks were also observed as Sludge B. The first peak occurred in the temperature range 120–240°C with a weight loss of 11.0%, while the second occurred in the temperature range 240–440°C with a weight loss of 14.5%. Moreover, another peak in the DTG curve started from 660°C and ended at 740°C with a weight loss of 6.5%.

3.5.2.2 Vaporization of Heavy Metals

Fig. 3.31 shows the relationship between the amount of phosphoric acid and the vaporization percentages of Zn, Pb, Cr, and Cu at 600 and 800°C. At 600°C, the vaporization percentage decreased rather slightly from 24% to 7% for Zn, from 21% to 12% for Pb, and from 17% to 11% for Cu when the phosphoric acid content increased from 0 to 30%, while the vaporization of Cr can be considered nearly unchanged at around 5%. At 800°C, the vaporization percentage decreased significantly from 43% to 11% for Zn,

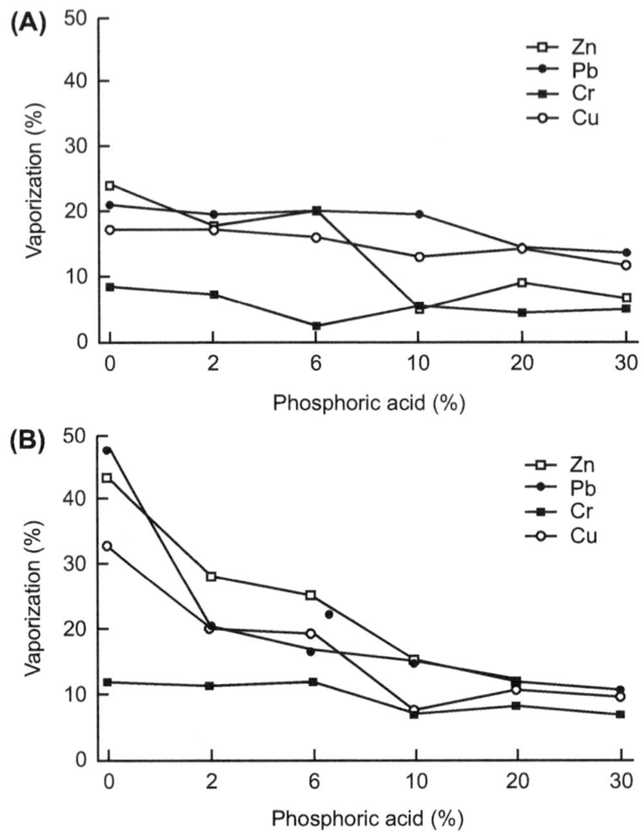

Figure 3.31 Vaporization percentages of heavy metal during thermal treatment of phosphated sludge at 600°C (A) and 800°C (B).

from 48% to 10% for Pb, and from 32% to 9% for Cu, while the vaporization percentage of Cr decreased slightly from 12% to 7%.

3.5.2.3 Crystalline Phase and Chemical Composition Analysis

The crystalline phases of the bottom ash of Sludge A, Sludge B, and Sludge C are shown in Fig. 3.32. The major crystalline phase of Sludge A at 600°C was $CaCO_3$, $Ca_5(CrO_4)_3O_{0.5}$, and $Ca_2Cr_2O_5$. As the content of H_3PO_4 increased, $Ca_2P_2O_7$, $Ca_{19}Cu_2(PO_4)_{14}$, $Ca_{19}Zn_2(PO_4)_{14}$, and $Ca_9Cr(PO_4)_7$ became the major crystalline phase.

Fig. 3.33 shows the XRD pattern of the bottom ash of Sludge A, Sludge B, and Sludge C at 800°C. The pattern consisted of strong intensity peaks corresponding to CaO and weak peaks of $Ca_2P_2O_7$, $Ca_{19}Cu_2(PO_4)_{14}$,

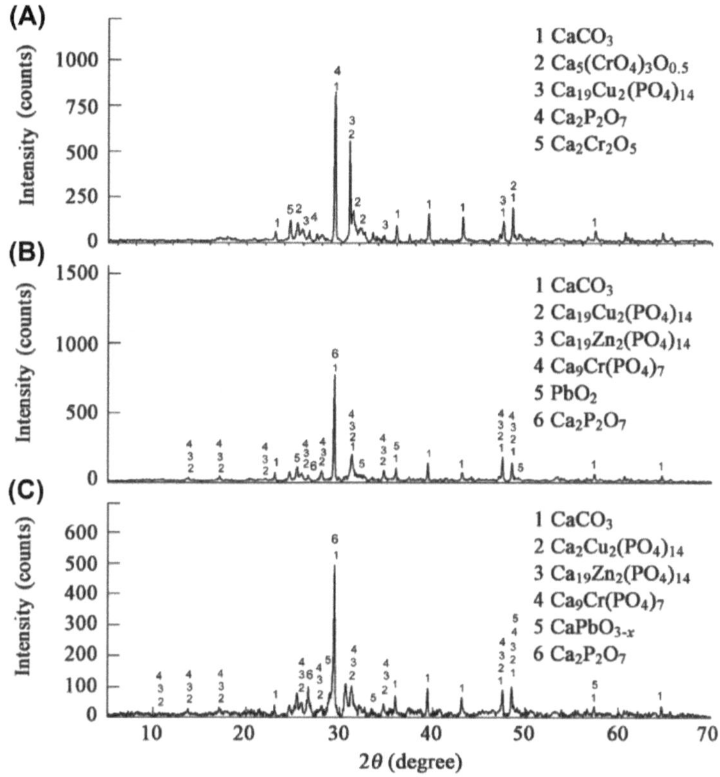

Figure 3.32 X-ray diffraction patterns of the bottom ashes of Sludge A (A), Sludge B (B), and Sludge C (C) at 600°C.

$Ca_{19}Zn_2(PO_4)_{14}$, and $Ca_9Cr(PO_4)_7$. When the content of H_3PO_4 rose, the peaks of $Ca_2P_2O_7$, $Ca_{19}Cu_2(PO_4)_{14}$, $Ca_{19}Zn_2(PO_4)_{14}$, and $Ca_9Cr(PO_4)_7$ became sharper and these crystals became the major phases in the bottom ash.

3.5.2.4 Scanning Electron Microscopy Analysis

Fig. 3.34 shows the morphology of bottom ash of sludge and Sludge A at 800°C. SEM analysis revealed the differences in the gross microstructure between the bottom ash of untreated sludge and pretreated sludge with phosphoric acid. The SEM micrograph of bottom ash of untreated sludge showed the presence of floc-like material. This floc-like material was believed to be the crystal grain of CaO. The micrograph of the bottom ash of pretreated tanner sludge (Sludge A) showed irregularly shaped mineral crystal grain covered on the CaO crystal (Fig. 3.34B), which was believed to

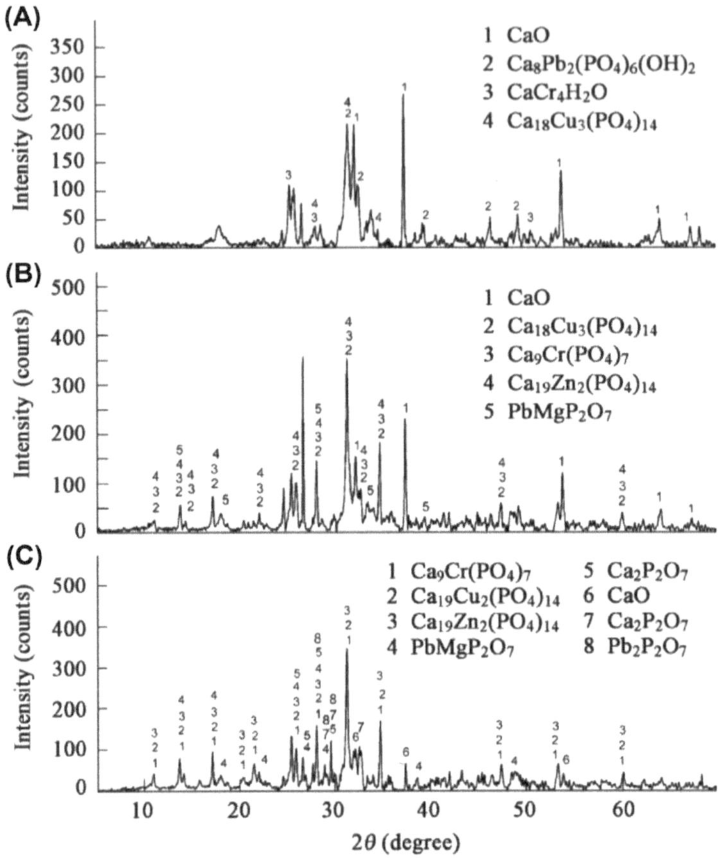

Figure 3.33 X-ray diffraction patterns of the bottom ashes of Sludge A (A), Sludge B (B), and Sludge C (C) at 800°C.

be the apatite-like mineral phase corresponding to $Ca_8Pb_2(PO_4)_6(OH)_2$ and $Ca_{18}Cu_3(PO_4)_{14}$ peaks in the XRD graph.

The reaction scheme of sludge pyrolysis is very complex. The following global reaction schemes for the pyrolysis of sewage sludge are as follows:

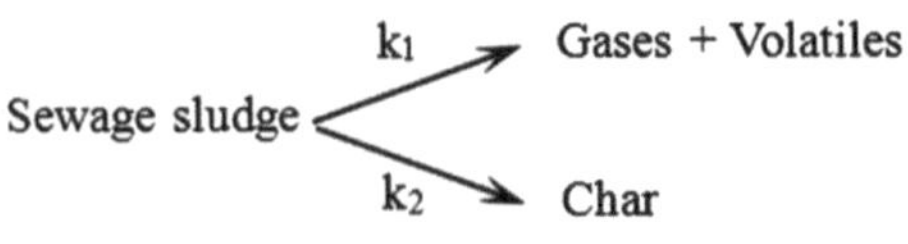

(3.12)

As seen in Fig. 3.30A, the pyrolysis of sludge took place in two stages. Prior to the first stage, the TG diagrams for sludge showed a slight decrease in weight at a temperature in the range of 50–220°C, owing to the

Figure 3.34 Scanning electron microscopy images showing the bottom ash of sludge (A) and Sludge A (B) at 800°C.

evaporation of adsorbed water. The most abundant release of volatile matter was observed in the first stage. In this stage, the peak of DTG curve had a shoulder at 220°C, which can be attributed to the decomposition and devolatilization of less complex organic structures, which was a small fraction, while the peak was caused by the decomposition of more complex organic structures corresponding to a larger fraction.

In the second stage of sludge pyrolysis, decomposition of inorganic materials mainly took place. Pure $CaCO_3$ shows an endothermic peak and the corresponding decrease in weight at 720°C. Thus, the weight loss in the second stage is mainly attributed to the decomposition of $CaCO_3$ via the following reaction:

$$CaCO_3 \leftrightarrow CaO + CO_2\uparrow \tag{3.13}$$

When sludge was mixed with phosphoric acid, $CaCO_3$ that contained in sludge can react with H_3PO_4 as

$$CaCO_3 + H_3PO_4 + H_2O \leftrightarrow CaHPO_4 \cdot H_2O + CO_2\uparrow \tag{3.14}$$

which leads to the formation of a foam.

Thus, for Sludge A, Sludge B, and Sludge C, the peak that started from 120°C and ended at 240°C was a result of the dehydration of crystal water of $CaHPO_4 \cdot 2H_2O$ and the evaporation of absorbed water. Compared with the DTG curves of Sludge A, Sludge B, and Sludge C, the more the amount of H_3PO_4 was mixed, the higher was the peak. Furthermore, the DTG curves of Sludge A, Sludge B, and Sludge C showed peaks at 360°C having a shoulder at 420°C. The peaks were a result of the decomposition of organic matter while the shoulders can be assigned to the decomposition of $CaHPO_4$ as

$$CaHPO_4 \leftrightarrow Ca_2P_2O_7 + H_2O \tag{3.15}$$

which can also be proved in the crystalline phase analysis (Figs. 3.32 and 3.33). Owing to the reaction with H_3PO_4, the amount of $CaCO_3$ decreased, which weakened the peaks of the DTG curves of Sludge A, Sludge B, and Sludge C at 720°C.

Although most of the phosphoric acid reacts with $CaCO_3$, addition of phosphoric acid also allows the formation of other insoluble metal phosphates as shown in Figs. 3.32 and 3.33, which makes heavy metals be trapped and fixed into stable crystal phases following the general expression as

$$Me^{2+} + H_3PO_4 + H_2O \leftrightarrow MeHPO_4 \cdot 2H_2O + H^+ \quad (3.16)$$

where Me: Pb, Cu, Zn, Mg, Fe.

In the process of incineration, the heavy metal crystal phases are also decomposed as shown in the following the reaction:

$$MeHPO_4 \leftrightarrow Me_2P_2O_7 + H_2O + H^+ \quad (3.17)$$

Therefore, $Mg_2P_2O_7$, PbP_2O_7 also can be identified by XRD as the minor crystalline phase, besides the major phase of $Ca_2P_2O_7$ and $CaCO_3$. Owing to the low content of phosphoric acid, most Cr was stabilized as the compound of $Ca_2Cr_2O_5$ and $CaCrO_4H_2O$.

Besides this, owing to the presence of excessive $CaCO_3$ in sludge, the following reaction also occurred, which was in agreement with the minor crystal phases in the bottom ash of Sludge A, Sludge B, and sludge C identified by XRD , such as $Ca_8Pb_2(PO_4)_6(OH)_2$, $Cu_5(PO_4)_2(OH)_4$, and $Ca_5Al_{12}(PO_4)_8(OH)_{22}$.

$$nMe(aq) + mH_2PO_4^- + kH_2O \leftrightarrow Me_n(P_2O_7)_m(OH)_k + (2n - m)H^+(aq) \quad (3.18)$$

where Me: Ca, Al, Mg, Cu, Cr, Pb, Zn, etc.

When the addition of phosphoric acid increased, the product of phosphate of metals in the sludge also increased. These phosphates of metals were apt to bond with each other as complex in the combustion process because of the following reaction:

$$Me_3(PO_4)_2 + Ca_3(PO_4)_2 \rightarrow Ca_nMe_m(PO_4)_k \quad (3.19)$$

Thus, heavy metals, including Cr, were fixed into more stable crystal as apatite-like minerals as shown in Figs. 3.32 and 3.33. Scanning electron microscope studies also provided conclusive proof on the presence of apatite-like mineral in the bottom ash, which can fix heavy metals to avoid transfer to the environment. However, most of Pb was stabilized as the compounds $PbMgP_2O_7$, and $Pb_2P_2O_7$ according to XRD analysis.

After being treated by phosphoric acid, the heavy metals showed low vaporization in general agreement with the formation of insoluble phases, with Pb showing the highest value of only 17%, which was probably because the phosphate of Pb was not easy to bond with the other phosphate to form the considerably stable complex. For Cu, Cr, and Zn, the vaporization percentages seemed to be independent of temperature and were less than 15% at 600 and 800°C, which showed a good correspondence with the XRD and SEM analysis. Meanwhile, 10% was the optimal dosage to control the heavy metal emissions. Moreover, the stabilization agent also reduced the leachability of several divalent cations (Pb^{2+}, Cu^{2+}, Zn^{2+}, and Cd^{2+}), even under aggressive leaching conditions.

The study deals with the detailed investigation on the thermal behaviors and heavy metal vaporization of sludge pretreated with 85% phosphoric acid. In the TGA experiments, three degradation steps were observed for pretreated sludge, which included the decomposition of $CaHPO_4 \cdot 2H_2O$, the pyrolysis of organic matter, and the decomposition of $CaCO_3$. In the presence of H_3PO_4, the vaporization percentage of heavy metal in sludge decreased efficiently with the increasing volumes of H_3PO_4. Pretreatment of sludge with proper amount of H_3PO_4 was considered among the measures of controlling heavy metal emissions in the process of sludge incineration.

3.6 PRODUCTION OF SLUDGE SOLIDIFIERS AND THEIR COMMERCIAL APPLICATIONS

3.6.1 General Description in Solidifier Production

A series of chemical solidifiers such as patented TNM1, TNS1, TNMC4, and TNA1 reagents had been produced and some of which have been widely applied in real-world sludge projects. Table 3.12 summarizes the basic properties of the reagents. The main starting materials for producing those solidifiers include magnesite, serpentine, limestone, bauxite, etc. Those materials, after being mixed in a preset ratio, are calcined at >1000°C for several hours to produce highly reactive solidifiers. The calcined products are then ground into the sizes of 80—120 μm and employed as cementitious media for the purpose of sludge solidification

Table 3.12 Chemical Components of the Major Solidifiers and Their Performances in Sludge Solidification/Stabilization

Number	Property	Name of Solidifiers			
		TNM1	TNS1	TNMC4	TNA1
1	Chemical components	$MgO-CaO-SiO_2$	$CaO-MgO-SiO_2$	$CaO-MgO-Al_2O_3-SiO_2$	$CaO\text{-}Al_2O_3\text{-}SiO_2$
2	Appearance	White, powder	Gray, powder	Gray, powder	White, powder
3	Particle size (mesh)	80–120	80–120	80–120	80–120
4	Degree of activity (s)	≤600	≤500	≤500	≤500
5	Insoluble matter (%)	≤3.5	≤5.5	≤3.5	≤3.5
6	Loss on ignition (%)	≤4.5	≤5.5	≤5.5	≤3.5
7	SO_3 (%)	≤4.0	≤5.0	≤5.0	≤3.0
8	Chloride (%)	≤0.1	≤0.1	≤0.1	≤0.1
9	Bulk density (g/m^3)	≤0.8	≤0.7	≤0.8	≤0.7
10	Dosage (% of wet sludge)	5–10	5–10	5–10	5–10
11	Stabilization efficiency	• water content: ≤60% • UCS: 50–100 kPa • Shearing strength: 30–50 kPa • pH: 9–10	• water content: ≤55% • UCS: 100–200 kPa • Shearing strength: 30–50 kPa • pH: 9–10	• water content: ≤55% • UCS: 100–200 kPa • Shearing strength: 30–50 kPa • pH: 9–10	• water content: ≤60% • UCS: 100–150 kPa • Shearing strength: 30–50 kPa • pH: 8–9
12	Advantages	• all types of sludge • low cost, low dosage	• Fast setting time • high strength • low alkalinity	• Fast setting time • high dewatering efficiency in high organic content sludge	• Fast setting time • low alkalinity

Figure 3.35 Main facilities installed for solidifier production and its products.

(Fig. 3.35). The main components of the final products are as follows (wt%): MgO 10–15%, Al_2O_3 15–20%, CaO 40–45%, and SiO_2 5–10%. In addition, to upgrade the pozzolanic reactivity of the reagents, various additives like Na_2SiO_3, phosphoric acid, SiO_2, Al_2O_3, $NaAlO_2$, $FeSO_4$, or oxalic acid are often added during sludge S/S.

The features and benefits of the newly developed solidifiers include:

- High availability of starting materials
- Ease of use and safe disposal
- Low addition dosage (<10 wt%)
- Faster hardening time (3–7 days)
- Substantial decrease in sludge volume compared with the traditional reagents such as PC, fly ash, bed ash, cement kiln dust, and lime kiln dust
- Costs saving (labor, transportation, and final disposal) due to faster processing time and highly reduced sludge volume
- Chelating ability (to fix heavy metals like Cu, Zn, and Mn while increasing the availability of micronutrients), allowing the green drying-incineration of DS and low SO_2 and NOx emissions
- Potential of odor control, reducing the impacts to a surrounding community.

Table 3.13 Several Different Solidifiers for River Silt Dewatering and Solidification/Stabilization

		Name of Solidifiers		
Number	**Property**	**TNDC1**	**TNDCS1**	**TNDCM4**
1	Chemical components	$MgO-CaO-SiO_2$	$CaO-MgO-SiO_2$	$CaO-MgO-Al_2O_3-SiO_2$
2	Appearance	Gray, powder	Gray, powder	Gray, powder
3	Particle size (mesh)	80–100	80–100	80–100
4	Degree of activity (s)	≤600	≤600	≤600
5	Insoluble matter (%)	≤6.5	≤5.5	≤5.5
6	Loss on ignition (%)	≤5.5	≤5.5	≤5.5
7	SO_3 (%)	≤5.0	≤3.0	≤3.0
8	Chloride (%)	≤0.1	≤0.1	≤0.1
9	Bulk density (g/m^3)	≤0.6	≤0.7	≤0.8
10	Dosage (%)	1–3	1–3	1–3
11	Stabilization efficiency	• water content: ≤50 • UCS: 100–300 kPa • Shearing strength: 30–60 kPa • pH: 8–9	• water content: ≤50 • UCS: 100–300 kPa • Shearing strength: 30–60 kPa • pH: 8–9	• water content: ≤50 • UCS: 100–300 kPa • Shearing strength: 30–60 kPa • pH: 7–9
12	Advantages	• all types of river silt • low cost	• high dewatering efficiency in high organic content rive silt	• high dewatering efficiency in high organic content rive silt • low alkalinity

Besides, on the basis of the experience from sludge solidifier production, several different solidifiers for river silt dewatering and S/S were also developed, as listed in Table 3.13.

3.6.2 Current Applications in Sludge Treatment Projects

Shanghai Laogang Waste Disposal Company, located in Pudong New Area of Shanghai, is one of the specialized companies in China for the treatment and disposal of different wastes. One of the typical input material is sludge

(A)
(B)
(C)

Figure 3.36 Sludge solidification/stabilization facility: (A) direct loading onto truck after mixing with reagent in a closed space; (B) natural curing for 3–7 days; and (C) soil-like solidified products.

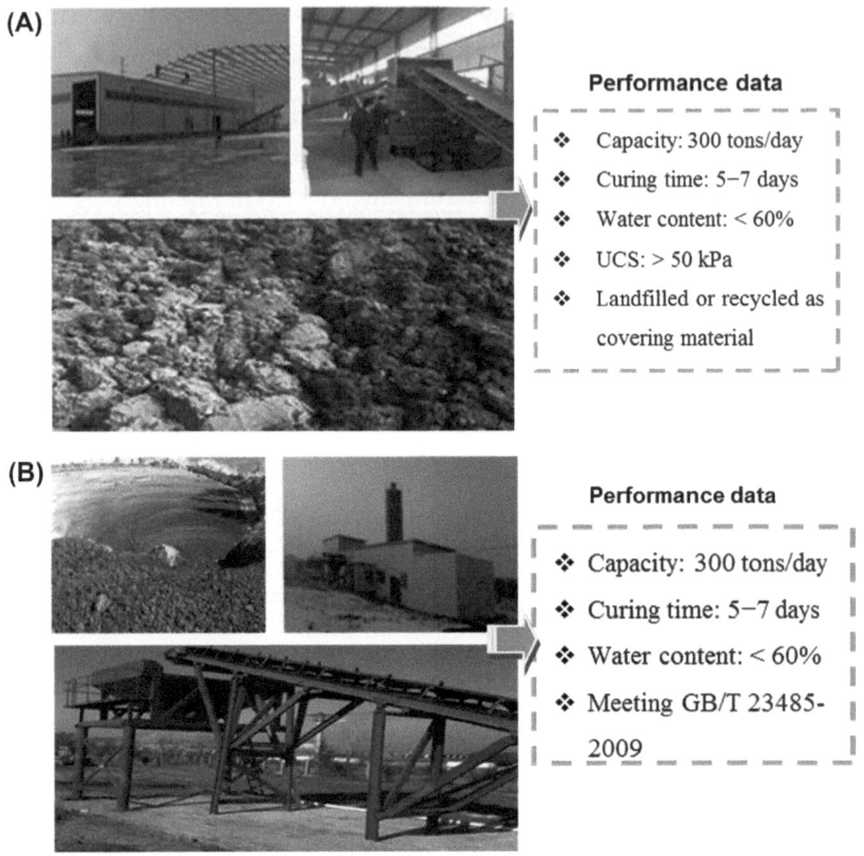

Figure 3.37 Sludge solidification/stabilization facilities, located in Binzhou City and Haerbin City, China (note: GB/T 23485-2009 < The disposal of sludge from municipal waste water treatment plant—quality of sludge for co-landfilling>).

from industrial and municipal WWTPs. Since 2009, the company has built and operates several urban sludge solidification plants (Figs. 3.36, 3.37 and 3.38), capable of handling 300-500 tons per day with TNM1 as the reagent. The detailed flowsheet can be found in Fig. 1.11 of Chapter 1.

Sludge was firstly mixed with reagent TNM1 in a closed space with automated H_2S/odor measuring system, then directly loaded onto truck, and taken to a special site for natural drying. After 3–7 days of curing, water content of the solidified sludge was decreased to $<60\%$, with the

Table 3.14 Physicochemical Properties of Solidified Sludge and Resultant Leachate Over Time

Curing Time (Day)	**1**	**2**	**3**	**4**	**5**	**6**
Solidified Sludge						
Water content (%)	67.5	63.2	57.5	53.3	51.3	41.2
Organic matter (%, dry basis)	23.88	26.89	33.58	28.67	23.69	28.10
Higher calorific value (kcal/kg)	1301.2	1298.6	1572.7	1463.6	1512.9	1413.2
Lower calorific value (kcal/kg)	18.79	99.68	323.63	363.91	430.39	583.92
Leachate						
pH	12.42	12.56	12.47	12.41	12.38	12.34
$NH_4^+ - N$ (mg/L)	190.1	147.7	39.0	49.0	20.0	13.6
COD (mg/L)	17,120	17,840	17,200	15,840	15,200	16,560
As (mg/L)	2.98	2.68	2.38	3.3	1.56	2.16
Zn (mg/L)	0	0	0	0	0	0
Pb (mg/L)	0.16	0.26	0.42	0.48	0.42	0.3
Cd (mg/L)	0.14	0.12	0.14	0.12	0.12	0.12
Ni (mg/L)	7.46	6.62	6.36	4.56	3.76	6.54
Cr (mg/L)	0.46	0.28	0.56	1.46	0.78	0.74
Mg (mg/L)	22.42	31.42	10.86	9.18	7.98	11.22
Ca (mg/L)	1643.4	1528.8	1576.6	933	1260.4	1073.4
Al (mg/L)	1.06	1.3	1.22	2.5	1.44	8.42

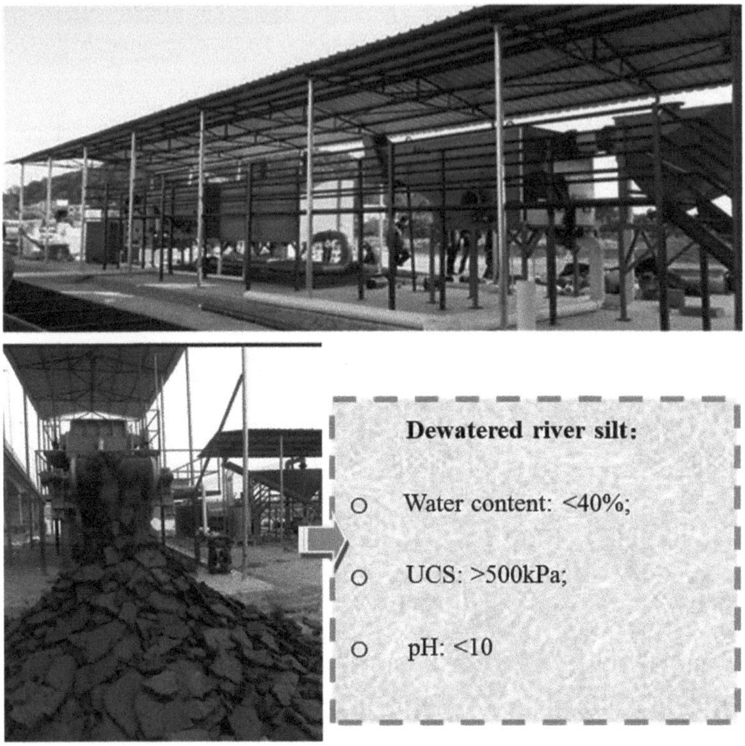

Figure 3.38 A river silt dewatering project, located in Dongguan City, China (after being conditioned with the reagents described in Table 3.13; the river silt was subject to an elastic plate-filter press for dewatering).

strength up to 100 kPa (Table 3.14). The final solidified products closely resembled soil and could be landfilled directly, coincinerated with other solid wastes or recycled as covering material of refuse landfills.

CHAPTER FOUR

Making of Sewage Sludge-Derived Controlled Low-Strength Materials (CLSMs)

4.1 PERFORMANCE APPRAISAL OF CONTROLLED LOW-STRENGTH MATERIAL USING DEWATERED SLUDGE AND MUNICIPAL SOLID-WASTE INCINERATION BOTTOM ASH (MSWI BA)

Recycling of wastes is a topic of growing importance in the field of waste management related to the environmental protection and cost-reductive advantages. Numerous studies have been focused on using sludge or refuse incineration bottom ash (BA) in construction materials such as cementitious binder, bricks, composite geomaterial, and secondary building material. Another feasible reuse of sludge and refuse incineration BA in bulk quantities is their conversion into controlled low-strength material (CLSM). The material is a self-compacting and flowable concrete and extensively used as a backfill in situ because of its better load-bearing capability. Because of the significant environmental benefits and sound properties of CLSM products, recycling of various wastes for the production of CLSM has attracted great interest.

CLSM refers to a cementitious material whose characteristics are similar to stabilized soils and that permits reexcavation afterward if necessary. The American Concrete Institute (ACI) Committee 229 has defined CLSM as a material resulting in a compressive strength of 8.3 MPa or less after hardening. CLSM, known as "self-compacting and flowable," allows perfect filling of any void and is used primarily for soil and structural filling in civil engineering works such as pipeline bedding, foundation backfill, abandoned mines, etc. The material is durable, economical, easy placement without vibration, and not labor-intensive and therefore seems more attractive and practical as compared to conventional backfills.

The utilizations of sewage sludge and refuse incineration BA in the construction industry remain quite limited as compared to other conventional by-products. According to our previous research, sewage sludge and BA contained typically substantial amounts of silica, calcium, and alumina,

Pollution Control and Resource Recovery: Sewage Sludge
ISBN: 978-0-12-811639-5
http://dx.doi.org/10.1016/B978-0-12-811639-5.00004-8

which could offer reasonably good strengths, and be beneficial to the making of CLSM. Furthermore, sewage sludge and BA, if reused in CLSM, will offer many advantages, namely, pollution control, natural resources recovery, and environmental sustainability, etc.

The objective of this work is to explore the possibility of using dewatered sludge (DS) together with municipal solid-waste incineration (MSWI) BA as raw materials in the production of CLSM to expand the use of these by-products for potential applications. The mechanical and microstructural properties of CLSM mixtures are characterized to provide deep insights into the cementitious mechanisms and optimize CLSM. The leaching behaviors of CLSM are also evaluated for assuring environmental protection.

4.1.1 Characterization of Calcium Sulfoaluminate Cement and Municipal Solid-Waste Incineration Bottom Ash (MSWI BA)

The X-ray diffraction (XRD) patterns of calcium sulfoaluminate $(C\overline{S}A)$ cement clinker and MSWI BA are illustrated in Fig. 4.1. As can be seen, $C\overline{S}A$ clinker was composed principally of yeelimite $(C_4A_3\overline{S})$, belite (C_2S), and anhydrate $(CaSO_4)$ corresponding to the rays at 23.7, 32.1, and 25.5 degrees (2θ). The characteristic weak peaks at 17.5 and 33.7 degrees (2θ) confirmed the presence of calcium fluoroaluminate $(C_{11}A_7 \cdot CaF_2)$. As to BA, moderate crystalline such as quartz (SiO_2) at 26.7 degrees (2θ), calcite

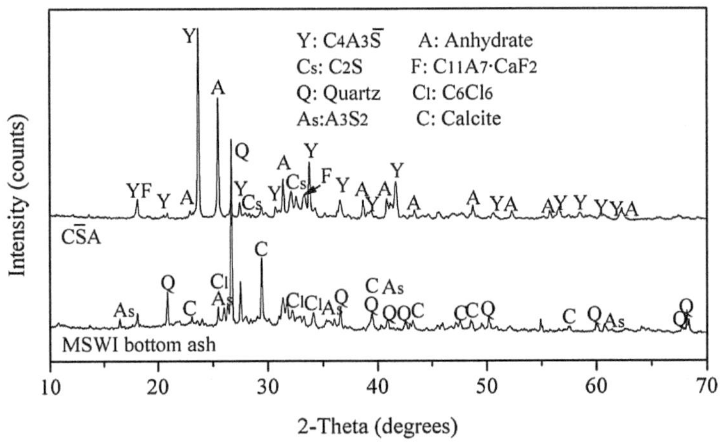

Figure. 4.1 X-ray diffractograms of calcium sulfoaluminate $(C\overline{S}A)$ cement clinker and MSWI BA.

($CaCO_3$) at 29.5 degrees (2θ), mullite ($3Al_2O_3 \cdot 2SiO_2$) at 16.5 degrees and 26.3 degrees (2θ) as well as carbon chloride (C_6Cl_6) at 26.3 degrees (2θ) could be identified, respectively.

4.1.2 Compressive Strength of Controlled Low-Strength Material

Six different CLSM mix designs were prepared in a definite ratio as presented in Table 4.1. A conventional mix, Mortar-Ref, was prepared for comparative purposes. Unconfined compressive strength (UCS) of CLSM mixes after 28 days of curing is presented in Fig. 4.2. As expected, greater the amount of $C\overline{S}A$ cement substituted by wastes, lower was the compressive strength of CLSM obtained. The highest strength of about 9.0 MPa was obtained for Mortar-Ref with DS:BA:$C\overline{S}A$ ratio of 1:0:4. Mortar-I with DS:BA:$C\overline{S}A$ ratio of 1:0.4:3.6, however, provided a 28-day strength of approximately 7.8 MPa with a 13% strength loss as compared to Mortar-Ref. When DS:BA:$C\overline{S}A$ ratio was further increased to 1:3.2:0.8, a strength of around 3.6 MPa was obtained for the Mortar-V sample. The drastic decline in strength might partially be ascribed to the weakened van der Waals forces between hydrated particles due to the deleterious effects of insoluble metal gels and organic materials from raw materials.

The CLSM with 28-day strength of 0.5−2.1 MPa is classified as the excavatable fills, which are suitable for use in nonstructural engineering such as backfilling, pipe trench refilling, etc. It was evident that the strength values of all CLSM specimens remarkably exceeded the upper excavatable limit. Therefore, CLSM mix designs manufactured herein seem more acceptable in structural and nonexcavation utilizations.

Table 4.1 Mixed Formulations of Specimen for the Six Controlled Low-Strength Material (CLSM) Mix Designs in a Definite Ratio

	Proportion by Weight					
Mix Id	**DS**	**BA**	**$C\overline{S}A$**	**Water**	**$C\overline{S}A$/(DS+$C\overline{S}A$+BA)**	**BA/$C\overline{S}A$**
Mortar-Ref	1	0	4	1.6	0.61	0/10
Mortar-I	1	0.4	3.6	1.6	0.55	1/9
Mortar-II	1	0.8	3.2	1.6	0.48	2/8
Mortar-III	1	1.6	2.4	1.6	0.36	4/6
Mortar-IV	1	2.4	1.6	1.6	0.24	6/4
Mortar-V	1	3.2	0.8	1.6	0.12	8/2

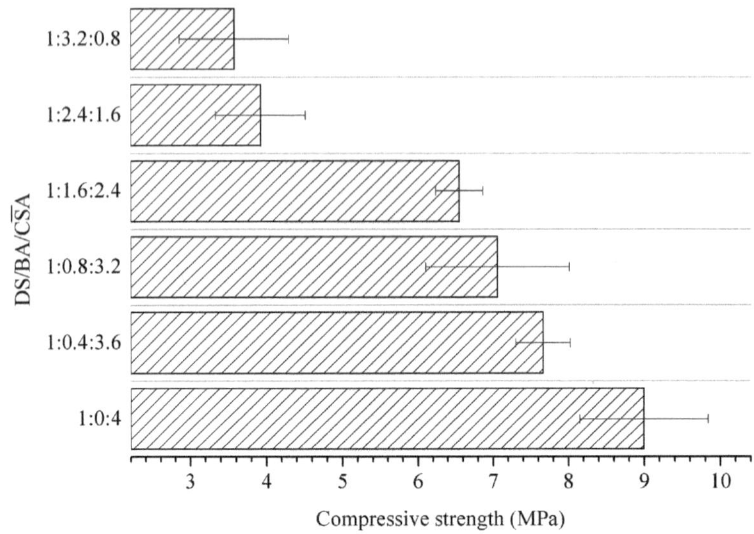

Figure 4.2 The effect of dewatered sludge (DS):BA:$C\overline{S}A$ ratios on 28-day strength of CLSM specimens.

4.1.3 X-Ray Diffraction Analysis of Controlled Low-Strength Material

XRD patterns of CLSM specimens after 28 days of hydration are shown in Fig. 4.3. The results revealed that sludge and BA gently modified the relative amount of minerals in the mortars depending on the ratio of DS:BA:$C\overline{S}A$. In the cases of lower ratios, the samples displayed a result similar to Mortar-Ref. Beyond the ratio of 1:1.6:2.3, however, a gradual decrease could be clearly seen in crystalline ettringite $(C_3A \cdot 3C\overline{S} \cdot H_{32}, \text{or AFt})$ and calcite ($CaCO_3$), in addition to a considerable increase in quartz (SiO_2). This was mainly attributable to the presence of hydrocarbons in raw materials, which hindered the normal reactions occurring in $C\overline{S}A$ clinker to a degree. Nevertheless, the positions of characteristic peaks of hydrates in all CLSM samples appeared exactly similar (Fig. 4.3), indicating that the retardation was not seriously significant, which offered a beneficial explanation for the good strength development of the designed materials (Fig. 4.2).

Furthermore, to deeply disclose the cementitious mechanisms by which the strength was enhanced, the variations of phases $C_4A_3\overline{S}$ and $CaSO_4(C\overline{S})$ present in CLSM systems during the period of hydration were detected. The strong consumption of $C_4A_3\overline{S}$ and $C\overline{S}$ at the later age (3–28 days) (Fig. 4.4) confirmed that the typical pozzolanic reaction

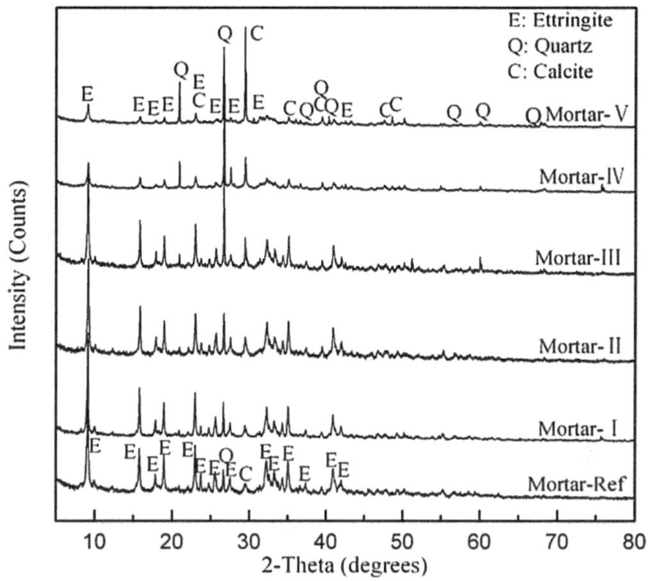

Figure 4.3 X-ray diffractograms of controlled low-strength material (CLSM) specimens on day 28.

$$C_4A_3\overline{S} + 2C\overline{S} + 34H \rightarrow C_3A \cdot 3C\overline{S} \cdot H_{32}(AFt) + 2AH_3 \tag{4.1}$$

took place, which subsequently resulted in the formation of AFt $(C_6A\overline{S}_3H_{32})$ and aluminum hydroxide (AH_3) gels to provide good binding properties. In reaction (4.1), shorthand notations of $C = CaO$; $A = Al_2O_3$; $\overline{S} = SO_3$; $H = H_2O$ were used. Meanwhile, the reaction of metallic aluminum present in BA with the alkalis from the hydration of $C\overline{S}A$ cement might promote the precipitation of AH_3 gels as well. The presence of AH_3 was not detectable as stated before (Fig. 4.3) possibly because of its poorly crystalline.

Additionally, sufficient $C_4A_3\overline{S}$ was considered to be the source of pozzolanic reaction (4.2) as

$$C_4A_3\overline{S} + 18H \rightarrow C_3A \cdot C\overline{S} \cdot H_{12}(AFm) + 2AH_3 \tag{4.2}$$

which directly produced monosulphoaluminate crystals $(C_3A \cdot C\overline{S}H_{12}$, or AFm$)$. The fact that XRD failed to identify AFm as presented in Fig. 4.3 suggested that the later consumption of $C_4A_3\overline{S}$ via reaction (4.1), might hinder reaction (4.2), consequently leading to the absence of AFm phases.

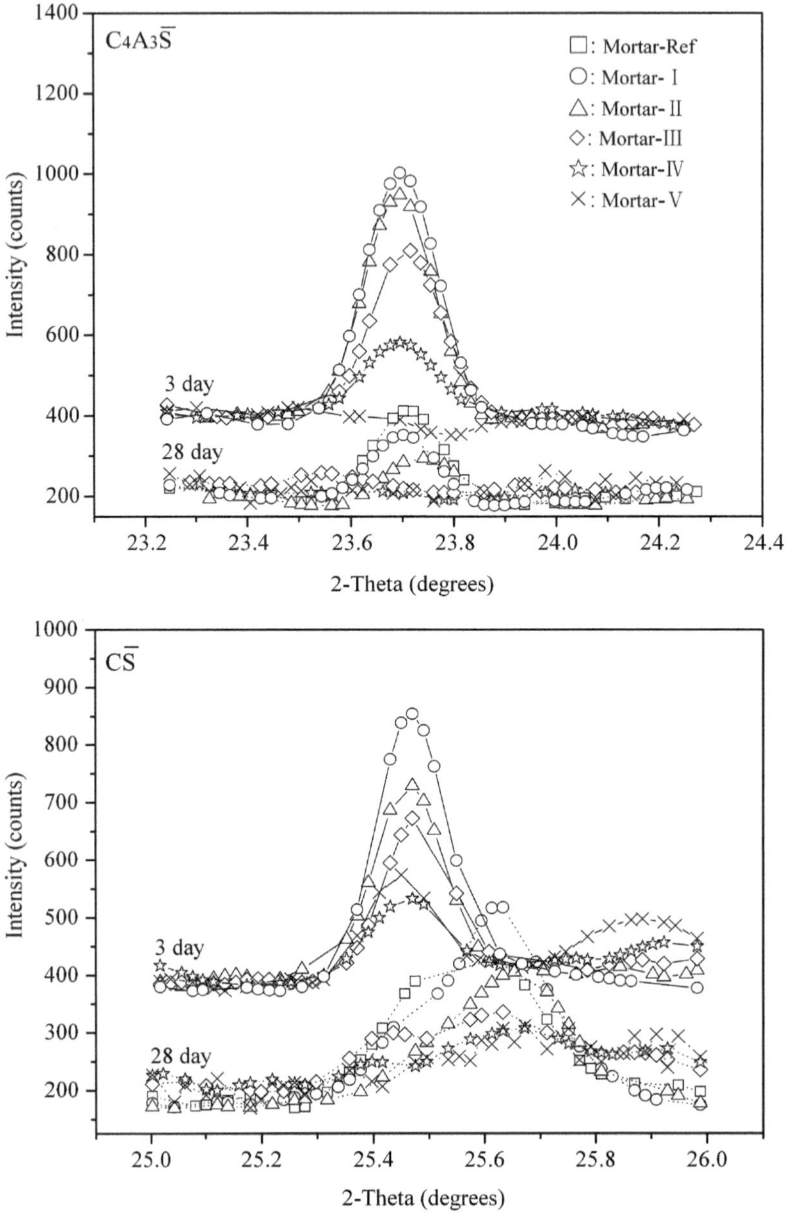

Figure 4.4 Variations of X-ray diffractograms of CLSM specimens with time, illustrating $C_4A_3\overline{S}$ and $C\overline{S}$.

4.1.4 Thermogravimetry—Differential Scanning Calorimetry Analysis of Controlled Low-Strength Material Specimens

XRD has failed to identify some poorly crystalline and amorphous phases, as described in Section 4.1.3. Thermogravimetry—differential scanning calorimetry (TG—DSC) analysis was subsequently employed to detect certain chemical species present. Fig. 4.5 displays the derivative TG (DTG)—DSC curves of all CLSM specimens after a 28-day hydration. The broad weight-loss peaks in the range 80—130°C in DTG curves were most probably attributed to the decomposition of AFt-related gel. The corresponding endothermic peaks could also be found in DSC curves. The considerable signal in the 170—280°C region was observed possibly because of the presence of AH_3.

On further increasing the temperature, a significant weight loss at around 650—700°C appeared, mainly thanks to the volatilization and destruction of organic compounds in sludge and BA. Also, up to 880°C or so, the peaks with various intensities were visible, relating to CO_2-loss of calcite. Apart from decarbonation of calcite, possible pyrolysis of certain inorganic minerals might also take place.

The intensity of peaks related to hydration process experienced variations depending on DS:BA:C$\overline{S}$A ratios, though all CLSM mixtures had a

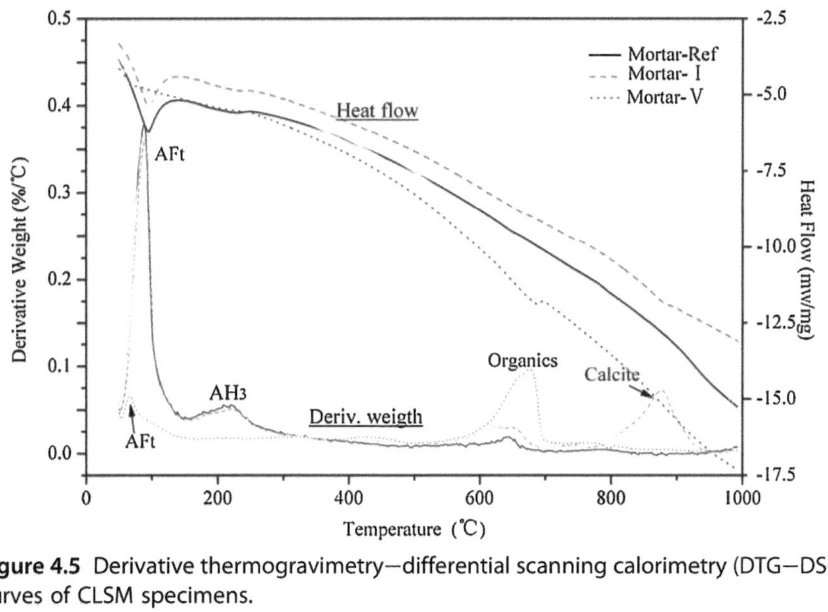

Figure 4.5 Derivative thermogravimetry—differential scanning calorimetry (DTG—DSC) curves of CLSM specimens.

similar thermal behavior, as illustrated in Fig. 4.5. It can be found that the weight-loss peaks for hydration products, that is, AFt, AH_3, and $CaCO_3$ became evidently weak as the ratio increased, indicating that the inert matters from raw materials might depress the normal pozzolanic reactions and the precipitation of the crystalline phases.

4.1.5 Fourier Transform Infrared Spectroscopy Analysis of Controlled Low-Strength Material Specimens

Fig. 4.6 depicts Fourier transform infraredspectroscopy (FT-IR) spectra of the hydrated CLSM pastes. The characteristic bonds at 3400–3600 cm^{-1} and 1655 cm^{-1} were attributed to O–H symmetric (ν_1) and asymmetric (ν_3) stretching vibrations and H–O–H in-plane bending vibration (ν_2) in molecular water, respectively. The O–H stretch bond of AH_3 appeared at around 3629 cm^{-1}. The typical bands of ν_2 and ν_3-CO_3^{2-} were detected at 1420 and 875 cm^{-1}, respectively, due to presence of calcite ($CaCO_3$). Also, the peaks at 850 and 1115 cm^{-1} were assigned to AlO_6 and ν_3-SO_4^{2-}, respectively, which were considered as fingerprint evidences for AFt crystal. Additionally, the band characteristics of the presence of organic loading were placed at 2852 and 2921 cm^{-1}.

FT-IR spectra depicted that the intensities of the absorption peaks corresponding to O–H of AH_3, AlO_6, ν_3-SO_4^{2-}, ν_2, and ν_3-CO_3^{2-} decreased

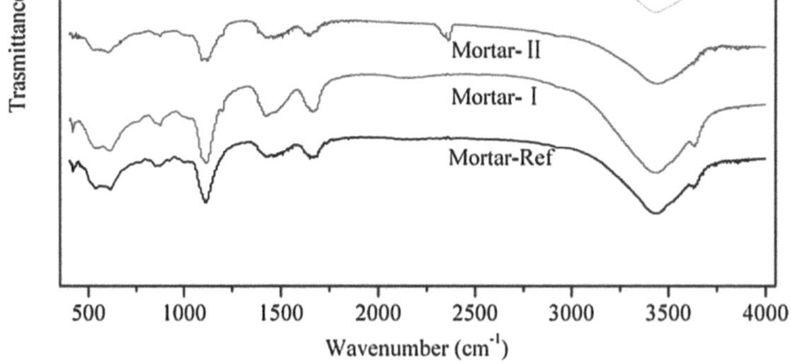

Figure 4.6 Fourier transform infraredspectroscopy (FT-IR) spectra of CLSM mortars cured for 28 days.

with increased DS:BA:C$\bar{S}$A ratio, revealing that pozzolanic reaction (4.1) occurred in a smaller extent by the addition of sludge and BA, thus leading to a lower resultant strength of specimens when compared to Mortar-Ref. These variations in the FT-IR spectra were exactly consistent with the results regarding XRD and TG-DSC analysis.

4.1.6 Scanning Electron Microscopy—Energy-Dispersive X-ray Spectroscopy Analysis of Controlled Low-Strength Material Specimens

The Mortar-Ref, Mortar-II, and Mortar-V were subjected to scanning electron microscopy—energy-dispersive x-ray spectroscopy (SEM-EDX) analysis to reliably investigate the effect of wastes presence on microstructures of CLSM matrixes. Visual observation of hydrated specimens obtained after 28 days in Fig. 4.7 reveals that the mortars exhibited similar microstructure and mineralogical compositions. The AFt crystal of needle-like morphology with lengths 10—20 μm was observed to extensively form in the available space offered by preexisting cracks and ash—paste interface. Simultaneously, the presence of AH_3 gels was very well confirmed with the help of EDX analysis with respect to a higher proportion of metallic aluminum in these areas (about 7.9, 9.1, and 8.9 wt.% for Mortar-Ref, Mortar-II, and Mortar-V, respectively). In addition to the identifiable hydration products discussed above, other abnormal compounds such as kalium-, magnesium-, and titanium-bearing phases were also recognized at the interface between cement paste and waste grains.

When comparing the SEM images of three CLSM mortars, the needle-shaped AFt crystals present in Mortar-Ref appeared clearly visible and regular, displaying a relatively high crystallinity. The crystals in Mortar-II, especially in Mortar-V, however, seemed more irregular with some exiguous tubers around them. The slight distinction in microstructure could be due to the fact that the available sludge and BA have participated into the hydration of C$\bar{S}$A clinker via Eq. (4.1), and became crystallochemically incorporated within the needle-like AFt phase, thus leading to the irregular morphology (Fig. 4.7A—B). While some coarse sludge and BA particles remained nearly unhydrated except of their surfaces as shown in Fig. 4.7A, they acted as aggregates of the sludge-/BA-amended materials and yielded the strength framework of the matrix. The hydration products that are AFt, AH_3, and $CaCO_3$ interluded among the different particles and enhanced the contacts of the aggregates that consequently contributed a lot to the material strength development (Fig. 4.2).

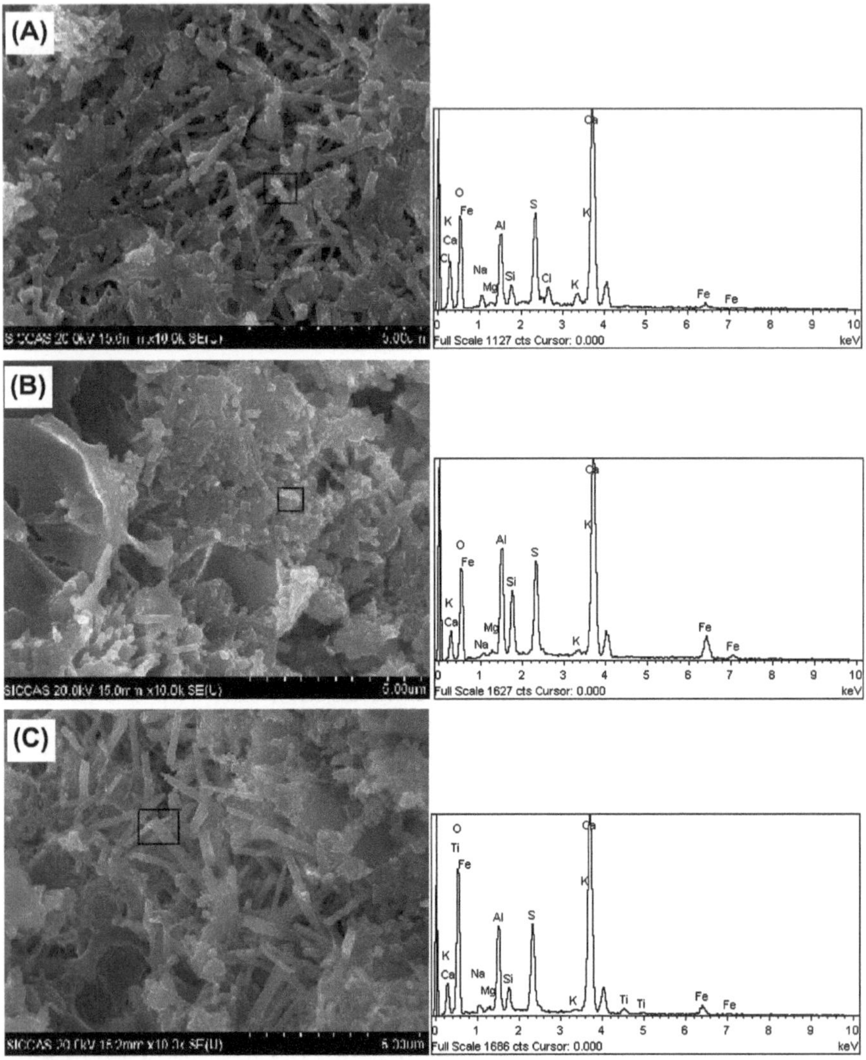

Figure 4.7 Scanning electron microscopy–energy-dispersive X-ray spectroscopy (SEM–EDX) spectrums of (A) Mortar-V, (B) Mortar-II, and (C) Mortar-Ref cured for 28 days in the CLSM.

From these findings, it could be rightly stated that sludge and BA were truly applied as additional ingredients in CLSM production. From the fact that smaller the particle size the higher the cementitious activity of materials, it could be verified that BA would be more beneficial in promoting the early strength of CLSM, provided that an excellent fineness comparable to that typical of cement is reached.

4.1.7 Leaching Behavior of Controlled Low-Strength Material Specimens

The experimental results for the targeted heavy metals and corresponding regulatory thresholds are presented respectively in Table 4.2. Cumulative amounts of metals leached from CLSM were considerably low, much below the regulatory thresholds. A slight increase in Cu from 0.08–0.13 to 2.22–2.42 mg/L was observed as the DS:BA:C$\bar{S}$A ratio increased, as shown in Table 4.2. In the case of As (arsenic), each specimen exhibited very low cumulative leachability of 0.02 mg/L. The similar behaviors were also observed for Ba, Mn, and Ni, regardless of DS:BA:C$\bar{S}$A ratio and leaching time, demonstrating very high immobility. Contrary to what were observed for the metals discussed, the effect of the increasing waste loading in CLSM mortars was more pronounced in preventing Cr and Pb leaching. As can be seen from Table 4.2, the cumulative leaching of Cr and Pb demonstrated a

Table 4.2 Cumulative Fractions of Leachable Substances in Toxicity Characteristic Leaching Procedure Test (mg/L) for CLSM

		CLSM Mortars						
Parameters	**Cumulative Time (h)**	**Ref**	**I**	**II**	**III**	**IV**	**V**	**Limit[a]**
As	24	–	0.01	–	0.01	0.01	–	5
	840	–	–	0.02	–	0.01	–	
Ba	24	0.08	0.08	0.07	0.08	0.11	0.14	100
	840	0.08	0.08	0.07	0.07	0.10	0.09	
Cd	24	–	–	–	–	–	–	1
	840	–	–	–	–	–	–	
Cr	24	0.17	0.15	0.16	0.14	0.09	0.06	5
	840	0.24	0.22	0.21	0.16	–	0.03	
Cu	24	0.08	0.25	0.52	1.04	2.26	2.22	100
	840	0.13	0.39	0.60	1.36	1.44	2.42	
Pb	24	0.03	0.03	0.01	–	–	–	5
	840	0.02	0.03	0.01	–	–	–	
Mn	24	–	–	–	–	0.01	0.06	50[b]
	840	–	–	–	–	–	0.03	
Zn	24	–	–	–	–	–	–	100
	840	–	–	–	–	–	–	
Ni	24	0.01	0.01	0.01	0.01	0.02	0.03	5
	840	0.02	0.01	0.01	0.01	0.02	0.02	

[a]Chinese standard: GB5085.3–2007.
[b]Adopted from Naganathan, S., Razak, H.A., Hamid S.N.A., 2010. Effect of kaolin addition on the performance of controlled low-strength material using industrial waste incineration bottom ash. Waste Manage. Res. 28, 848–860.

constant decrease that reached 0.03−0.06 and 0 mg/L, respectively, at Mortar-V with DS:BA:C$\bar{S}$A ratio of 1:3.2:0.8. In particular, no Cd and Zn were detected in any of the six CLSM specimens. Hence, it was very clear that the sludge-/BA-bearing CLSM would not pose any environmental risk when applied in structural engineering.

The low leachability characteristics of metals concerned in this investigation were possibly because dissolved metal iron substituted the parent irons (Ca^{2+}, Al^{3+}, etc.) and were encapsulated within the AFt-related crystals. Additionally, the adsorption from van der Waals force also favored the stability of metals-containing phases.

Furthermore, total organic carbon (TOC) concentrations in the extracts were measured. It was found that all CLSM dosages exhibited low TOC leaching ranging from 365 to 635 mg/L. These concentrations nearly could not pose health harm since leachate would eventually get diluted and attenuated if discharged into the nature. The porosity decrease and sinuousness of the porous system due to the formation of hydration crystals were presumably responsible for the low leachability.

4.2 MECHANICAL AND MICROSTRUCTURAL PERSPECTIVES OF CONTROLLED LOW-STRENGTH MATERIAL CURED FOR 1 YEAR

4.2.1 Compressive Strength of Controlled Low-Strength Material

Seven groups of CLSM mixtures were created in a definite ratio (Table 4.3). Fig. 4.8 shows that UCS of CLSM specimens strongly depended on their mix designs and the mineral constituents of BA. The control, that is, Mortar-Ref in the absence of BA, produced an UCS of around 2.5 MPa. In respect of the mortars with milled BA, it was apparent that the addition of BA caused a significant enhancement in the compressive strength (Fig. 4.8A). Mortar-I_{65} corresponding to DS:BA:C$\bar{S}$A ratio of 1.0:0.1:0.9 displayed the 1-year strength gain of 6.2 MPa, increase of roughly 150% if compared to Mortar-Ref. When DS:BA:C$\bar{S}$A ratio was increased to 1.0: 0.4:0.6, the UCS of Mortar-II_{65} declined to 3.5 MPa but was still 40% higher than that of Mortar-Ref. The further rise of DS:BA:C$\bar{S}$A ratio to 1.0:0.8:0.2 (i.e., Mortar-III_{65}), however, was found to be more deleterious than beneficial to the mechanical properties. The UCS of Mortar-III_{65} decreased to roughly 2.0 MPa, approximately 20% lower than that of Mortar-Ref. This was probably due to the presence of the excess metal

Table 4.3 Mixed Formulations of CLSM Specimens

	Proportion by Weight				
Mix Id	**DS**	**BA**	**C$\bar{S}$A**	**C$\bar{S}$A/(DS + BA + C$\bar{S}$A)**	**BA/C$\bar{S}$A**
Mortar-Ref	1.0	0	1.0	1.0	0/10
Mortar-I_{65}	1.0	0.1	0.9	0.45	1/9
Mortar-II_{65}	1.0	0.4	0.6	0.30	4/6
Mortar-III_{65}	1.0	0.8	0.2	0.10	2/8
Mortar-I_{900}	1.0	0.1	0.9	0.45	1/9
Mortar-II_{900}	1.0	0.4	0.6	0.30	4/6
Mortar-III_{900}	1.0	0.8	0.2	0.10	2/8

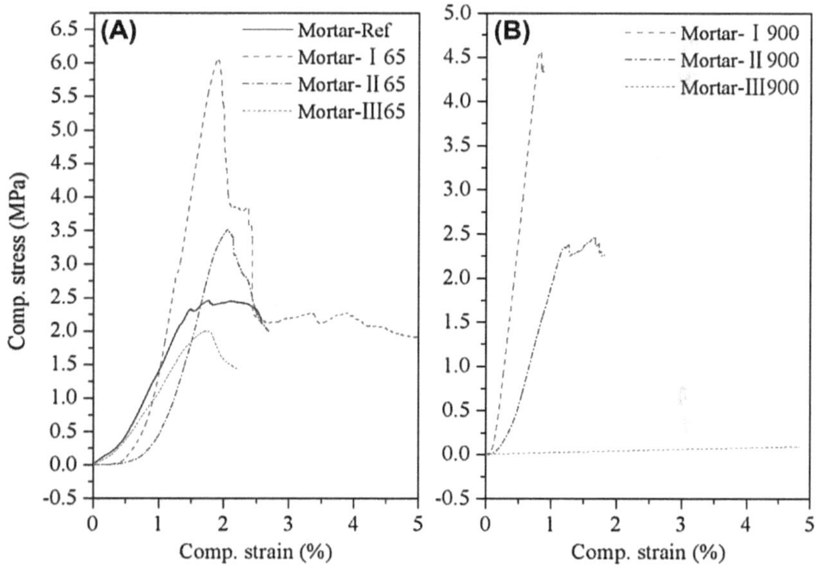

Figure 4.8 Effect of DS:BA:C$\bar{S}$A ratios on the unconfined compressive strength (UCS) of CLSM specimens (A) Mortar-Ref, Mortar-I_{65}, Mortar-II_{65}, and Mortar-III_{65} and (B) Mortar-I_{900}, Mortar-II_{900}, and Mortar-III_{900} cured for 1 year.

gels and organic matters in BA, which retarded the normal hydration occurring in C$\bar{S}$A cement.

Comparatively, the UCS of CLSM containing BA thermally treated at 90°C (Fig. 4.8B) were observed to be 4.6 (Mortar-I_{900}), 2.5 (Mortar-II_{900}), and 0.7 MPa (Mortar-III_{900}), respectively, decreasing up to 26%, 29%, and 65% in comparison with the corresponding CLSM mortars formulated with milled BA. This phenomenon was linked to the highly denser surfaces of BA caused by the melting of minerals at 900°C, which reduced and blocked water and iron transport required for BA hydration. The retardation of hydration thus caused the strength to decrease remarkably. These findings evidently implied that at 1.0:0.1:0.9 ≤ DS:BA:C$\bar{S}$A ≤ 1.0:0.8:0.2, milled BA combined with sludge and C$\bar{S}$A cement could produce CLSM with desired mechanical properties, and thermal treatment of BA was not necessarily required prior to reuse during CLSM production.

The crystalline phases for the above CLSM specimens were investigated by XRD analysis (Fig. 4.9). The mineralogical compositions of CLSM mortars containing milled BA with DS:BA:C$\bar{S}$A of 1.0:0.1:0.9 and 1.0:0.4:0.6 (i.e., Mortar-I_{65} and Mortar-II_{65}, respectively) were similar to Mortar-Ref. The major crystalline phases recognized in CLSM mortars consisted of quartz (SiO_2), ettringite $\left(C_3A\cdot 3C\bar{S}\cdot H_{32}, \text{ or AFt}\right)$, and calcite ($CaCO_3$) with small amounts of anhydrite ($CaSO_4$). Comparison of the strength development with the XRD results showed that the excellent UCS was primarily due to the formation of AFt gels. In addition to such an effect, the higher UCS for CLSM might also be explained considering the effect of frictional resistance induced by BA, which was expected to enhance the contacts of the aggregates and the subsequent mechanical performance. As the DS:BA:C$\bar{S}$A ratio was increased from 1.0:0.4:0.6 to 1.0:0.8:0.2, the intensity of XRD peaks of SiO_2 increased drastically whereas

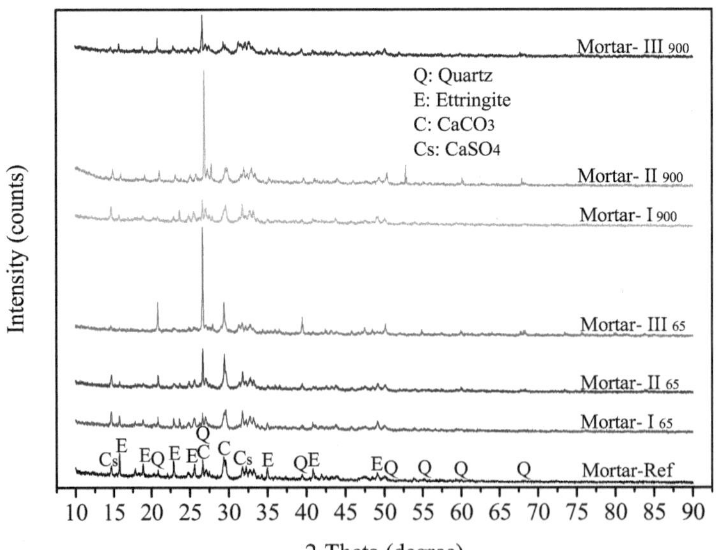

Figure 4.9 X-ray diffractograms of various CLSM mortars cured for 1 year.

the peaks of AFt and $CaCO_3$ crystals tended to decrease. Higher BA dosage will introduce more salts of Zn and Pb (Table 4.3) with the resultant increase in quantity of calcium hydroxyzincate ($CaZn_2(OH)_6 \cdot 2H_2O$) and Pb-containing phases. Those compounds can cover the surface of cement particles and hinder the normal reactions of $C\overline{S}A$ clinker. As a result of that, the strength of Mortar-III_{65} body was deteriorated (Fig. 4.9). Notably, DS:BA:$C\overline{S}A$ ratio exerts a crucial role in the formation of crystals and simultaneously the strength performance of CLSM.

Fig. 4.9 also presents the phases' evolution of CLSM samples containing thermally treated BA. Similarly, the pronounced crystalline phases of SiO_2, AFt, $CaCO_3$, and $CaSO_4$ were identified in XRD patterns for the mortars with DS:BA:$C\overline{S}A$ of 1.0:0.1:0.9 and 1.0:0.4:0.6 (i.e., Mortar-I_{900} and Mortar-II_{900}, respectively). As compared to the CLSM manufactured with milled BA, the peak intensities of phases detected in CLSM with thermally treated BA became typically narrower and lower. In particular, when DS:BA:$C\overline{S}A$ ratio was 1.0:0.8:0.2 (Mortar-III_{900}), the peaks related to AFt and $CaCO_3$ crystals nearly disappeared while the peak of SiO_2 increased. The formation of new complex crystalline phases observed before (Fig. 4.9) in thermally treated BA, which reduced the chemical reactivity of BA and retarded the pozzolanic cementing action induced by $C\overline{S}A$ cement, might be the core reason for the decrease in the strength of CLSM mortars.

4.2.2 Fourier Transform InfraredSpectroscopy (FT-IR) Analysis of Various Controlled Low-Strength Material Specimens Cured for 1 Year

The evolution of functional groups in the crystalline phases was also characterized by FT-IR analysis. As depicted in Fig. 4.10, the characteristic bond at about 3448 cm^{-1} was O—H symmetric (ν_1) and asymmetric (ν_3) stretching vibrations of hydration water, whereas the band present at 1650 cm^{-1} was assigned to H—O—H in-plane bending vibration (ν_2) of interlayer water. The band located at around 3629 cm^{-1}, linked to O—H stretch bond, was a sound proof of $Al(OH)_3$ gels' presence. The two bands at 1482 and 876 cm^{-1} corresponded to ν_2 and ν_3-CO_3^{2-} resulting from $CaCO_3$ phases. Besides, the bands at 857 and 1113 cm^{-1} were mainly attributable to AlO_6 and ν_3-SO_4^{2-}, respectively, supporting the presence of AFt crystals. With the addition of milled BA, the band 857 cm^{-1}, related to AlO_6 in AFt gels, became higher and increased. Although the peaks of AFt gels somewhat overlapped with those of $CaCO_3$ in XRD patterns (Fig. 4.9), it tended to

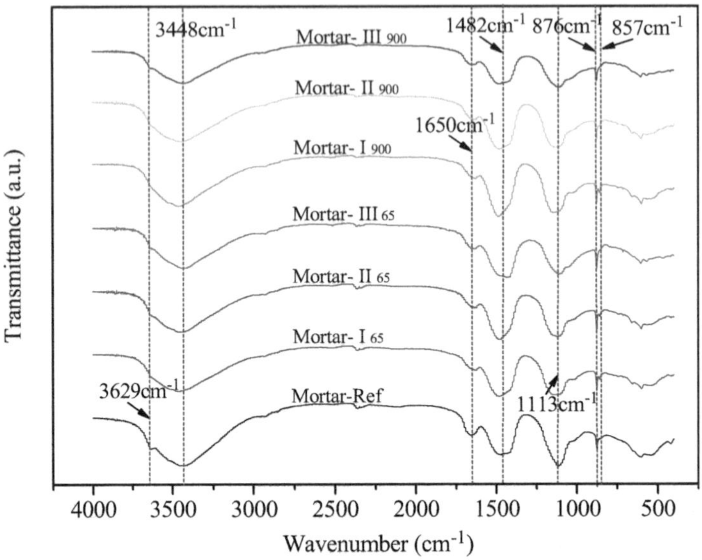

Figure 4.10 FT-IR spectra of various CLSM mortars cured for 1 year.

form clearly with increasing DS:BA:C$\bar{S}$A ratio up to 1.0:0.1:0.9 (Mortar-I_{65}) according to FT-IR spectra in Fig. 4.10. This result agreed with the increase in the UCS when BA was used to replace the part of C$\bar{S}$A cement. The highest UCS value of CLSM at 6.2 MPa was obtained in Mortar-I_{65} mixture, which revealed that the addition of milled BA did benefit the AFt gels' formation and improve the mechanical properties of CLSM mortars. As compared with the FT-IR spectra of CLSM fabricated with milled BA, the band intensities of the above mentioned functional groups in the specimens containing thermally treated BA decreased significantly, in accordance with the results regarding XRD analysis (Fig. 4.9). This further explained why the addition of thermally treated BA brought about the perceptible degradation of strength performance of CLSM products.

4.2.3 Morphological Structures (SEM-EDX) Analysis of Controlled Low-Strength Material Specimens Cured for 1 Year

To understand more about the surface morphology and crystalline phases of CLSM mortars, SEM-EDX analysis was conducted. Fig. 4.11 illuminates the SEM images of samples Mortar-Ref, Mortar-I_{65}, Mortar-II_{65}, Mortar-I_{900}, and Mortar-II_{900}. It can be seen from the SEM image of Mortar-Ref that the majority of fibrous- and needle-like crystals between 5 and 10 μm in size

Figure 4.11 SEM-EDX spectrums of various CLSM mortars cured for 1 year.

were formed predominantly in the available space offered by preexisting cracks and aggregate–paste interface. According to the EDX results, XRD, and FT-IR analysis (Figs. 4.9 and 4.10), the crystals should be AFt gels. The AFt crystals are known to hold a critical role in diminishing the porosity of system and leading to a denser microstructure. Mortar-I_{65} with milled MSWI BA (DS:BA:$C\overline{S}A$ = 1.0:0.1:0.9) showed the presence of newly well-formed AFt crystals with polygonal morphology on the crystalline surface. This was possibly because of the crystallochemical incorporation of sludge and BA into AFt phases, which might cause the solid wastes to change from their amorphous to crystalline forms. The new AFt phases, with a large amount of intercrystalline pores, formed among the different particles and enhanced the contacts of the solid phases, which densified the matrix and contributed a lot to the material's strength and chemical durability. With respect to Mortar-II_{65} (DS:BA:$C\bar{S}A$ ratio = 1.0:0.4:0.6), the coarse sludge and BA particles, which remained almost unchanged except of their surfaces, were easily distinguishable in loose contact with the unhydrated clinker relics. The major hydration products, that is, AFt crystals and $CaCO_3$ gels, were identified in relatively minor amounts, implying that the cement hydration was substantially inhibited. The irregular meshed structure with abundant big pores in the interior of CLSM due to the retardation of $C\bar{S}A$ cement possibly accounted for the negative effect of the overdose of milled MSWI BA on the strength development.

Fig. 4.11 also displays the microstructure of CLSM containing BA thermally treated at 900°C. It is evident that the SEM analysis of Mortar-I_{900} with DS:BA:$C\bar{S}A$ ratio of 1.0:0.1:0.9 was similar to that of Mortar-II_{900} with DS:BA:$C\bar{S}A$ ratio of 1.0:0.4:0.6. It was quite loose and rough as well, and there tended to be more approximately slit-shaped pores present compared with Mortar-II_{65}. Neither highly crystalline AFt nor $CaCO_3$ crystals were detected, which was unfavorable for the compaction of CLSM mortars. The difference in the microstructure of CLSM prepared with milled and thermally treated MSWI BA was embodied by the difference of surface characteristics of BA particles. Firing at 900°C might enhance the formation of vitreous and impermeable layer covering the surface of BA, which prevented the water and ion transport required for the hydration. This was considered to further increase the porosity of the system and subsequently to debase the mechanical performance of CLSM. The SEM observations in general agreed well with the evolution trends of the strength of solidified samples.

4.2.4 Assessment of Leaching Behavior of Controlled Low-Strength Material Specimens

Organic substances leached from various CLSM forms were analyzed by using three-dimensional excitation–emission matrix (EEM) fluorescence spectroscopy to evaluate the durability performance of CLSM mortars upon exposure to aqueous solution environments. Typical EEM fluorescence spectra of organic substances in leachate are depicted in Fig. 4.12. Two main peaks could be readily identified from the fluorescence spectra. The first peak was present at the Ex/Em of 240–250/395–405 nm (Peak A), whereas the second peak was identified at the Ex/Em of around 290–300/380–390 nm (Peak B). The two fluorescence peaks have been

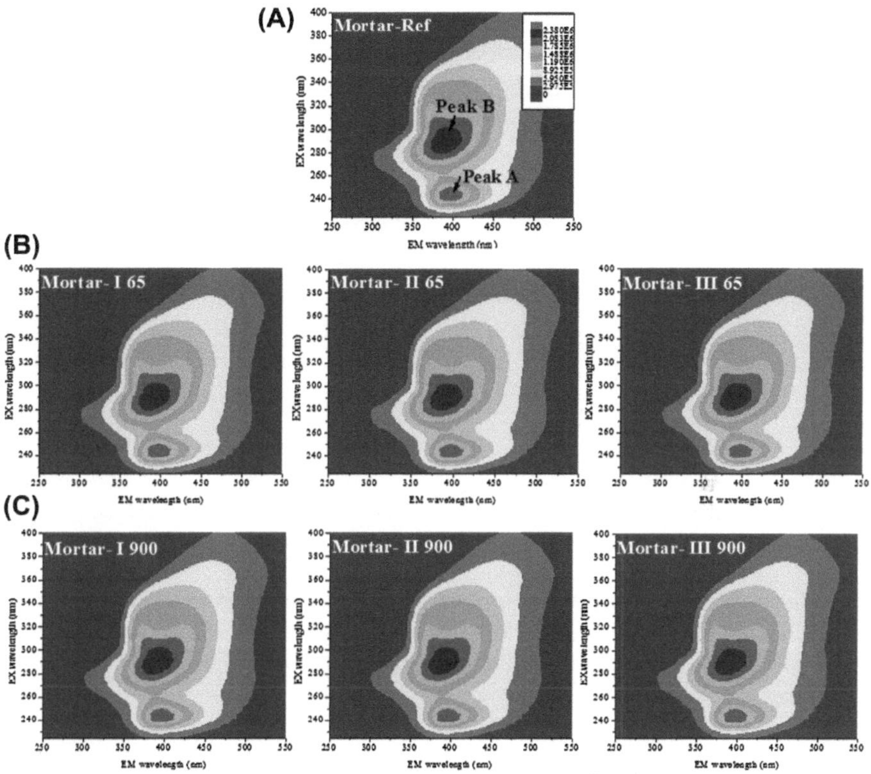

Figure 4.12 Excitation–emission matrix (EEM) fluorescence spectra of organic substances leaching from various CLSM mortars without (A) or with milled (65°C) (B) and thermally (900°C) treated (C) bottom ash (BA) after 1 year of curing.

described as the humic-like substances. Protein-like substances were not observed herein, indicating that the curing process of CLSM mortars was characterized by a decrease in protein-like matters and an increase in humic-like substances. Note that all leachate samples showed higher similarity in EEM fluorescence spectra, which revealed that the leaching and transforming behaviors of organic matters present in solidified CLSM pastes were not substantially influenced by DS:BA:C$\bar{S}$A ratio and properties of BA. In addition, it is worth mentioning that the concentrations of all heavy metals for selected CLSM mixtures were lower than the stated limits in China (GB 5085.3–2007). The low leachability of heavy metals might be related to the strong fixing of AFt and monosulphate since Cr(III), Pb, and Zn were strongly encapsulated into calcium silicate matrix by replacing Ca. In addition, the increase of the aluminum (hydr)oxides during the weathering process might keep heavy metals such as Cu and Zn from leaching. Therefore, the good mechanical strength and low leachability observed after 1 year of curing were strongly encouraging for the recycling and reuse of DS and BA for CLSM production.

CHAPTER FIVE

Harvest of Bioenergy From Sewage Sludge by Anaerobic Digestion

5.1 OVERVIEW ON CURRENT ADVANCES IN SLUDGE PRETREATMENT TO IMPROVE ANAEROBIC BIODEGRADABILITY

As per IEA's latest estimates, the world's primary energy consumption reached 13,371.0 million tons of oil equivalent in 2012, growing by 52.3% over the last 20 years. The excess and useless use of fossil fuels has accelerated the depletion of global nonrenewable resources and climate change. The world is facing a severe, critical shortage of energy, as well as an unprecedented challenge. Developing sustainable technologies have become of the upmost importance in response to the rising energy demands. One alternative to fossil fuels is renewable methane-rich biofuel produced through anaerobic digestion (AD) of biomass such as SL. AD is of great promise for sewage-sludge treatment as it removes odors and pathogens, stabilizes sludge and, more importantly, produces renewable energy in the form of methane. This can either cover part of the energy requirements for SL treatment or, to a certain degree, alleviate human's dependence on fossil fuels (Fig. 5.1).

AD comprises several successive biochemical processes (i.e., hydrolysis, acidogenesis, acetogenesis, and methanogenesis) involving different groups of microbes (Fig. 5.2). In the first step, complex organic matters such as proteins, polysaccharides, and lipids are solubilized and hydrolyzed into simple soluble components (e.g., amino acids, long-chain fatty acids, sugars, and alcohols) under the assistance of extracellular enzymes. The hydrolyzed molecules in the second step are converted by acidogenic (or fermentative) bacteria to short-chain volatile fatty acids (VFAs) and other minor by-products such as ammonia (NH_3), H_2, and CO_2. During acetogenesis process, acetogens further decompose the higher organic acids (e.g., propionic and butyric acids) to form primarily acetic acid, and H_2 via β-oxidation; however, the conversions are only favorable thermodynamically under particularly low

Pollution Control and Resource Recovery: Sewage Sludge
ISBN: 978-0-12-811639-5
http://dx.doi.org/10.1016/B978-0-12-811639-5.00005-X

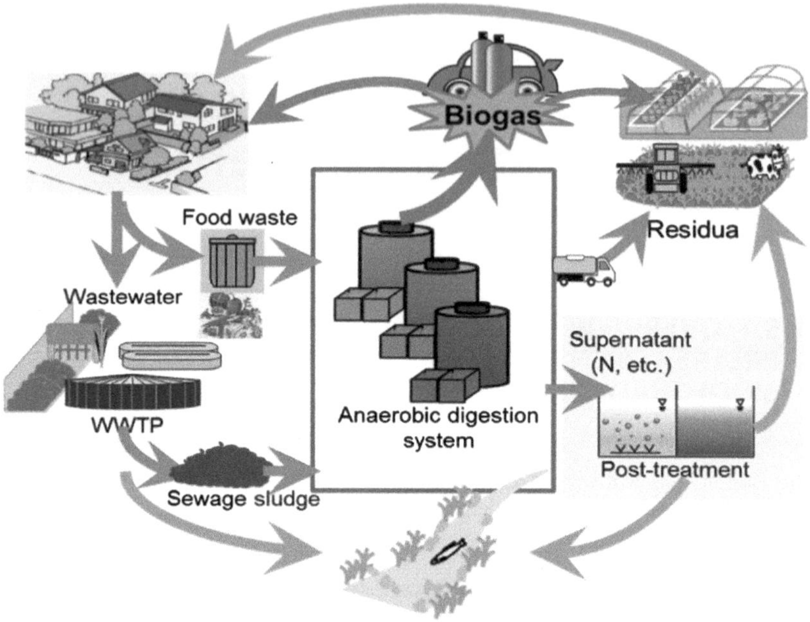

Figure 5.1 Schematic overview of organic wastes [e.g., sewage sludge (SL), food waste, etc.] reduction, and renewable and recyclable supplies of resources.

concentrations of the reaction products (acetate and H_2) (i.e., acetate concentration: 10^{-4}—10^{-1} mol/L; H_2 partial pressure required for propionate: 10^{-6}—10^{-4} atm; and for butyrate: (1.0—7.0) $\times 10^{-3}$ atm). Finally, a myriad of methanogenic *archaea* metabolize acetate as well as H_2 and CO_2 to produce methane. Acetate is the most preferred methanogen substrate, producing 60—70% of the methane with the remaining 30—40% from the redox reaction of H_2 and CO_2.

Hydrolysis of particulate organics associated with microbial cells and extracellular polymeric substance (EPS) has been proven as the rate-limiting step for the start-up of the anaerobic systems while methanogenesis is considered rate-limiting for further fermentation of the released soluble substrates. To accelerate the hydrolysis and enhance subsequent methane productivity, a variety of sludge pretreatment options such as mechanical, thermal, chemical, biological processes, or integrations of these ones have been developed at laboratory or pilot level so far with various levels of success. If properly designed, pretreatments can facilitate the release of intracellular substances by rupturing the cell walls and making them more accessible to subsequent microbial actions.

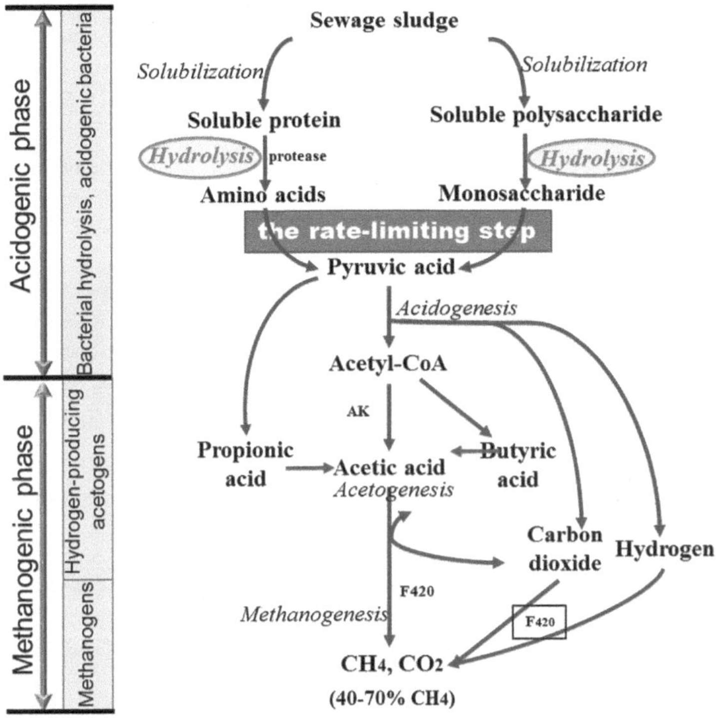

Figure 5.2 Proposed metabolic pathway for methane production from SL anaerobic digestion.

Slow hydrolysis, because of the complex and rigid floc structure of sludge, gives rise to long retention time and low degradation efficiencies. In this sense, numerous strategies have been previously studied on laboratory and pilot levels to enhance the hydrolytic limiting stage and to improve the performance of the global process (Table 5.1). The main goal of pretreatment is to disrupt EPS matrix and cells wall and to make the available nutrients accessible to microbes, thus speeding up the conversion of organic solids and methane productivity. Currently, many different approaches including mechanical, thermal, chemical, and biological as well as several combinations of different processes have been developed for this purpose.

5.1.1 Mechanical Pretreatment

5.1.1.1 Ultrasonic Pretreatment

Ultrasonication is a well-established mechanical technology for sludge disintegration. Ultrasound waves cause the periodical compression and rarefaction when propagating through the medium (Fig. 5.3A). The

Table 5.1 Summary of the Advantages and Disadvantages of Different Sludge Pretreatment Methods

Pretreatment	Mechanisms	Control Parameters	Advantages	Disadvantages	Commercial Technologies
Ultrasonication	Cavitation (hydro-mechanical shear forces, oxidizing effect)	Power input, exposure time	Increased methane yield	High energy demand	Biosonator, Sonix, Iwe.Tec, Smart DMS, Sonolyzer, Hiescher
Microwave	Thermal effect, athermal effect, etc.	Power input, frequency, exposure time	Quick and uniform heating, increased methane yield, pathogen removal	High energy demand	Aspal SLUDGE, Praxair Lyso
Electrokinetic disintegration	High-voltage field	Voltage, frequency, exposure time	Increased biogas production, economic benefits	High energy demand, complex operation and maintenance	BioCrack, OpenCEL, PowerMod
High-pressure homogenization	Turbulence, cavitation, shearing forces	Pressure, exposure time, cycle	Easy operation, low cost, increased methane yield, popularity on large scale	High energy demand, complex operation and maintenance	MicroSludge, Crown, Cellruptor
Thermal	Heating in the range of 100∼ °C	Temperature, exposure time, pressure	Increased methane yield, sanitation, odor removal, sludge reduction, improved dewaterability	High energy demand, risk of recalcitrant compounds formation, release of ammonia	CambiTHP, Biothelys, Exelys, Turbotec, Lysotherm, Biorefinex

Acidic	Hydrolysis of hemicellulose	Dose, exposure time	Increased methane yield, simple device, easy operation, low cost, effective for lignocellulosic biomass	Chemical cost, corrosion, special materials for reactor construction, neutralization before digestion	
Alkali	Solvation, saphonication	Dose, exposure time	Increased methane yield, simple device, easy operation, low cost, suitable for lignin breakdown	Chemical cost, corrosion, special materials for reactor, toxicity of Na^+, neutralization before digestion	
Ozonation	Formation of hydroxyl radicals, etc.	Dose, exposure time	Increased methane yield, pathogens removal, sludge reduction, flexible operation	Risk of liberated cellular materials mineralization	Full-scale operations in excess sludge reduction of WWTP

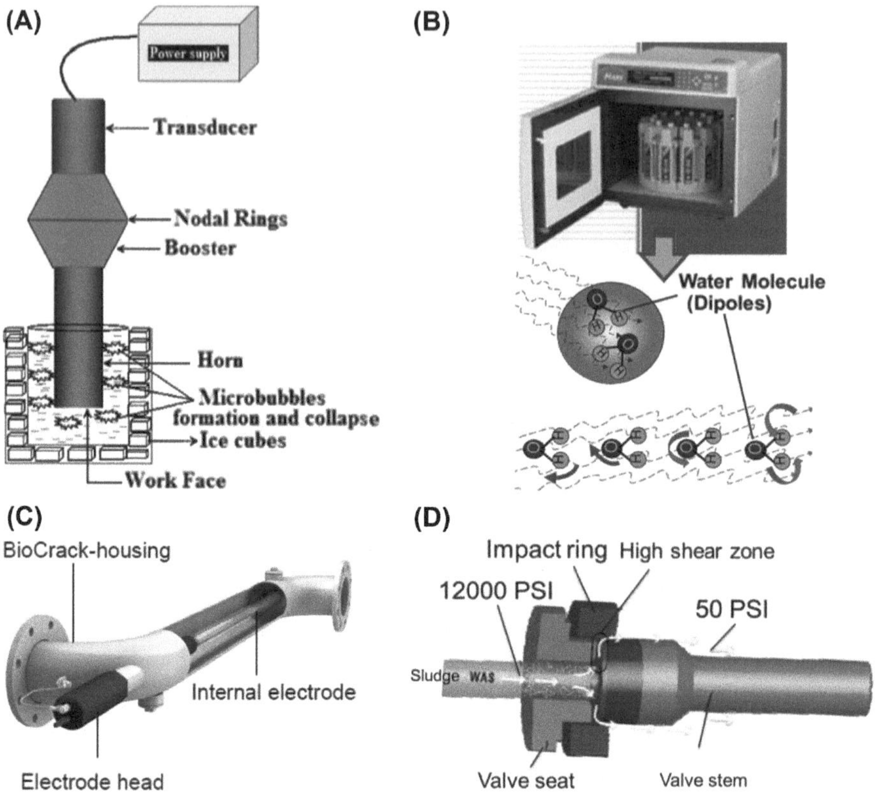

Figure 5.3 Configurations of ultrasonication [50] (A), microwave irradiation [20,68] (B), electrokinetic disintegration (the BioCrack module) (C), and high-pressure homogenization (HPH) (the MicroSludge module) [54] (D).

microbubbles formed during this process violently collapse within a few microseconds after reaching a critical size, inducing the occurrence of cavitation. The sudden and violent collapse leads to extreme conditions (a local temperature of around 5000K and a pressure above 500 bars) and initiates powerful hydro-mechanical shear forces and highly reactive radicals (H• and •OH). Both the hydro-mechanical shear forces and the oxidizing effect of H• and •OH contribute to the break-up of sludge flocs and the liberation of intercellular material. In comparison, hydro-mechanical shear forces, nonetheless, are stronger in sludge rupture than radicals.

Low-frequency (20—40 kHz) ultrasound, first applied in lab-scale cell lysis study in the 1960s, has been frequently adopted on the lab, pilot, and full-scale levels to sludge disintegration. Sludge solubilization efficiency relies on specific energy (SE) input $\left(SE = \frac{P_{US} \times t}{V \times TS}, \text{J/kg TS}\right)$. The threshold of SE usually reported in literature is between 1000 and 16,000 kJ/kg TS. It is worth noting that for a given SE, higher power was more efficient than a longer treatment time but lower power. It is reported that ultrawave system has at least 20 full-scale and 17 pilot-scale installations for the pretreatment of sludge, mostly in Germany. These ultrasound installations may result in a 15–35% increase in volatile-solids destruction and a 15–35% increase in biogas production. Besides the use in sewage-sludge AD, ultrasound also has been used for enhancing the sludge dewaterability.

5.1.1.2 Microwave Irradiation

Like ultrasonication, microwave (MW) irradiation is considered as a popular alternative to the conditional heating (CH) technology. In the electromagnetic spectrum, MW irradiation operates in wavelengths of 1 mm^{-1} m with corresponding oscillation frequencies 0.3–300 GHz. In industry, a shorter frequency either close to 900 MHz or at 2450 MHz is often adopted. Damages to sludge cells with MW irradiation may occur in two ways (Fig. 5.3B): (1) thermal effect that is generated through the rotation of dipoles under oscillating electromagnetic fields, which heats the intracellular liquor to boiling point and brings out the break-up of bacterial cells; and (2) athermal effect that is induced by the changing dipole orientation of polar molecules, giving rise to the possible breakage of hydrogen bonds and unfolding and denaturing of complex biological molecules, which kills microorganisms at lower temperatures.

MW irradiation is applicable in sewage-sludge disintegration before AD with high efficiency. Microwave pretreatment resulted in an effective solubilization of the organic fraction in the sludge and a 50% increase in biogas production. Single microwave pretreatment resulted in 50% reduction of *Clostridium perfringens* while the quantity of *C. perfringens* present in the digested sludge without MW pretreatment was almost the same as in raw sludge. Additionally, several lines of evidences also confirmed the favorable potential of MW irradiation in improving the dewaterability of digested sludge.

Nevertheless, there are conflicting conclusions on the athermal effect of MW irradiation in the literature. MW and CH pretreatments under identical heating profiles indicated that although MW athermal effect does not exert discernable effect on the chemical oxygen demand (COD) solubilization, it substantially upgraded the mesophilic anaerobic biodegradability of activated sludge and biogas production. Heating method (CH versus MW) had no statistically significant effect ($p > .05$) on biosolids solubilization and methane production. There was no pattern of microwave-heated digesters showing enhanced performance over conventionally heated digesters due to the athermal effect. This creates the need for more efforts to determine whether and/or to what extent the athermal effect influences sludge physicochemical characteristics as well as biodegradability.

5.1.1.3 Electrokinetic Disintegration

Electrokinetic disintegration (or pulsed electric field) is one of the high-voltage electric field methods. During disintegration process, the charges created by the high-voltage field induce the sudden disruption of rigid sludge flocs and cellular membranes, thereby making the nutrients easily available to the fermenting bacteria.

Electrokinetic disintegration, as a newly developed sludge pretreatment technology, has been implemented extensively in industry. Electrokinetic pretreatment of 63% of the input waste sludge increased biogas production by over 40% and reduced the biosolids requiring disposal by 30%. The electrokinetic disintegration was able to drastically reduce the amount of biological sludge produced by the plant, without affecting its treatment efficiency, which gave rise to a considerable net cost saving for the company. In another full-scale implementation, it similarly confirmed the net positive economic benefit as a result of the energy offsets from the increase in biogas (60%) and reduction in biosolids disposal (40%). Besides, German Vogelsang is one of the representatives for electrokinetic disintegration device producers (called as "BioCrack module"). The BioCrack module is composed of a system of pipes with alongside electrodes that applied a voltage of around 30—100 kV (Fig. 5.3C). The company claims that the application of BioCrack module to pretreat sludge increases biogas yields up to 20% at poised power of 35 W per module while offering roughly 30% downstream energy savings (pumping, mixing, etc.).

5.1.1.4 High-Pressure Homogenization (HPH)

The major principle of high-pressure homogenization (HPH) relies on abrupt pressure gradient, high turbulence, cavitation as well as strong shearing forces, which are aroused under strong depressurization of highly compressed sludge suspensions (up to 900 bar) (Fig. 5.3D). During this process, sludge flocs break and cell membrane ruptures releasing the intracellular substances. Hence, HPH can bring improvement in sludge disintegration and biodegradation performance.

There are many lab-scale investigations or full-scale demonstrations nowadays reporting the effect of HPH pretreatment on sludge solubilization and biogas production. Homogenizing pressure was found as the most significant factor ($p < .05$) affecting solubilization of particulate COD and biopolymers in sludge. At 12,000 psi and 0.009 g NaOH/g TS, HPH + 2PAD system achieved the maximum methane production (0.61−1.32 L CH_4/L digester-d) and VS removals (43−64%), as well as significant pathogen removals and positive energy balance. HPH option is characterized by easy operation, low investment, and high energy-efficiency and accordingly has been in high popularity in large-scale implementations over the past years. In a demonstration project, a modified homogenizer at 150 bar with flow rate of 2.7 m^3/h was employed to pretreat concentrated sludge; the results indicated about 23% sludge reduction with more than 30% increased biogas production, leading to enormous savings in sludge disposal costs.

5.1.2 Thermal Hydrolysis

Thermal hydrolysis (TH) is a well-established and commercially implemented pretreatment technology that, originally used to enhance sludge dewaterability, has been extensively studied in an effort to improve digestibility of SL. The performance of the TH process heavily relies on the treatment temperature and time used. Sludge solubilization increased with increasing temperature and the TH effect impacted more on carbohydrates and proteins than on lipids. For an anaerobic sludge treatment system comprised of TH, the optimal temperature for TH was 170°C under treatment time of 30−60 min, reducing the solid retention time (SRT) by 5 d and increasing the biogas production. No inhibitory effect appeared at the temperature studied. Hereafter, TH pretreatment of sludge has increasingly drawn the attention of environmental researchers at large. TH process (60−210°C) was applied to pretreat six different waste sludges and found that sludge solubilization increased with treatment temperature up to 190°C. In the studied temperature range, biodegradability enhancement or methane production increase by TH showed a function of sludge COD solubilization. Further increasing the TH temperature (i.e., to 210°C) reversely weakened biodegradability. They attributed this to the formation of recalcitrant compounds in the *Maillard* reactions, that is, melanoidins (high molecular-weight heterogeneous polymers) which are particularly difficult to degrade and even inhibit the biodegradation of other organics. They further concluded that the thermal treatment time had less impact on the sludge solubilization in comparison with temperature. Sludge solubilization did not seem to be different even if its treatment was longer (4 h to 30 min). On the contrary, if temperatures are not high enough, hours to several days of heating period may be required. In consequence, most of the studies in literature agree that the optimal conditions for TH pretreatment are temperatures of 160–180°C and times in the range 30–60 min. The major benefits of TH pretreatment include pathogen's destruction, odor removal, reduction of sludge volume, and improved dewaterability with subsequent enhancement of sludge handling as well as positive energy balance.

Owing to the inherent advantages, several kinds of full-scale TH processes such as CambiTHP and Biothelys have been commercialized worldwide. The main operating parameters are 150–165°C, 20–30 min, and 8–9 bar for Cambi, and 165°C, 30 min, and 9 bar for Biothelys. Fig. 5.4 illustrates the TH flow diagrams of CambiTHP and Biothelys. The CambiTHP is a proven and reliable process that has been used in projects

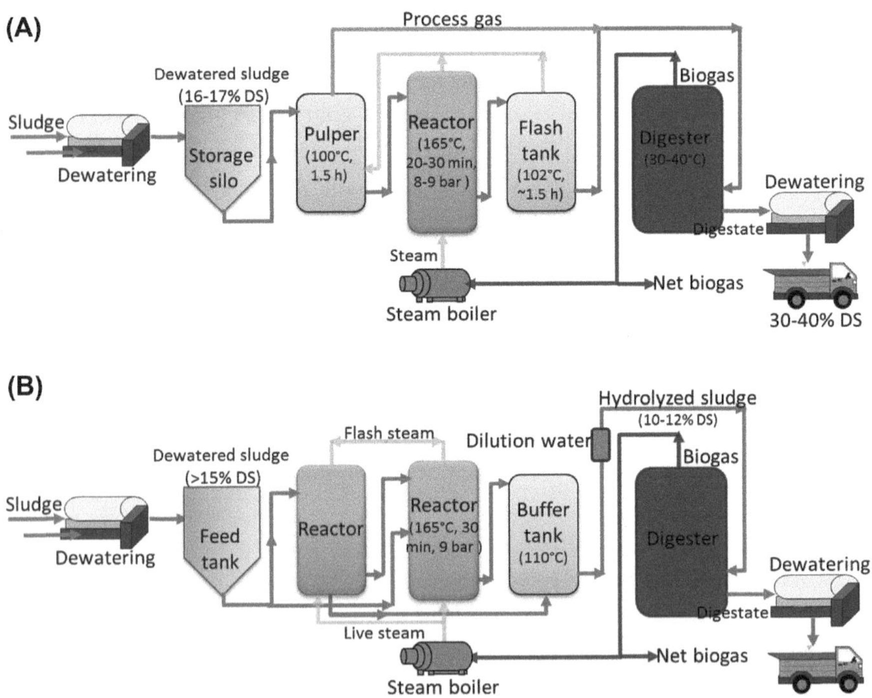

Figure 5.4 Thermal hydrolysis process flow diagram: CambiTHP (A) and Biothelys (B).

around the world since 1995. The main merits of this patented technology, as provided in Cambi homepage, include: (1) increased sludge biodegradability and biogas production, (2) significant sludge cake volume reduction, (3) 2—3 times increase in digester capacity, (4) eliminated foaming problems, (5) improved sludge dewaterability (up to 40% dry solids after being dewatered), and (6) pathogen-free biosolids (class A) with no regrowth or reactivation of bacteria.

5.1.3 Chemical Pretreatment

Chemical pretreatment employs strong reagents to deform the cell wall and membrane, favoring the availability of sludge organic matter for enzymatic attacks. The major reagents employed in the literature include acids, alkali, and oxidants (ozonation and peroxidation).

5.1.3.1 Acidic and Alkali Pretreatment

Acidic and alkali pretreatments have shown of great promise in biomass solubilization because of their multiple advantages, for example, a simple device, ease of operation, high methane conversion efficiency and low cost. Acidic hydrolysis is performed using acids such as HCl, H_2SO_4, H_3PO_4, and HNO_3 while alkali pretreatment usually employs several alkaline solutions, inducing NaOH, KOH, $Ca(OH)_2$, $Mg(OH)_2$, CaO, and ammonia. The addition of acid or base avoids the necessity of high temperature and thus can be operated at ambient or moderate temperatures. The effectiveness of acidic or alkali pretreatment may vary with the types and characteristics of the studied substrates because of their distinct affinity to organic components. Acidic pretreatment is indicated to be more effective for lignocellulosic biomass.

The main reaction that occurs in this process is the hydrolysis of hemicellulose, which releases the monomeric sugars and soluble oligomers from the cell wall matrix into the hydrolyzate thereby improving the enzymatic digestibility. The method offers good performance in hemicellulose removal but hardly has impact on lignin hydrolysis, and the lignin condensates and precipitates. Besides, it may induce the formation of toxic by-products, such as furfural and hydroxymethyl furfural , strong inhibitors to microbial fermentation. Other drawbacks associated with acidic method include great toxicity and strong corrosivity because of extremely low levels of pH, therefore special materials are required for the reactor construction. Hence, to solve the above problem, the integration of acidic-thermal pretreatment has been presented in the literature.

Alkali pretreatment is comparatively suitable for lignin breakdown. The basic principles of the alkali hydrolysis are based on solvation and saphonication, which induce depolymerization and cleavage of lignin-carbohydrate linkages, accordingly rendering the uneasily biodegradable substances more accessible to the extracellular enzyme. In addition, it solubilizes the xylan hemicellulose by saponifying the intermolecular ester bonds (e.g., acetyl and uronic acid substitutions, etc.), though to a less degree than the acidic pretreatment. Alkali method gains higher popularity in sludge disintegration before being sent to the digesters when considering its unique benefits in providing additional alkalinity that increases the buffer capacity of systems, specific methanogenic activity, and process stability. Among the alkaline reagents, NaOH is the most effective in sludge solubilization and enhancing biogas production, with a ranking of efficacy being NaOH > KOH > $Mg(OH)_2$ and $Ca(OH)_2$. However, results tend to underline that the alkali-pretreated sludge with overloading NaOH is not compatible with anaerobic biological digestion. Too high dose of Na^+ might interfere with the metabolic pathway of anaerobic microflora and deteriorate the methane output. Moreover, the formation of refractory molecules as a consequence of pH modifications can also bring about the limitation. Other drawbacks of single alkali hydrolysis include necessity of treated sludge reneutralization (also for acidic pretreatment), increased mineral content of digested sludge, etc.

Of late, alkali pretreatment has been combined with other sludge disintegration methods such as ultrasound, microwave, thermal, HPH, and electrolysis, with the purpose of reducing alkali consumption and maximizing the methane recovery. In addition, alkali conditions (especially pH 10) have also exhibited particularly high attractiveness in promoting sludge fermentation and production of short-chain fatty acids, which can be recovered as the organic reactants necessary for biological nutrient removal (BNR) in wastewater treatment plants (WWTPs).

5.1.3.2 Ozonation

Of the oxidation techniques that have been referred to in the literature, ozonation is the most widely used peroxidation process. It can damage the cell membrane, disintegrate the zoogloea structure, and has been successfully put into practice in excess sludge solubilization and reduction, as well as recirculation of valuable products for improved BNR in the activated sludge process.

Ozonation has been also combined with anaerobic sludge digestion to bypass the hydrolysis step and upgrade biogas production. Sludge solubilization efficiency is dose-dependent and linearly correlates with the amount of applied ozone in a moderate range. On the other hand, the efficiency of the ozonation process is closely related to both the mass transfer and reaction kinetics of ozone. Kinetic reactions taking place between the dissolved ozone and sludge mixed liquor have a lower average rate, and as a consequence, in spite of the increase in applied zone dose, it is not possible to observe an ever-increasing COD solubilization. Moreover, too-high ozone dose could cause the partial or even complete mineralization of the liberated cellular materials, impacting ultimate methane productivity. The optimal dose of ozone for sludge solubilization varies from 0.05 to 0.5 g O_3/g-TS, depending on the characteristics of sludge and the pretreatment conditions employed. In addition, sludge ozonation is energy-intensive because it requires high energy for ozone production (12.5 kWh/kg O_3), transfer to the sludge (2.5 kWh/kg O_3), and energy consumption to produce liquid oxygen (0.5 kWh/Nm^3 O_2). In order to achieve high performances for reducing the costs involved in sludge ozonation, a microbubble ozonation process was developed; sludge solubilization efficiency rose from 15 to 30% for the bubble contactor to 25–40% at ozone doses of 0.06–0.16 g O_3/g-TSS, with ozone utilization improving from 72% to 99%. The application of microbubble ozonation accelerates the formation of hydroxyl radicals and speeds up sludge solubilization, thus reducing the impact of high capital requirements.

Several ozone systems for sludge solubilization and reduction are commercially available, such as (1) Aspal SLUDGE (Air Liquide) that offers high solid content of excess sludge, improved dewaterability, and low energy consumption; and (2) Praxair *Lyso* (Praxair Technology, Inc.) that allows up to 80% sludge reduction, 75% reduction in ozone use, as well as an improvement in the settling and dewatering characteristics of sludge.

5.1.3.3 Fenton Oxidation

Another well-established oxidation technology is Fenton process, which involves reactions of hydrogen peroxide (H_2O_2) with catalyst iron ions (Fe^{2+}) to produce highly active hydroxyl radicals (•OH). Hydroxyl radicals have a higher oxidation potential of +2.80 V (in acidic conditions) than H_2O_2 (+1.36 V) and ozone (+2.07 V) and are particularly effective for the disintegration of sludge EPS and the cell lysis of microorganisms, resulting in the release of both intracellular materials and bound water. In this context,

Fenton oxidation has been intensively applied for enhancing the sludge dewatering.

Additional to its application in sludge dewatering, Fenton process enhances biogas production and minimizes sludge weight that needs to be disposed of. The effectiveness of this process depends on several variables, namely reagents concentrations (H_2O_2 and Fe^{2+}), Fe^{2+}/H_2O_2 ratio, treatment time, as well as initial pH and temperature. A secondary sludge treated with 60 g H_2O_2/kg-TS, 0.07 g Fe^{2+}/g H_2O_2 and pH 3, resulted in 15% increase in methane yield (from 430 to 496 m^3 CH_4/Mg-$VS_{degraded}$), 3.1 times increase in net energy as well as considerably reduced GHG emissions (0.128 Mg CO_2/Mg). Under the optimal conditions of 4 g $Fe^{2+/0}$/kg-TS, 40 g H_2O_2/kg-TS and pH 3 within 1-h oxidation period, CFP and FTP enhanced the methane production by 26.9% and 38.0% respectively, relative to the untreated reactor. A major drawback for Fenton reagent is the necessity of low pH values (pH < 4.0) to avoid hydrolysis and precipitation of Fe^{3+}; moreover, treated sludge needs neutralization before digestion. Apart from this, excess H_2O_2 or Fe^{2+} dose may also scavenge hydroxyl radicals, lowering the concentration of •OH radicals.

5.1.4 Biological Pretreatment

5.1.4.1 Temperature-Phased Anaerobic Digestion (TPAD)

Biological pretreatment comprises the approaches such as, but not limited to, enzymatic hydrolysis, predigestion, as well as the use of fungi or biosurfactants. Predigestion (e.g., TPAD) that combines a thermophilic pretreatment unit prior to mesophilic AD is an effective strategy. The combination of both temperature regimes in TPAD promotes the hydrolysis of feedstocks and acidogenesis in thermophilic range and ensures higher syntrophic acetogenesis and methanogenesis in the subsequent mesophilic stage. As a result, this technology possesses several advantages over single-stage mesophilic digestion process, including higher methane production, better solids-destruction, use of low-quality thermal energy, low energy input, and sterilization of pathogens. There have been a majority of studies evaluating TPAD systems so far. The highest specific methane yield of 370 mL/g-VS_{added} can be obtained in TPAD (70°C, HRT 2–3 d, and 37°C), 30–50% higher than the mesophilic and thermophilic single-stage tests, with methane production of up to 300 mL/g-VS_{added} (HRT 14 d, 35°C) after thermophilic pretreatment at 1–2 days, pH 6–7, and 65°C. Besides increasing the methane production (0.62 L/g-VS), TPAD (55°C, HRT 3 d, and 35°C, 15 d) helped obtain

Class A biosolids with fecal coliform 10^3 MPN/g-TS, and *Salmonella* spp. 1 MPN/4 g-TS.

In addition, TPAD has been also integrated with other pretreatment methods such as microwave, ultrasound prior to thermophilic phase, or ozonation between thermophilic and mesophilic stages. Ten semicontinuous reactors were established to test the influence of MW pretreatment, phase separation, digestion temperature, and SRT on process performance (Fig. 5.5). Microwaving coupled with staged digestion enhanced the solids destruction, upgraded biogas production, shortened HRT, eliminated pathogens indicators completely, and/or improved dewaterability of digestate. Despite the promising perspectives outlined above, it should be noted that TPAD process is still in its infancy. Works related to parameter optimization, analysis of energy balance, especially system upscaling, etc. need to be further conducted before considering its large-scale operations. Attention should be also paid to the composition and functional diversity of resident bacterial communities involved in TPAD process, aiming to provide biokinetic parameters for design and optimization of the digesters.

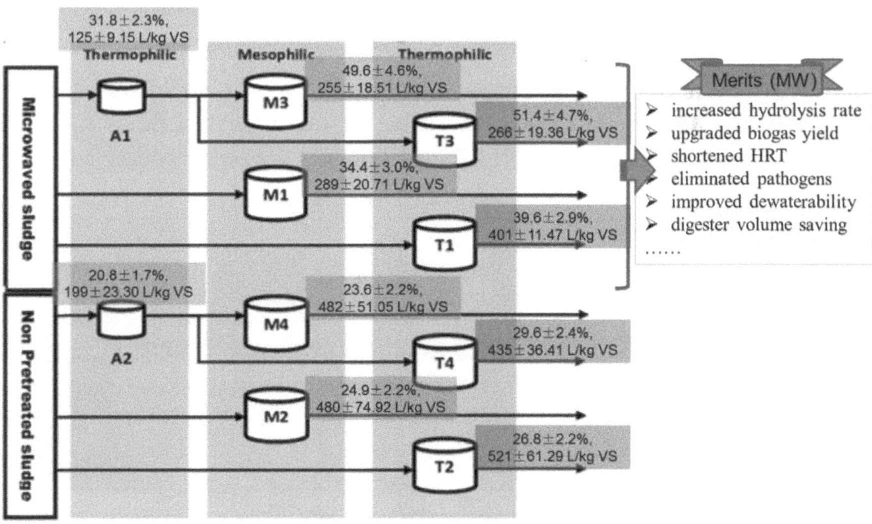

Figure 5.5 Schematic diagram of ten digester configurations and the corresponding performances at total solid retention time 5 d: VS removal (%) and biogas production (L/kg-$VS_{removed}$). *Further modified from Coelho NMG, Droste RL, Kennedy KJ. Evaluation of continuous mesophilic, thermophilic and temperature phased anaerobic digestion of microwaved activated sludge. Water Res 2011;45:2822-34.*

5.1.4.2 Microbial Electrolysis Cell (MEC)

Microbial electrolysis cell (MEC) that is modified from microbial fuel cell (MFC) is a recent emerging technique for methane generation via electromethanogenesis. In an MEC process, exoelectrogenic bacteria in anode consumes organic matters anaerobically while releasing electrons to the anode and protons into solution; with a small voltage input, the electroactive microorganisms (i.e., electrotrophs) in cathode can accept electrons transferring to cathode through the external circuit, or alternatively use cathodic H_2 as electron carriers to drive methane formation (Fig. 5.6). Moreover, it has been demonstrated that the intergradation of MEC with anaerobic digesters (or called "bioelectrochemical reactor, BER") can substantially enhance the production of methane and system stability, and even remediate AD systems that exhibit process failure. Very recently, several technical attempts have been conducted to couple MEC with AD systems for methane production; indeed, the relatively positive results have been achieved. A "BER" containing the carbon fiber fabric (CFF) (BER + CFF) for methane fermentation of thickened sludge was configured. Biogas production in BER + CFF was 3.57 L/L d at HRT of 4 d, increasing by 3.3 times when compared with the control (0.83 L/L d). In a single-chamber MEC fed with sludge, 1.7−5.2 and 11.4- to 13.6-fold increment in hydrogen and methane productions were noticed at 1.4 and 1.8 V, respectively, from a bioelectrochemical enhancement in hydrogen and methane production. In addition to methane, MEC has been also introduced into bioelectrocatalytic reduction of CO_2 toward the production of other commodities such as formate, acetate, bioalcohols, etc. Regarding the external power supply,

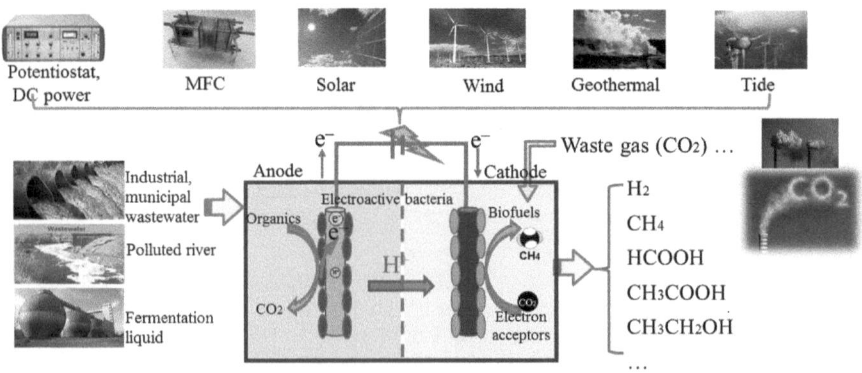

Figure 5.6 Schematic diagram of a typical two-chamber microbial electrolysis cell and its applications.

besides the commonly used DC power, other alternative power sources such as solar and wind can be also used in this unique anaerobic system, via which both solar/wind and bioenergy is stored simultaneously.

In spite of the promising perspectives outlined above, unlike the other forms of biological processes, there are still many critical challenges associated with MEC-based option that need to be addressed, including slow start-up, pH issues, electrode corrosion/deterioration, high ohmic loss and high overpotentials. Moreover, the performances of this system may differ greatly due to the variations in cells configuration, electrode materials, inoculum sources, substrate compositions, and operational conditions. All of these factors have discounted the repeatability and reliability of experimental results, which ultimately restricts the commercial implementation and even upscaling of this system. Thus, how to virtually use the MEC for real-world application remains a vital challenge. Additionally, though there is a general recognition for electron transfer mechanisms occurring between bacterial cells and electrodes, a more detailed exploration in microbial community dynamics as well as syntrophic interactions between fermenting bacteria and electroactive bacteria (e.g., exoelectrogens in anode and electrotrophs in cathode) is yet to be performed.

5.1.5 Challenges and Future Perspectives

Anaerobic digestion holds the numerous potential as a substantial approach to harvest bio-methane from SL. A number of pretreatments as mentioned before have been proposed to enhance this process. At present, mechanical, thermal, and chemical pretreatments have been intensively investigated. There are a few patented technologies that have been applied in practical sludge treatment: CambiTHP, Biothelys (TH), Biosonator (ultrasonication), Aspal SLUDGE, Praxair Lyso (microwave), BioCrack (electrokinetic disintegration), MicroSludge, and Cellruptor (HPH) (Table 5.1). However, the work on biological techniques (TPAD, MEC) is not exhaustive and has still been undergoing lab-scale experiments (Fig. 5.7A and B). System upscaling is extremely necessary in order to make the biological processes practically applicable for industrial applications. Besides, to maximize resources recovery from SL, the digestate can be posttreated to produce value-added by-products [i.e., struvite ($MgNH_4PO_4 \cdot 6H_2O$) by chemical precipitation (Fig. 5.7D) or pyrochar and activated carbon via pyrolysis (Fig. 5.7E)]. MEC can also be employed to reduce CO_2 to CH_4 by means of electromethanogenesis, and digestate-derived pyrochar to adsorb the volatile toxic contaminants in biogas (i.e., siloxanes) for biogas purification (Fig. 5.7C).

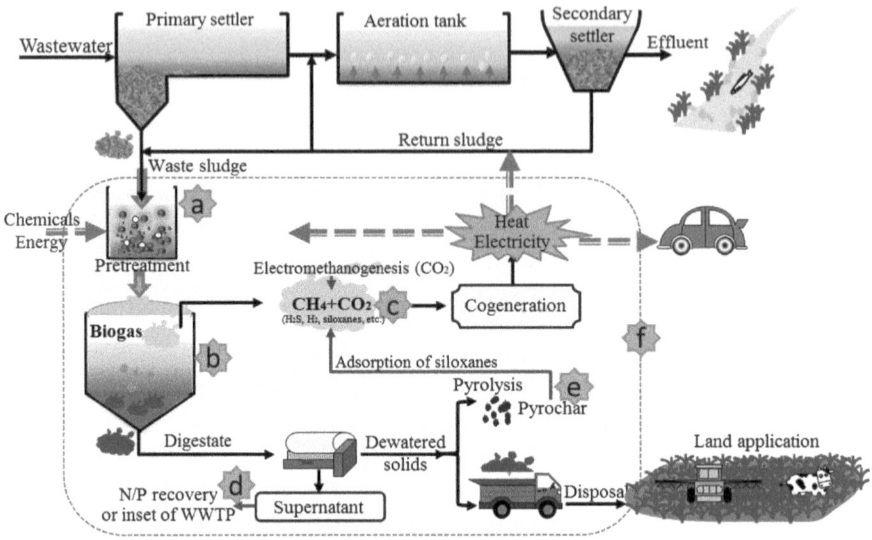

Figure 5.7 Process flowchart of the sludge processing steps.

Furthermore, pretreatments have the potential to enhance the methane productivity; however, it does not always, strictly speaking. It had claimed that an increment in organic matter solubilization cannot be directly related to an enhancement of the AD of sludge in terms of biogas production. Likewise, it was also noticed that soluble organics [soluble chemical oxygen demand (SCOD), soluble carbohydrates, and soluble proteins] have not influenced on the biochemical methane potential (BMP) in the observed region. Concomitant liberation of unwanted substances (i.e., biorefractory components, excess ammonia) impairs the system stability and inhibits methane conversion. The opposite observations reflect that for an objective assessment of pretreatment effectiveness, in addition to the commonly used disintegration degree reflected by DD_{COD}, other parameters such as the feature of pretreatment methods provided, the chemical compositions of liberated soluble organics, as well as the microbial communities of inoculums added (i.e., methanogenesis activity) have to be considered (Fig. 5.7A and B).

In addition, a systematic assessment of different pretreatment options is quite necessary for deciding which one would be the most suitable from an industrial point of view. Making a comparison, however, is a very hard task since multiple influencing factors should be considered (Table 5.1). The technical feasibility of one method is dependent upon not only the degree of sludge disintegration and methane conversion efficiency, but also energetic and environmental benefits. Besides to increase methane conversion efficiency, pretreatment can also effect the energy required for AD, dewatering, transportation, and ultimate disposal (landfill, incineration, compost, or land application) as well as corresponding greenhouse gas emission (GHG) emissions. Unfortunately, until yet most of previous studies dealing with sludge pretreatment have mainly focused on the former with hardly considering energetic and environmental issues. To simplify the evaluation procedure, researchers often ignore some factors, such as dewatering, etc., that are also important costs for the WWTP and could influence the final outcome. The lack of standardization on the effectiveness of digestion systems remains a huge barrier to a meaningful comparison, though an approach to evaluate the effectiveness of pretreatment methods were proposed, which involved the energy input for pretreatment, AD, dewatering of the digestate, transporting the dewatered solids from the WWTP to the disposal site, and during final disposal (Fig. 5.7F).

The general parametric values considered for the calculations are summarized in Table 5.2. An example is also given to evaluate the mass-energy balance and GHG emissions in the integrated process of thermal pretreatment (CambiTHP) and mesophilic digestion (Fig. 5.8). The resultant information would help the industry to determine the performance of one cotreatment route for achieving the optimal sludge conversion and energy recovery. The emerging approach shows a potential to alleviate such problem (i.e., the lack of standard cost-benefit optimization tool), but further researches are still required to confirm their real potentials. Besides, it is important to admit that for the same pretreatment option, the performance may vary considerably with sludge characteristics, pretreatment conditions, and AD process parameters increasing the difficulty in comparison. Even worse is that, nowadays, the active attempt in the combination of different pretreatment processes make the comparison more difficult. Other factors that will impact the evaluation results include the operation skill of operator, the maintenance frequency of devices, the local circumstance of labor, land price, the market for renewable energy exchange, etc., in "real world" scenario. Apparently, for a reliable comparison, supplementary information (e.g., "standard cost-benefit optimization tool") is urgently needed to evaluate such technologies from the energetic, economic, and environmental perspectives of view (Fig. 5.7F).

Table 5.2 General Parameters Used in Mass-Energy Balance and Greenhouse Gas Emissions

Parameters	Values	References
Volume of sludge to be treated	100000 m^3/d	
Step (i): Pretreatment		
Sludge temperature	12°C	Gassara et al. [22]
Influent DS concentrations	15%	
VS (% of DS)	75%	Turovskiy and Mathai [67]
Specific gravity	1.023	
Pretreatment conditions	150–165°C, 20–30 min, 8–9 bar	CambiTHP
Specific heat of sludge	4200 kJ/m^3 °C	Metcalf [45]
CO_2 emission due to energy utilization	530 g CO_2/kWh	Pilli et al. [52]
Step (ii): Anaerobic digestion		
Mesophilic	35°C	
SRT	20 d	
Energy consumed for mixing	0.005 kW/m^3, 20 min mixing/h	Lubken et al. [40]
VS $removal_{control}$	55%	
VS $removal_{pretreated}$	~65%	CambiTHP
CH_4 yield	0.67 m^3 CH_4/kg $VS_{degraded}$	Turovskiy and Mathai [67]
Heating value of CH_4	35.8 MJ/m^3 CH_4	Tervahauta et al. [66]
CO_2 emission during CH_4 combustion	1964 g/m^3 CH_4	
Heat loss during operation	150.84 kJ/d m^3	Taricska et al. [64]
Step (iii): Dewatering		
Energy consumed for dewatering	101.4 $kWh/10^3$ kg DS	Brown et al. [9]
DS $content_{control}$	20–25%	CambiTHP
DS $content_{pretreated}$	40%	CambiTHP
Step (iv): Transportation		
Distance (WWTP → land application site)	50 km	Gassara et al. [22]
Diesel consumed for transportation	35 L/100 km (3 axle semi-trailer vehicles), 14.1 tons of capacity	Gassara et al. [22]
CO_2 emission	2730 g CO_2/L-diesel	Gassara et al. [22]
CH_4 emission	12×10^{-2} g CH_4/L-diesel	Gassara et al. [22]
N_2O emission	8×10^{-2} g N_2O/L-diesel	Gassara et al. [22]

Table 5.2 General Parameters Used in Mass-Energy Balance and Greenhouse Gas Emissions—cont'd

Parameters	Values	References
Step (v): Land application		
Energy consumed for land application	351.68×10^{-3} kWh/kg DS	Wang et al. [73]
CO_2 emission	17.20 g CO_2/kg DS	Hospido et al. [27]
CH_4 emission	3.18 g CH_4/kg DS	
N_2O emission	30 g CO_2/kg DS	

Note: the heat-trapping potential of CH_4 and N_2O are about 21 and 310 times of CO_2 on a per molecule basis, respectively [22].

5.2 ANAEROBIC DIGESTION TECHNIQUE AND COMBINED ELECTRICALALKALI PRETREATMENT TO INCREASE THE ANAEROBIC HYDROLYSIS RATE OF SLUDGE

Very recently, electrolysis has shown to be an efficient alternative for sludge modification prior to aerobic digestion. The principle of electrolysis pretreatment relies on the ohmic heating, electrophoresis, and electro-osmosis, resulting in particle disintegration and microbial cell lysis. The pretreatment especially if assisted by chemical methods will become more energy-efficient for sludge treatment as they solubilize sludge and obviate the need for high voltages. Furthermore, the combined treatment can not only reduce the required quantity of chemicals and energy, but also greatly enhance the efficiency of sludge AD by the synergistic functions and thus is more economically attractive.

In this section, the combined electricalalkali process is applied to pretreat sludge, aiming to enhance sludge disintegration and subsequent AD. Firstly, pretreatment effectiveness is analyzed by comparing the SCOD, disintegration degree (DD_{SCOD}) of sludge, total suspended solid (TSS), volatile suspended solid (VSS), protein (PN), and polysaccharide (PS) in pretreated and nonpretreated sludges to propose the best conditions to achieve the effective sludge solubilization, and also to explore the responsible reasons for enhancing the anaerobic biodegradability of sludge. Secondly, the performances of AD of different pretreated sludges were assessed in terms of methane yields via biochemical methane potential (BMP) tests in batch mesophilic anaerobic digesters. Finally, preliminary testing of dewatering effectiveness before and after AD was also performed to clarify the exact role of AD on sludge dewaterability.

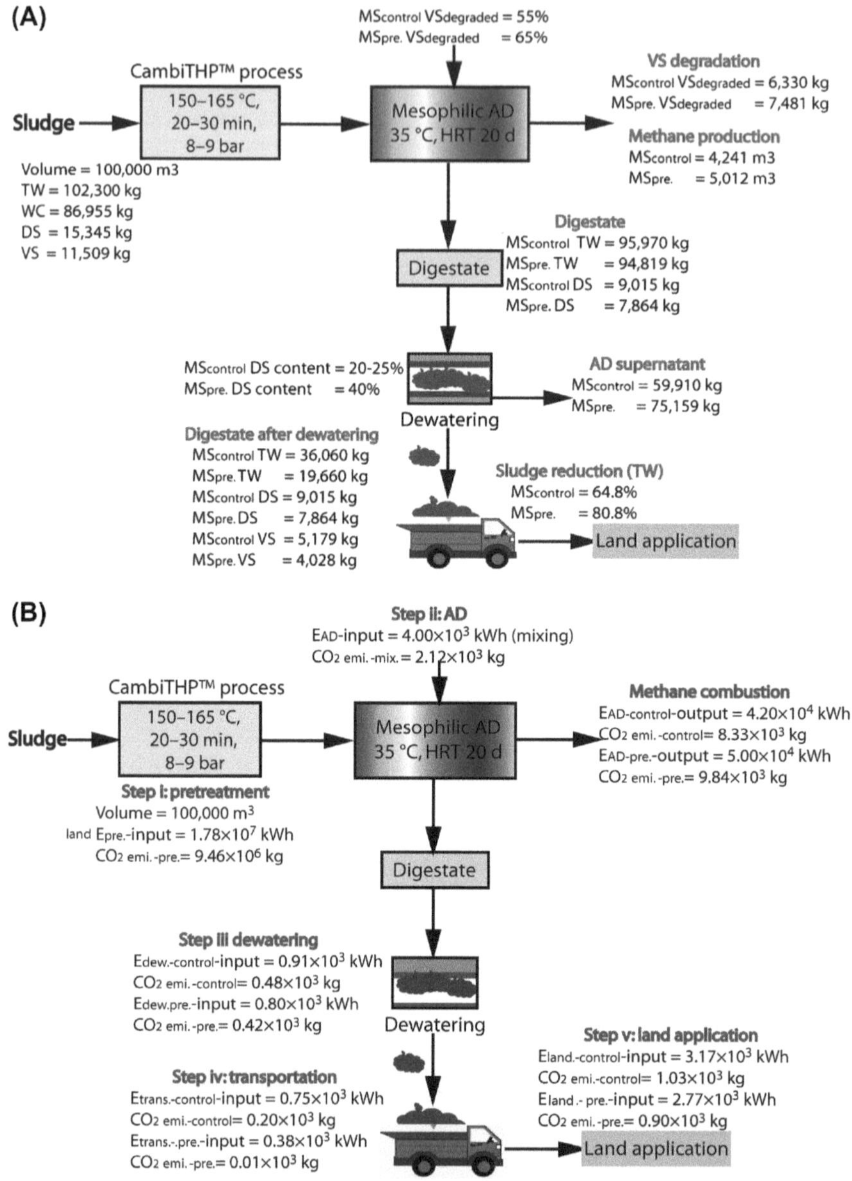

Figure 5.8 Mass-energy balance and greenhouse gas (GHG) emissions for the control and thermal pretreated sludge at mesophilic digestion: (A) mass balance and (B) energy balance and GHG emissions (*TW*, total weight; *WC*, water content).

5.2.1 Test Materials and Experimental Procedures

5.2.1.1 Electrical-Alkali Modification Procedure

Sludge sample was withdrawn from a full-scale WWTP in Sendai, Japan (Table 5.3). A series of batch-scale experiments were conducted under constant electrical voltage electrolysis conditions in a 500-mL undivided single-compartment glass cell by using a pair of Ti/RuO_2 meshes as the anode and cathode. Different operating conditions of applied voltages (5, 10, 15, and 20 V) and pH values (no adjustment, 9.2, 10.2, 11.0, and 12.2) were conducted to identify the optimum conditions. Sludge disintegration efficiency was represented by disintegration degree (DD_{SCOD}), which can be calculated by Eq. (5.1).

$$DD_{SCOD}(\%) = \frac{SCOD_S - SCOD_{S0}}{SCOD_{NaOH} - SCOD_{S0}} \times 100 \tag{5.1}$$

where $SCOD_S$ and $SCOD_{S0}$ are for pretreated and untreated sludge, respectively. $SCOD_{NaOH}$ was obtained through alkaline hydrolysis, in which 0.5 M NaOH was added to the sludge suspension and mixed for 24 h at the room temperature.

5.2.1.2 Biochemical Methane Potential (BMP) Assay

BMP assay was batch-performed for the various treatment conditions at mesophilic conditions in series of 120-mL serum bottles with 75 mL of working volume (Fig. 5.9). The bottles were filled with 50 mL of inoculum and 25 mL of sludge samples. A blank assay, only filled with 75 mL of inoculum, was conducted in order to determine the biogas production from endogenous inspiration of the inoculum. Later on and before the start-up of reactors, the pH of all digesters were neutralized to around 7.0. Once the reactors were loaded, the headspace was purged with high-purity nitrogen gas at 0.5 L/min for 1−2 min to remove oxygen. After sealed with rubber stoppers secured by aluminum crimp, the bottles were placed in a water bath ($35 \pm 1°C$) under continuous shaking (100 ± 1 rpm).

The volume of biogas was measured periodically by releasing the pressure in the bottles using 10−20 mL glass syringes and its composition ($CH_4 + CO_2$) by gas chromatography equipped with a thermal conductivity detector (GC-8A, Shimadzu, Japan). The digestion period was 42 days by which stage the methane production nearly ceased. Finally, a first-order kinetic model was used for analysis of the BMP data. First-order hydrolysis rate, k_{hyd} (/day), and degradability, f_d (mL CH_4/g-COD_{added}), were

Table 5.3 General Characteristics of Sludge and Seed Sludge Used in This Study

	pH	Moisture Content (%)	TSS (g/L)	VSS (g/L)	TCOD (mg/L)	SCOD (mg/L)	Total PN (mg/L)	Soluble PN (mg/L)	Total PS (mg/L)	Soluble PS (mg/L)
Sludge	7.5 ± 0.0	99.1 ± 0.0	7.4 ± 0.3	6.4 ± 0.1	$12,664.5 \pm 86.9$	1331.0 ± 4.4	2535.8 ± 8.1	149.5 ± 0.6	804.0 ± 9.2	55.0 ± 0.2
Seed sludge	7.2 ± 0.1	98.3 ± 0.0	15.3 ± 0.2	11.1 ± 0.2	$21,209.9 \pm 289.8$	622.5 ± 10.9	5433.6 ± 5.9	146.4 ± 5.1	2473.4 ± 2.3	59.9 ± 0.0

PN, protein; *PS*, polysaccharide; *SCOD*, soluble chemical demand oxygen; *TCOD*, total chemical demand oxygen; *TSS*, total suspended solids; *VSS*, volatile suspended solids.

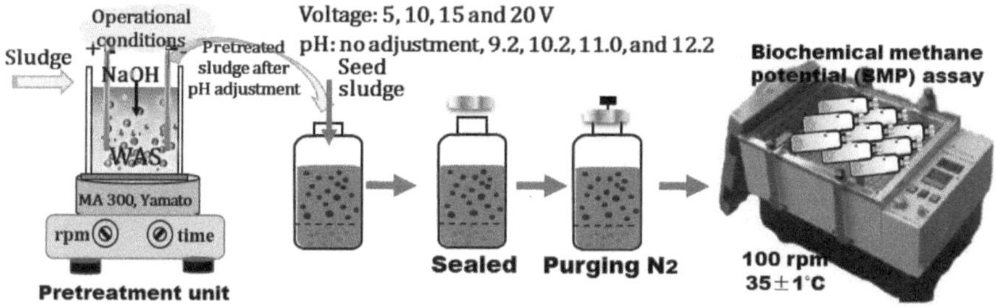

Figure 5.9 Experimental procedures of biochemical methane potential assay.

estimated by fitting the data to Eq. (5.2). Parameter uncertainty levels were assessed through parameter uncertainty analysis.

$$Y_{CH_4} = \frac{V_{CH_4}}{COD_{added}} = f_d(1 - \exp(-k_{hyd}t)) \tag{5.2}$$

where Y_{CH_4} is the specific methane yield at a given time (mL CH_4/g-COD_{added}), V_{CH_4} is the volume of methane produced (mL CH_4), COD_{added} is the mass of COD present in the BMP reactors (g-COD_{added}), and t is the time since the setup of the BMP (days).

5.2.2 ElectricalAlkali Pretreatment for Sludge

5.2.2.1 COD Solubilization and SCOD Release

SCOD is regarded as the main parameter for evaluation of the bioavailability of sludge particulate material. Increased SCOD can be readily used as the substrate to produce methane during AD. Fig. 5.10 indicates the effects of electrolysis and electricalalkali pretreatments on SCOD and degree of sludge disintegration (i.e., DD_{SCOD}). It is evident from the figure that, for the electrolysis pretreatment alone, the rising voltage from 5 to 20 V led to a very slight increase in SCOD (from 940.6 ± 17.39 to 1401.7 ± 34.8 mg/L) (Fig. 5.10A), and the DD_{SCOD} values were calculated as in the range of 0.1–8.2% (Fig. 5.10B). The low DD_{SCOD} demonstrated the limited enhancement of electrolysis on sludge disruption. Compared with the electrolysis pretreatment alone, the combination of electrolysis and alkaline pretreatments plays a vital role in improving sludge hydrolysis. The transfer of COD from the particulates to the soluble fractions was profoundly improved by increasing the alkaline dosages. For instance, at 5 V without pH adjustment (pH 7.5), the SCOD was 940.6 ± 17.4 mg/L. This value reached up to 1029.8 ± 50.7 mg/L for the sludge pretreated at pH 9.2, to about 1534.4 ± 18.1 mg/L for the sludge pretreated at pH 10.2, to

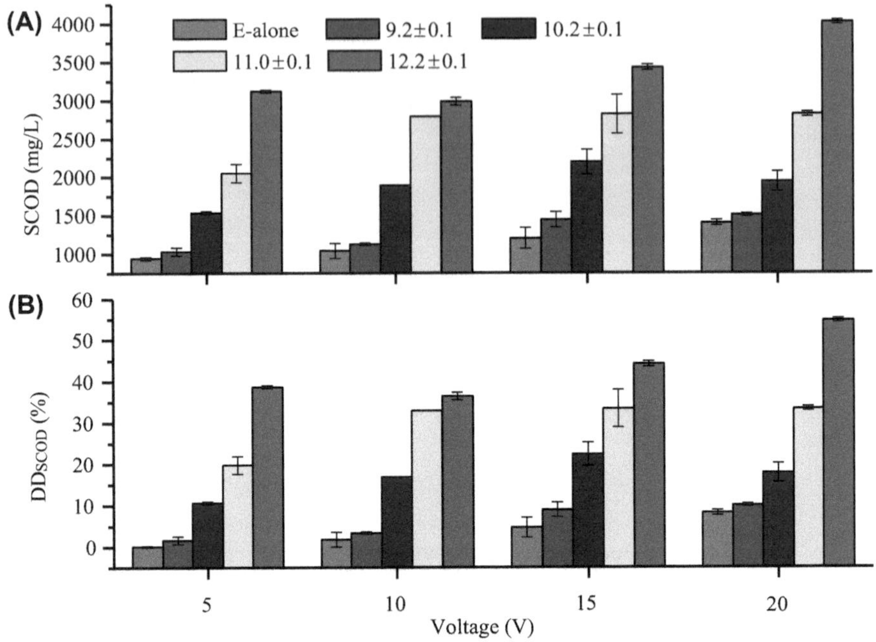

Figure 5.10 Effects of electricalalkali pretreatment on soluble chemical oxygen demand (SCOD) (A) and DD_{SCOD} (B) for 40 min (*E-alone*, electrolysis alone). Error bars represent standard deviations of duplicate tests.

2051.8 ± 119.5 mg/L for the sludge pretreated at pH 11.0, and to 3117.4 ± 18.1 mg/L for the sludge pretreated at pH 12.2. For these sludge samples, the DD_{SCOD} progressively jumped from 0.1 ± 0.2% (electrolysis alone) to roughly 1.6 ± 0.9%, 10.6 ± 0.3%, 19.8 ± 2.1% and 38.7 ± 0.3% at pH 9.2, 10.2, 11.0 and 12.2, respectively. Furthermore, higher operational voltages can emphasize the alkaline impact. The higher the applied voltage, the better sludge hydrolysis can be found at a fixed pH value.

As depicted in Fig. 5.10B, when at pH 12.2, DD_{SCOD} went up to 54.7 ± 0.5% at 20 V, which was 43.3% higher compared with that (38.7 ± 0.3%) at 5 V ($F_{observed} = 1674.38$, $F_{crit} = 18.51$, $p_{(.05)} = 5.97 \times 10^{-4} < .05$). The remarkable increase in DD_{SCOD} achieved by the combined method indicated a clear synergistic disintegration effect. Single electrolysis (5–20 V) could disrupt the partial extracellular biopolymer surrounding the microbial cells, but was not able to release and crack the cells. In contrast, electrolysis in combination with alkali could effectively accelerate the destruction of sludge polymeric network and subsequent exposure of the cells trapped. The exposed cells could not withstand the harsh conditions induced by the combined process, and were further totally broken down by destroying the chemical bonds in the cell wall/membrane. Disruption of sludge cells provoked the leakage of more extra- (EPS) and intracellular polymeric substances into the liquid of sludge, as a result of which the disintegration/solubilization of sludge was enhanced radically.

5.2.2.2 Total Suspended Solids and Volatile Suspended Solid Reduction

Fig. 5.11 presents the variations of TSS and VSS along with rising voltages and alkali dosages. With the electrolysis pretreatment alone, it was found that TSS and VSS after pretreatments at the applied voltages of 5−10 V were approximately the same as in the original raw sludge (Table 5.3), and the removal efficiencies were negligible; while the TSS and VSS removal efficiencies drastically increased to 16.2% and 26.6% and then to

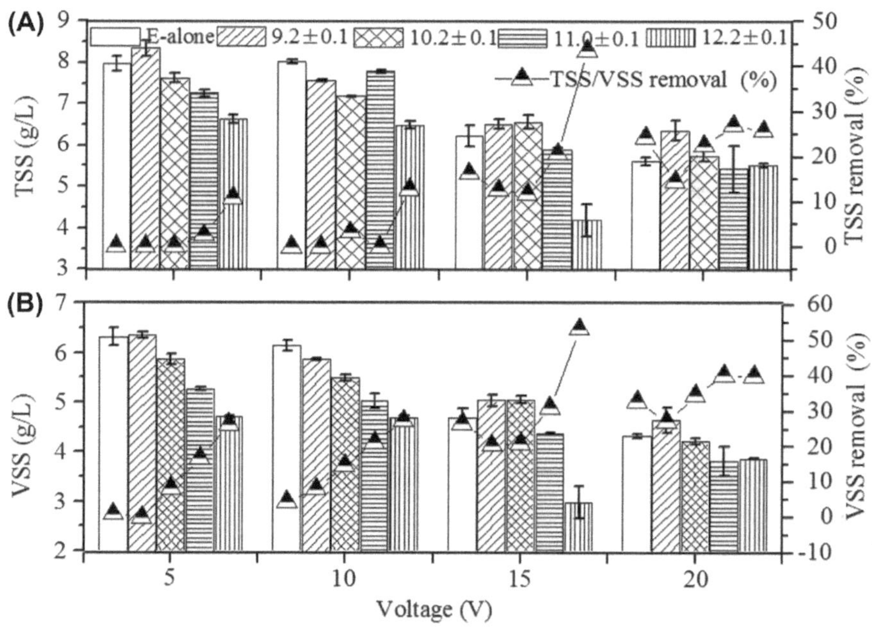

Figure 5.11 Effects of electricalalkali pretreatment on total suspended solid (TSS) (A) and volatile suspended solid (VSS) (B) for 40 min (Column: TSS and VSS; Line: TSS and VSS removal efficiencies). Error bars represent standard deviations of duplicate tests.

24.4% and 32.1%, respectively, as the voltage tested was raised to 15 V and to 20 V. Contrary to single electrolysis, combined pretreatment, particularly at high alkali loadings (pH > 10.2), resulted in a much higher TSS and VSS removals. In the case of 5 V, the choice of using lower pH (9.2) did not cause a perceptible enhancement in the removal efficiencies in comparison with alkali-free treatment. However, when pH was 11.0, TSS and VSS removal efficiencies increased to 2.6% and 16.9% or so ($p_{(.05)} < .05$), which was further elevated to 10.8% and 26.2% ($p_{(.05)} < .05$), respectively, as pH increased to 12.2. Obviously, to achieve desired removal efficiencies, high pH is required in the combined pretreatment process. With this analysis, pH 11.0 and pH 12.2 were chosen to be more effective in terms of sludge disintegration and suspended solids removal.

In addition, the statistical analysis performed by determining Pearson's correlation coefficients (R_p) implied that both TSS and VSS exhibited a statistically significant negative correlation with SCOD ($R_p = -0.600$, $p_{(.05)} = 5.15 \times 10^{-3} < .05$; and $R_p = -0.732$, $p_{(.05)} = 2.45 \times 10^{-4} < .05$, respectively) and DD_{SCOD} ($R_p = -0.600$, $p_{(.05)} = 5.17 \times 10^{-3} < .05$; and $R_p = -0.732$, $p_{(.05)} = 2.46 \times 10^{-4} < .05$, respectively), respectively, apparently revealing that the removal of suspended solids was achieved mainly through solubilizing the organic compounds from the particulate to soluble parts. The experimental results described in Figs. 5.10 and 5.11 verified that the electricalalkali pretreatment did not only reduce the particulate fractions by converting them to soluble organics, but also reduced the solid content of the sludge, which was considered to be particularly important for reducing the amount of sludge and downstream costs. Besides, most of complex organic compounds could also be degraded into monomers due to the swelling of sludge colloidal structure caused by pretreatments, which would make it much easier to be decomposed by microorganisms in the stage of AD. Therefore, the process can be considered as a competitive pretreatment process and a part of potential solution of sludge AD.

5.2.2.3 Soluble Protein and Polysaccharide Releases

The effects of combined electrolysis and electricalalkali pretreatments to disrupt the complex floc structure of sludge and to solubilize the extra- (EPS) and intracellular polymeric substances (PN, PS, etc.) were also visualized through analyzing PN and PS present in the soluble fractions of sludge. As seen in Fig. 5.12, by using a combination of electrolysis and alkali, the releases of soluble PN and PS could be enhanced significantly. Moreover, a higher pH led to the releases of more PN and PS, agreeing well with

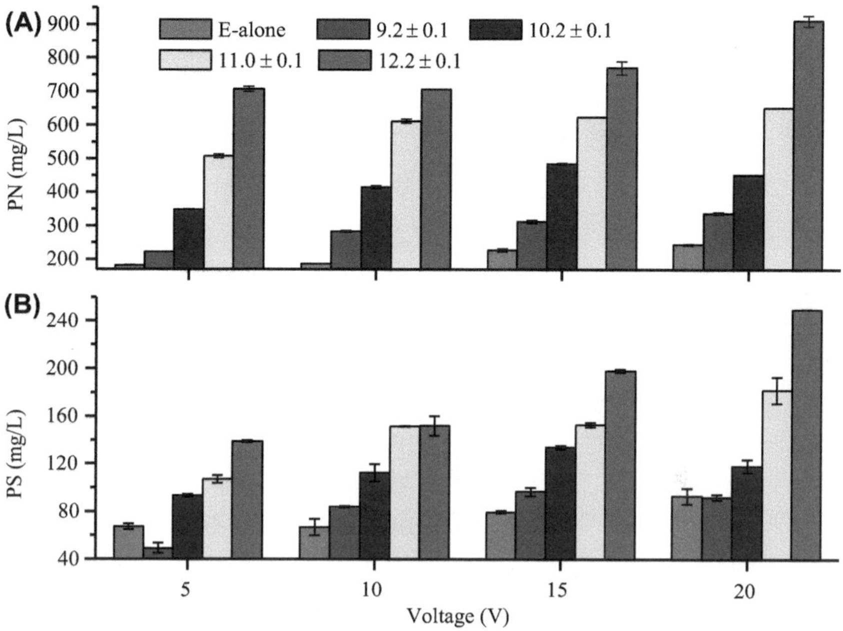

Figure 5.12 Effect of electricalalkali pretreatment on soluble protein (PN) (A) and polysaccharide (PS) (B) for 40 min. Error bars represent standard deviations of duplicate tests.

SCOD behavior given before in (Fig. 5.10). It is noticeable from the figure that soluble PN and PS concentrations, for instance, were 182.3 ± 1.6 mg/L and 67.2 ± 2.5 mg/L, respectively, at the tested voltage of 5 V without the presence of alkali. They were found to be 348.4 ± 1.2 mg/L and 93.4 ± 1.1 mg/L at pH 10.2, which were further increased to 706.9 ± 7.0 mg/L and 139.0 ± 0.8 mg/L, respectively, when pH was elevated to 12.2. The main reason for the dramatic increase of the concentrations of those compounds in the soluble phase of sludge after the combined pretreatment in this study was presumably due to the dissociation of acidic groups contained in the EPS of sludge and the repulsions between the negatively charged EPS. Because of this deterioration, numerous microbial cells trapped within sludge gels were liberated and exposed to OH^- ions. At alkaline pHs, the cells could not maintain an appropriate turgor pressure on account of saponification of lipids and hydrolysis of RNA in the cell walls/membrane and, as a consequence, disrupted completely. This possibility can be confirmed through the experimental results obtained by the scanning electron microscopy (SEM) analysis as illustrated in

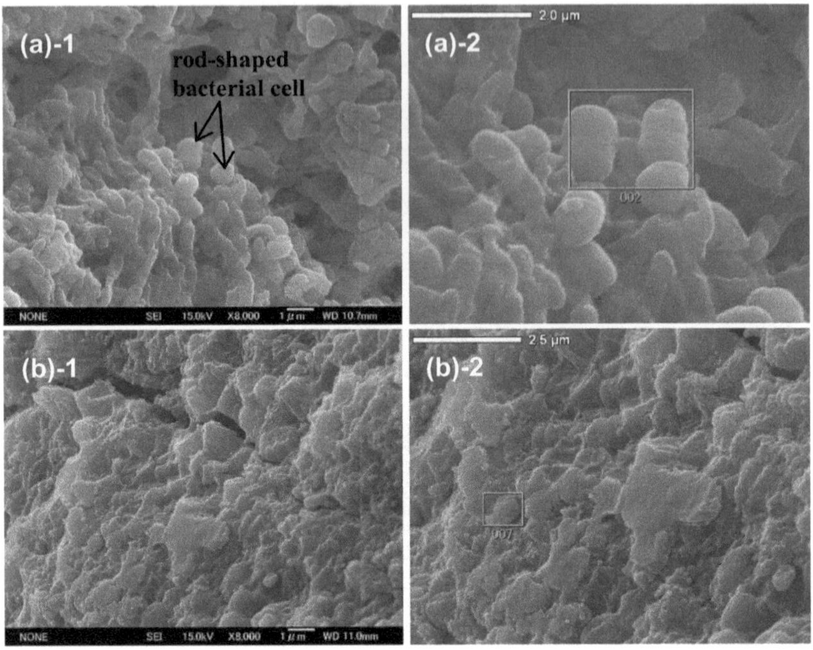

Figure 5.13 Scanning electron microscope (SEM) of sludge flocs pretreated by electrolysis alone (5 V) (a1-2) and by combined electrical-alkali process (5 V plus pH 9.2) (b1-2). a1 and b1, and a2 and b2 represent the surface and interior of the two pretreated sludges.

Fig. 5.13. The sludge flocs pretreated with electrolysis alone were rounded and regular, and large numbers of rod-shaped bacterial cells entrapped inside the sludge matrix were clearly detected (Fig. 5.13A). In the course of electrolysis integrated with alkaline solubilization, however, no sign of cells was identified after pretreatment, suggesting that the combined process not only had the capability to denature the EPS, but also to break cells apart (Fig. 5.13B), thus disabling the intracellular defense system and completely rupturing the floc's structure. Break-up and lysis of microbial cells accompanied the substantial release of intercellular compounds, so improving the increase of soluble PN and PS.

Relatively speaking, during the sludge pretreatment, much more PN than PS was released as presented in Fig. 5.12. This is mainly attributable to the fact that PN is the most principal constituent of the microbial cells, and the cells' disruption will cause more PN release out of the cells. For this reason, soluble PN concentrations in the liquid phase are usually higher than soluble PS.

5.2.2.4 UV–Vis Spectroscopic Properties of the DOM in Sludge Supernatant

The absorption of dissolved organic matters (DOM) at 254 nm (UV-254), used as a rough indicator of the concentrations of conjugated aromatic molecules, was measured and the results are depicted in Fig. 5.14. UV–Vis absorbance data clearly confirmed the presence of aromatic organics in the extracted DOM from sludge, composed of aromatic proteins, and a small amount of polycyclic aromatic hydrocarbons (PAHs), polychlorinated biphenyls (PCBs), etc. As expected, UV absorbance of aromatic molecules present in DOM from the sludge samples treated by electrical-alkali process continued to increase as the applied voltage elevated from 5 to 20 V (0.4–1.2-fold increase). It also radically increased with the rise in the tested pHs, on an average by 118.7% between 7.5 (electrolysis alone) and 10.2 ($p_{(.05)} = 4.32 \times 10^{-3} < .05$) and 112.9% between 10.2 and 12.2 ($p_{(.05)} = 2.02 \times 10^{-3} < .05$). Higher UV-254 values after combined pretreatment were compelling evidence of sludge flocs (e.g., EPS) and microbial cells being disrupted and solubilized (Fig. 5.13), which were a potential trigger for the release of aromatic protein-related organic matters from sludge solids into supernatant. Indeed, the lysis of microbial cells coincided with the release of not merely aromatic proteins but also hydrophilic

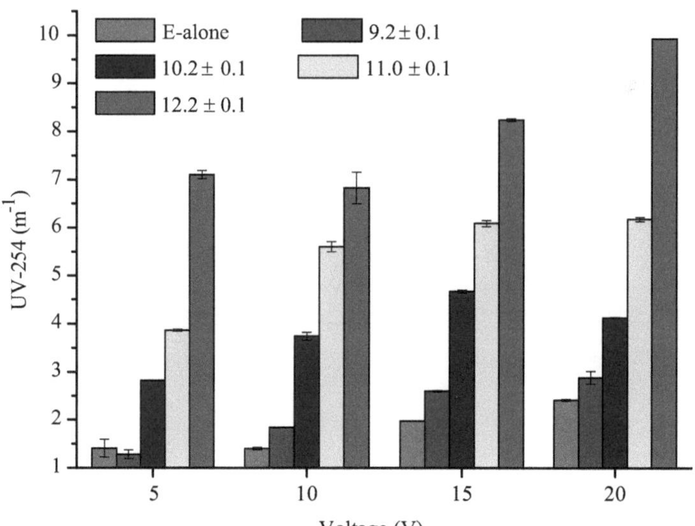

Figure 5.14 UV–Vis absorbance of the dissolved organic matters (DOMs) in the sludge supernatant after the electricalalkali pretreatment. Error bars represent standard deviations of duplicate tests.

UV-inactive DOM (e.g., polysaccharides, aliphatic components, carboxylic components, and unsaturated fatty acids), which do not absorb in this portion of the UV region, as suggested by the temporal enhancement of PS concentrations (Fig. 5.12B). The increased DOM by pretreatment usually warrants higher AD performance. Based on the aforementioned observations, it can thus be said that the electricalalkali pretreatment is an effective approach for promoting the sludge hydrolysis.

5.2.3 Batch Anaerobic Digestion Tests for Sludge

5.2.3.1 Effect of Pretreatment on Methane Production

The BMP assays were carried out to reveal the feasibility of using combined electricalalkali pretreatment to improve the anaerobic biodegradability of the sludge. The cumulative methane yields in terms of mL CH_4 per gram COD added for the different pretreatment conditions were followed in duplicate tests. The methane production from the experiment without the sludge addition was subtracted. Results of the BMP tests with model simulations are given in Fig. 5.15. Model lines presented in this figure are based on the best fit of first-order hydrolysis rate (k_{hyd}, /day) and ultimate degradability (f_d, mL CH_4/g-COD_{added}) with standard deviations (Table 5.4). The first week of incubation is critical because the maximum substrate utilization generally occurs during this period. Anaerobic hydrolysis rates of

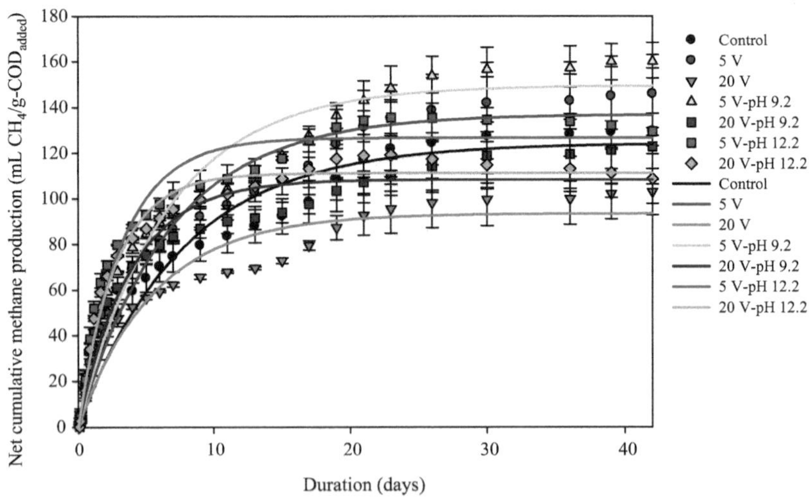

Figure 5.15 Net cumulative methane production (CMP) for different pretreated sludge samples. Error bars represent standard deviations of duplicate tests.

Table 5.4 Parameter Estimation for First-Order Production Rate Coefficient (k_{hyd}) and Ultimate Degradability (f_d) for Sludge AD

Sample	k_{hyd} (/day)	f_d (mL CH_4/g-COD_{added})	R^2
Control (nonpretreated)	0.13 ± 0.01	124.39 ± 3.54	0.9659
5 V	0.15 ± 0.01	136.93 ± 3.72	0.9618
20 V	0.18 ± 0.02	93.44 ± 3.23	0.9215
5 V-pH9.2 ± 0.1	0.15 ± 0.02	149.72 ± 4.63	0.9474
20 V-pH9.2 ± 0.1	0.26 ± 0.03	108.35 ± 2.88	0.9378
5 V-pH12.2 ± 0.1	0.31 ± 0.03	126.73 ± 2.54	0.9607
20 V-pH12.2 ± 0.1	0.36 ± 0.02	111.38 ± 1.63	0.9775

non- or pretreated sludge samples at the beginning of BMP assays would thus provide valuable information on the potential effects of electricalalkali process at different pretreatment conditions corresponding to different degrees of flocs solubilization and disruption. From the BMP results in Fig. 5.15, it was clear that the electricalalkali pretreatment consistently increased anaerobic hydrolysis rate (k_{hyd}) of sludge. The pretreated sludge samples contained higher levels of soluble organic matters (PN, PS, etc.) with respect to the raw sludge, which were metabolized more rapidly, and therefore produced fairly high k_{hyd} constants (Table 5.4). The significantly positive correlations between k_{hyd} and SCOD, DD_{SCOD}, PN as well as PS ($R_p = 0.922$, $p_{(.05)} = .003$; $R_p = 0.945$, $p_{(.05)} = .001$; $R_p = 0.957$, $p_{(.05)} = 7.32 \times 10^{-4}$; and $R_p = -0.600$, $p_{(.05)} = .005$, respectively) as illustrated in Fig. 5.16A also strongly verified this hypothesis.

Interestingly, the electricalalkali pretreatment resulted in an increase in the AD rate but had no or very slight impact on methane potential. Ultimate methane yields from the BMP tests implied that only the 5 V-pH 9.2 sample achieved discernible improvement in methane yields (149.72 mL CH_4/g-COD_{added}) at the 42 days of reactor operation, which was approximately 20.3% higher than that of the nonpretreated sludge (control, 124.39 mL CH_4/g-COD_{added}) (Fig. 5.15 and Table 5.4). In the rest of the reactors fed with the pretreated sludge, however, methane productions were found to be comparable and even slightly lower as compared to the control. For example, at around day 15, the cumulative methane yield of nonpretreated sludge surpassed the sludge pretreated at 20 V-pH 12.2, although the coupled pretreatment process had disrupted sludge flocs and solubilized a higher fraction of COD.

After 42 days of operation, the control had produced more methane with respect to the one obtained with the pretreated sludge. Obviously,

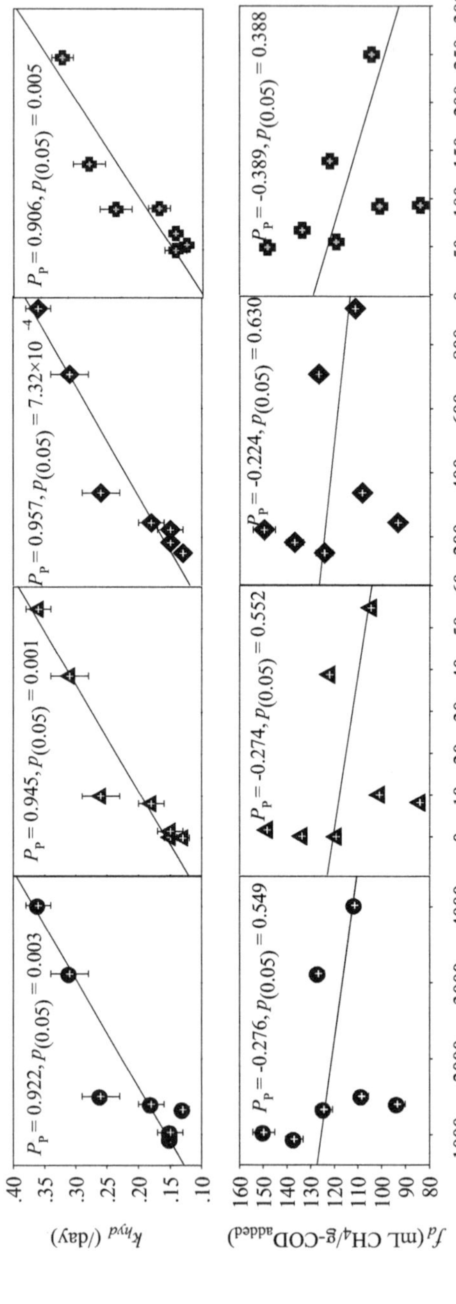

Figure 5.16 Pearson's correlation between first-order production rate coefficient (k_{hyd}), degradability (f_d) and physical characteristics of different pretreated sludge samples. Error bars represent standard deviations of duplicate tests.

the improvement was independent of the tested voltages and NaOH loadings. To provide deep insight into the influence of solubilization degree on sludge AD, the statistical analysis in terms of Pearson's correlation coefficient (R_p) and p-values was thus performed. The results presented in Fig. 5.16B showed that there were no clear correlations existing between ultimate methane yields (ultimate degradability, f_d) and SCOD, DD_{SCOD}, PN or PS ($R_p = -0.276$, $p_{(.05)} = .549$; $R_p = -0.274$, $p_{(.05)} = .552$; $R_p = -0.224$, $p_{(.05)} = .630$; and $R_p = -0.389$, $p_{(.05)} = .388$, respectively), supporting the finding that increased solubilization of sludge from a pretreatment does not always induce an improved AD efficiency. The reason for this phenomenon might be ascribed to different causes (Fig. 5.17)—as for example, the leakage of refractory and inhibitory soluble organics from sludge to liquid phase or partial chemical mineralization of solubilized compounds during the severe pretreatment conditions.

In addition, an overdose of added chemicals such as sodium might inhibit the microbial activity, interfere with their metabolic pathway for methane production, and cause the slightly lower ultimate methane output from the pretreated sludge. From these results, it seems that DD_{SCOD}, SCOD, PN, or PS contents achieved are not the only indicators that need to be taken into account for an ideal comparison of pretreatment effectiveness in AD of the sludge. Other influencing factors, such as the type and dose of chemical applied, compositions of soluble organics released after pretreatment etc., should also be very important.

5.2.3.2 Humification Process During Anaerobic Digestion of Sludge

Fig. 5.18A presents typical UV—Vis spectra of the organic matters extracted from the different digested sludge samples after 42 days of operation. The

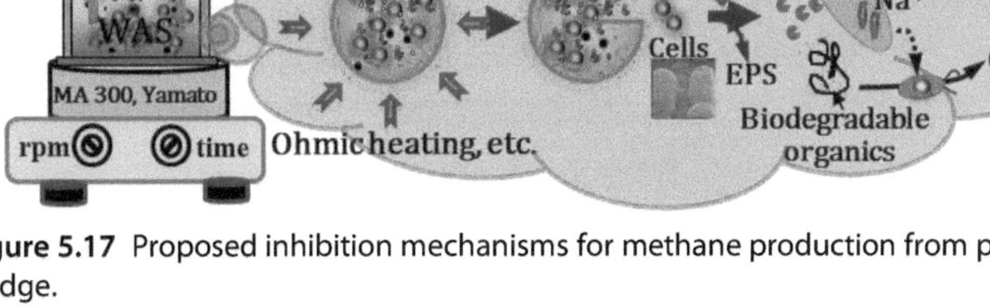

Figure 5.17 Proposed inhibition mechanisms for methane production from pretreated sludge.

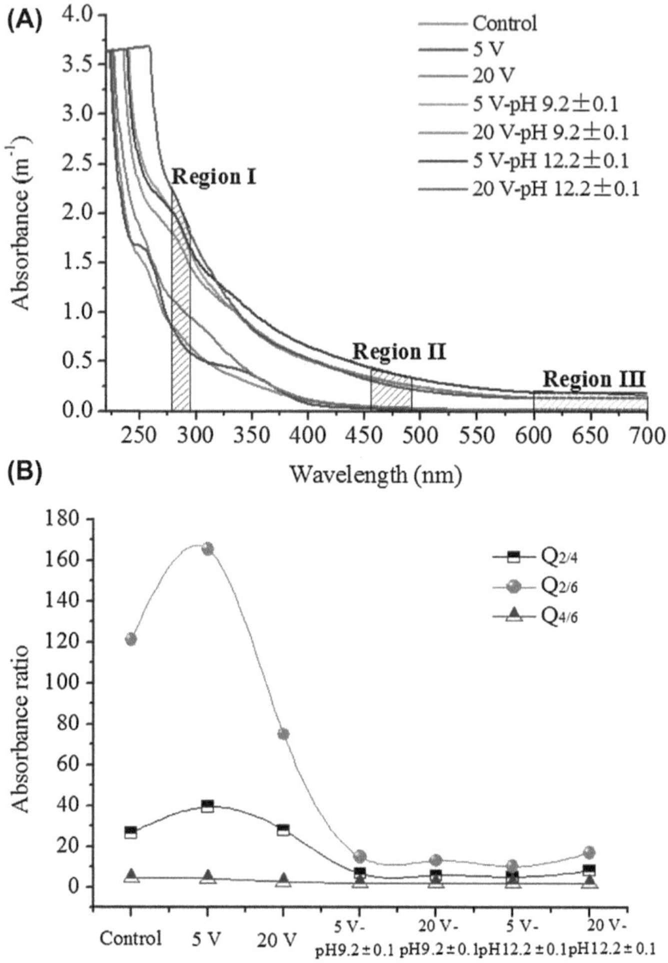

Figure 5.18 Evolution of the UV–Vis spectra (A) and absorbance ratios (B) of 0.5 M NaOH extracts from the different digested sludge samples after 42 days of digestion.

UV–Vis absorbance showed a monotonic decrease as the wavelength increased. The presence of absorbance between 250 and 300 nm are mainly related to $n \rightarrow \pi^{\star}$ and $\pi \rightarrow \pi^{\star}$ electron transition of C=O and C=C double bonds in unsaturated systems, particularly aromatic groups. The variation of UV–Vis spectrum shape in this range is thus a reflection of the transformation of aromatic components, configuration, or heterogeneity. The potential of the technique in characterizing the contributions of EPS to sludge dewaterability has been highlighted in the previous research.

To better understand the digestion performance of different sludge samples, the absorbance ratios indicating the degree of humification were determined to analyze the three absorption regions in the spectrum. In general, the absorbance at shorter wavelengths (260–280 nm) is related to organic matters at the very beginning of transformation, such as lignin and quinone (Region I). The absorbance at intermediate wavelengths (460–480 nm) denotes the organic matters at the beginning of humification (Region II) while the absorbance located at longer wavelengths (>600 nm) is regarded as the indicative of strongly humified materials with a high degree of aromatic and condensed groups (Region III). Then, the absorbance ratios, which are employed to assess the degree of organic matter maturity in digested sludge, were calculated as follows: $Q_{2/4} = A_{280}/A_{472}$, $Q_{2/6} = A_{280}/A_{664}$, and $Q_{4/6} = A_{472}/A_{664}$. The $Q_{2/4}$ reflects the proportion between the lignin and other materials at the beginning of humification, and the contents of materials at the beginning of transformation. The $Q_{2/6}$ represents the relation between nonhumified and strongly humified materials and the $Q_{4/6}$ often serves as the humification index. On the basis of these absorbance ratios, the degree of humification of digested sludge might be visualized.

Fig. 5.18B exhibits a rapid decline of three absorbance ratios ($Q_{2/4}$, $Q_{2/6}$, and $Q_{4/6}$) as the electricalalkali pretreatment was utilized. At the end of 42-day digestion, the $Q_{2/4}$, $Q_{2/6}$, and $Q_{4/6}$ for the sludge sample pretreated at 5 V-pH 9.2 ± 0.1, for example, were found to be 6.8, 15.2, and 2.2, decreases of approximately 74.3%, 87.5%, and 52.2%, respectively, in comparison with the nonpretreated sludge (i.e., the control). Low $Q_{2/6}$ or $Q_{4/6}$ reflects a high degree of aromatic condensation and an increased level of organic material humification. The electricalalkali process led to a higher content of soluble PN, PS, and also SCOD in liquid phase, which are easily degradable compounds. Once fed into reactors, rapider mineralization of soluble PN, PS etc. released during electricalalkali pretreatment and subsequent conversion into humic substances occurred through the degradative pathway (lignin theory) and the condensation route (polyphenol theory). Thus, the pretreated samples produced relatively lower absorbance ratios than that of the control sample. However, it is also interesting to find that the humification index $Q_{4/6}$ values (around 2.0–4.6) for different samples were below 5.0, which is core characteristic of matured humic material, indicating that all the digested sludge reached maturity at the end of digestion irrespective of experimental conditions. From these results, we can note that the electricalalkali process could accelerate the maturity process of sludge, however, has a very slight contribution to enhancing the final degree of humification.

5.2.4 Dewaterability Assessment for Digested Sludge

The sludge dewaterability was assessed by means of centrifuge (15,000 g for 10 min) and reported in Fig. 5.19. The electrolysis or combined with low-alkali pretreatment markedly upgraded the sludge dewaterability ($p_{(.05)} < .05$), whereas in the case of high-alkali only a slight decrease of moisture content was observed ($p_{(.05)} > .05$) possibly due to the progressive cell lysis, releasing the intracellular substances that have a high affinity for water. $Fe(II)/S_2O_8^{2-}$ increases the sludge dewaterability through a mechanism that involves the nonselective ruptures of sulfate radicals $(SO_4^{-}\cdot)$ on EPS and intracellular substances contained in sludge and subsequently diminishing their water retention properties. On the contrary, NaOH has negligible redox potential and could not further oxidize the released intracellular substances. Thus, the released bound water during electricalalkali pretreatment was reabsorbed by the intracellular substances, which finally offset this improvement effect in sludge dewatering.

The digested sludge was reported to be harder to dewater as a result of the large release of colloidal particles and biopolymers during hydrolysis step. The biological hydrolysis disrupted sludge flocs and caused an evident release of fine charged biocolloids, elevating the net surface charge of the flocs and eventually impairing its compressibility and dewaterability.

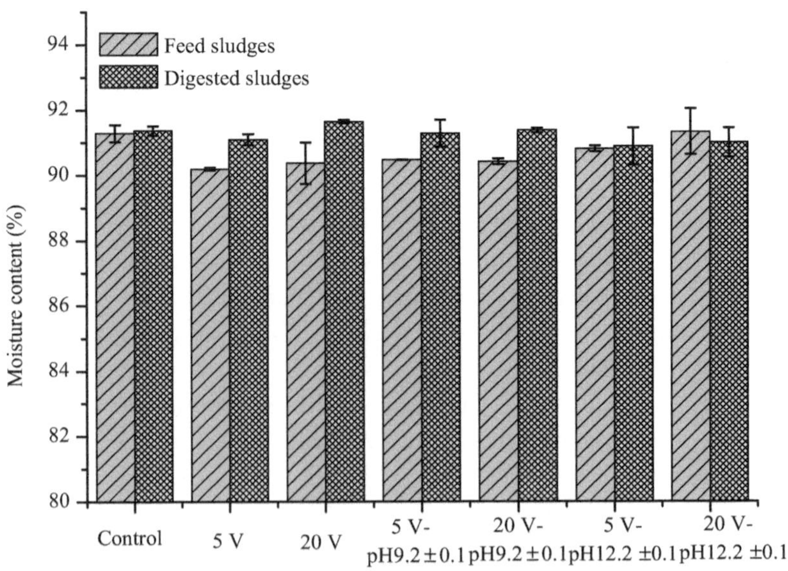

Figure 5.19 Dewaterability of feed and digested sludge samples. Error bars represent standard deviations of duplicate tests.

The moisture content of different pretreated sludge samples increased after AD, suggesting the worsen dewaterability. Electricalalkali pretreatment initially promoted the sludge dewatering characteristics to some extent as noticed before; however, subsequent AD attenuated the positive effects of pretreatment and the dewaterability deteriorated. Sludge, even after anaerobically digested, still has relatively high moisture content. Therefore, posttreatment processes, especially dewatering, should be highlighted as an integral part of sludge AD. Additional efforts should be focused on the study and implementation of dewaterability enhancement of digested sludge in the future studies, which is crucial for the utilities with limited site spaces.

5.3 INFLUENCE OF ZERO-VALENT SCRAP IRON (ZVSI) SUPPLY ON METHANE PRODUCTION

Zero-valent iron (ZVI), a strong reducing reagent ($E^0 = -0.447$ V), has been universally used in contaminated site remediation, decolorization of Rhodamine B, degradation of azo dye, and microbial removal of perchlorate. Most recently, the promotion for AD using ZVI has also been documented. The preliminary idea of using ZVI is based on the release of hydrogen through electron transfer during ZVI corrosion/oxidation, which can serve as an electron-donor source for methanogens, sulfate-reducing bacteria (SRB) or dechlorinating bacteria.

Unlike the exciting results achieved before, slight stimulation and even potential inhibition of ZVI on methanogens and SRB have been observed. The toxicity principles of nano ZVI may include: (1) fast H_2 production and accumulation, which impedes the thermodynamical conversions of propionic and butyric acids into acetic acid and methanogenesis (H_2 threshold of 23–75 nM for methanogens); and (2) disruption of cell membranes via reductive decomposition of ZVI, which inhibits methanogenesis. Up to date, the behaviors of ZVI in anaerobic systems are still inconsistent and sometimes contradictory. The questions, therefore, remain as to whether and how ZVI will affect hydrolysis, acidogenesis, and methanogenesis in anaerobic process. Besides, most previous studies were undertaken mainly with ZVI, and very limited research referred to ZVSI has been conducted. ZVSI is a by-product produced from machinery industries and easily available at local level; moreover, it has excellent reactivity similar to ZVI but of very low price. Provided that ZVSI can substitute ZVI, this will not only lower the overall project costs but also stimulate the waste-scrap conservation. Owing to the aforementioned factors, a series of batch experiments with six different dosages of ZVSI are thus performed herein to exploit the exact roles of ZVSI in sludge digestion, with the primary aim of developing a sustainable strategy to drive sludge stabilization while accomplishing bioenergy recovery.

5.3.1 Zero-Valent Scrap Iron (ZVSI) and Sewage Sludge

Sludge sample was obtained from a WWTP in Sendai, Japan (Table 5.5). Batch-scale experiments of influence of ZVSI (approx. 2.0 mm in width and 0.35 mm in thickness) on the hydrolysis, acidogenesis, and methanogenesis of sludge were carried out through BMP tests. Hydrolysis (%H), acidogenesis (%A), methanogenesis (%M), and biodegradability (%B) were calculated according to Eqs. (5.3)–(5.6):

$$\%H = \frac{(SCOD - SCOD_0 + COD_{CH_4})}{TCOD_0 - SCOD_0} \cdot 100 \tag{5.3}$$

$$\%A = \frac{(COD_{VFA} - COD_{VFA_0} + COD_{CH_4})}{TCOD_0} \cdot 100 \tag{5.4}$$

Table 5.5 Characteristics of Sewage Sludge Used

Parameter	Unit	Value[a]	SD[b]
pH		6.2	0.0
TS (total solids)	g/L	27.3	0.6
VS (volatile solids)	g/L	22.4	0.4
TSS (total suspended solids)	g/L	26.5	0.3
VSS (volatile suspended solids)	g/L	21.9	0.2
TCOD (total chemical oxygen demand)	mg/L	30,141	1,003
SCOD (soluble chemical oxygen demand)	mg/L	86	2
Total protein	mg/L	8881.1	563.9
Total carbohydrate	mg/L	2292.9	42.0
Soluble protein	mg/L	11.7	0.7
Soluble carbohydrate	mg/L	12.3	0.6
Ammonia nitrogen ($NH_4^+ - N$)	mg/L	13.2	0.2
Sulfate (SO_4^{2-})	mg/L	2.6	0.2

[a]All values are expressed in average data.
[b]*SD*, standard deviation.

$$\%M = \frac{COD_{CH_4}}{TCOD_0} \cdot 100 \tag{5.5}$$

$$\%B = \%A + \frac{Y_A}{1 - Y_A} \cdot \left(\%A - 100 \cdot \frac{COD_{VFA_0}}{TCOD_0}\right) + \frac{Y_M}{1 - Y_M} \cdot \%M \tag{5.6}$$

where Y_A is the yield of acidogenic microorganisms (0.050 g COD/g-COD) and Y_M is the yield of methanogenic archaea (0.029 g COD/g-COD).

5.3.2 Effect of ZVSI on Biogas Production

Fig. 5.20 illustrates cumulative methane yield profiles under different temperatures and ZVSI dosages. As can be seen, temperature, as a key operational parameter, played a decisive role in methane production and controlled the overall biokinetics of digestion process. During the first stage (20°C), methane yield showed negligible discrepancy irrespective of ZVSI dosages ($p > .05$), and the average values were lower than 1.5 ± 0.1 mL/g-VSS_{fed} attributable to the relatively low anaerobes activities. In contrast, increasing temperature to 35°C could profoundly promote methane production. Moreover, the higher the ZVSI was added the more the methane generated. For example, with an increase of ZVSI from 0 (ZVSI-free) to 0.05 g/g-VSS, methane yield exhibited a progressive climb from 126.4 ± 1.0 to 132.4 ± 3.7 mL/g-VSS_{fed} ($p = .16$), which was further enhanced ($p < .05$) as ZVSI increased to 0.1 g/g-VSS and above.

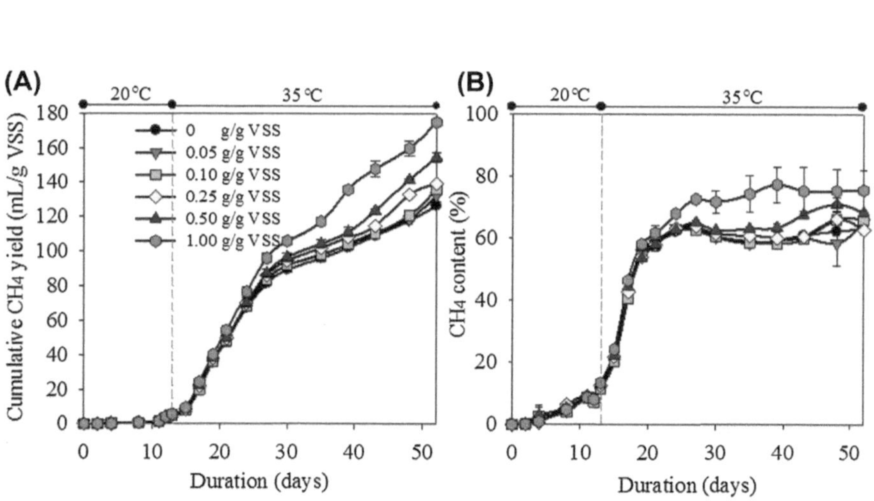

Figure 5.20 Effect of zero-valent scrap iron (ZVSI) dosages on CH_4 production: (A) CH_4 yield and (B) CH_4 content from sludge.

The highest methane production of 174.9 ± 1.5 mL/g-VSS$_{fed}$ was obtained at 1.0 g-ZVSI/g-VSS, 38.3% higher relative to the ZVSI-free ($F_{observed} = 1489.5$, $F_{crit} = 18.5$, $p = .0007 < .05$). It suggested that mesophilic condition (35°C) was always required for simulating the function of ZVSI and maintaining the superior stability of AD process. As with the methane content, the distinct difference was also observed under mesophilic condition. From Fig. 5.20B, methane content at 0 g-ZVSI/g-VSS was quite low and stable in 62.9–63.2% during 27–52 days, which further rose to 65.1–68.3% and then to 72.6–75.5% ($p < .05$) as ZVSI elevated to 0.5 and 1.0 g/g-VSS.

The commonly accepted mechanism of electron transfer from ZVSI to microorganisms is via the microbial corrosion and surface oxidation of ZVSI. Under anaerobic conditions, ZVSI is oxidized to ferrous iron (Fe^{2+}), producing cathodic hydrogen (H_2/[H]) via Eq. (5.7). Hence, H_2 production can be used as a direct measure to reflect the reactivity and corrosion extent of ZVSI. Fig. 5.21 shows the H_2 evolution in ZVSI-sludge system. As can be seen, fast H_2 production and accumulation took place during the initial 2 days at 20°C. With the increase of ZVSI, the yield of H_2 increased and

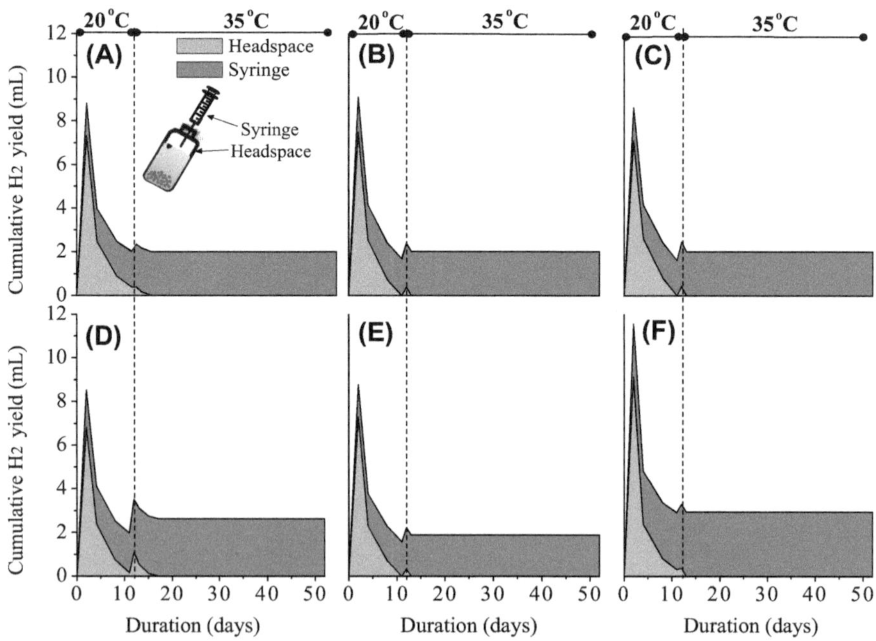

Figure 5.21 Production of H_2 during ZVSI corrosion/oxidation: (A) 0, (B) 0.05, (C) 0.10, (D) 0.25, (E) 0.50 and (F) 1.00 g-ZVSI/g-VSS in sludge.

reached the greatest of 12.3 ± 1.5 mL at 1.0 g-ZVSI/g-VSS. High H_2 content (or partial pressure) was detrimental to methane production because of the associated CO production and/or the impeded thermodynamical conversions of propionate and butyrate into acetate, which inhibited acetoclastic methanogenesis (*Methanosaeta* and *Methanosarcina*). Notably, H_2 did not accumulate continuously herein even at the highest dosage of ZVSI 1.0 g/g-VSS, but was quickly consumed during the following biological process (especially at 35°C), during which no H_2 was detected. The results confirmed that H_2 produced during both ZVSI corrosions could be effectively utilized by hydrogenotrophic methanogens (e.g., *Methanobacterials*, *Methanococci* and *Methanomicrobials*) via Eq. (5.8) to produce CH_4. The addition of ZVSI contributed to the production of more H_2, and thereby methane yield was also expected to increase (Fig. 5.20A). Meanwhile the autotrophic methanogenesis via using H_2 as energy source could not only form extra CH_4 but also consume part of CO_2 (Eq. 5.8) and, as a consequence, the percentage of methane increased with the applied ZVSI dosages (Fig. 5.20B). Likewise, homoacetogenic bacteria could also use CO_2/H_2 as sole energy source for the formation of methanogenic precursor acetate (Eq. 5.9), which further reduced CO_2/H_2 contents and raised methane percentage (Eq. 5.10). Other authors similarly noted that iron could be involved in energy metabolism as a cytochrome and ferredoxin in methylotrophic methanogens.

$$Fe^0 + 2H_2O \rightarrow Fe^{2+} + H_2 + 2OH^- \quad \Delta G^{0\prime} = -5.02\ KJ/mol\ Fe^0 \tag{5.7}$$

$$4H_2 + CO_2 \rightarrow CH_4 + 2H_2O \quad \Delta G^{0\prime} = -131\ KJ/mol \tag{5.8}$$

$$4H_2 + 2CO_2 \rightarrow CH_3COO^- + H^+ + 2H_2O \quad \Delta G^{0\prime} = -95\ KJ/mol \tag{5.9}$$

$$CH_3COO^- + H^+ \rightarrow CH_4 + CO_2 \quad \Delta G^{0\prime} = -75.7\ KJ/mol \tag{5.10}$$

Except for methane production, the prompt consumption of H_2 also is the prerequisite of continuous microbial-assisted corrosion of ZVSI. Because of high Gibbs free energy ($\Delta G^{0\prime} = -5.02$ kJ/mol Fe^0, Eq. 5.7), the formation of cathodic hydrogen through ZVSI corrosion is thermodynamically slow. The accumulation of H_2 is unfavorable for the corrosion reaction, and even halts the generation of cathodic hydrogen. H_2 removal by hydrogen-utilizing microorganisms via Eqs. (5.8)−(5.10) can effectively

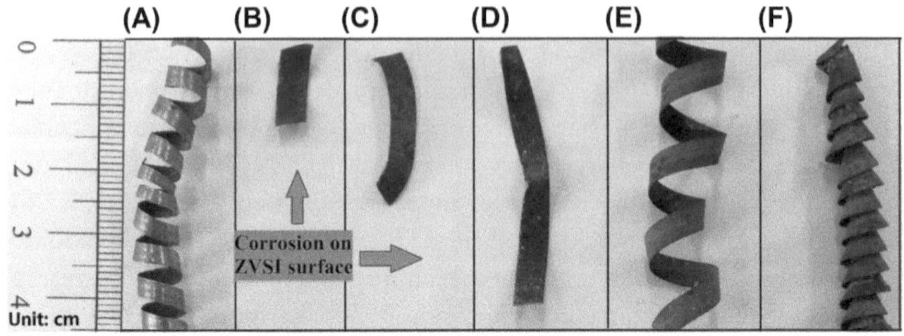

Figure 5.22 Initial ZVSI (A), and their corrosion during the overall digestion process: (B) 0.05, (C) 0.10, (D) 0.25, (E) 0.50, and (F) 1.00 g-ZVSI/g-VSS.

decline the amount of H_2 and drive the ZVSI bio-corrosion (Fig. 5.22). Thus, the otherwise unfavorable ZVSI bio-corrosion by cathodic depolarization occurred by continual microbial conversion of H_2 so that its pressure was kept low. In this case, methane formation was enhanced. Besides, methanogens could also convert CO_2 to methane with ZVSI as the direct electron donor via Eq. (5.11), which was necessarily beneficial for accelerating ZVSI corrosion. The above-named results clearly demonstrated that ZVSI can serve as an electron donor for methanogenesis (i.e., methanogenic *archaea*) and favored the improvement of methane production.

$$8H^+ + 4Fe^0 + CO_2 \rightarrow CH_4 + 4Fe^{2+} + 2H_2O \quad \Delta G^{0\prime} = -150.5\ kJ/mol\ CH_4 \tag{5.11}$$

5.3.3 Effect of ZVSI on Sludge Solubilization

The solubilization of sludge is a key step before hydrolysis. The dynamic changes of SCOD, protein, and carbohydrate during AD process were thus determined (Fig. 5.23). On the overall, the average SCOD, protein, and carbohydrate concentrations in different sludges rapidly rose during the initial 12 days (20°C) regardless of ZVSI, presumably due to the continued endogenous biomass decay and slow rate of methanogenesis (Fig. 5.20A). The former made a large amount of soluble compounds release while the latter retarded the further conversion of these organics into methane. Also, as for ZVSI, at a certain time, SCOD, protein, and carbohydrate had an ascending trend with ZVSI dosages (Fig. 5.23). As an example, with the fermentation time of 12 days at 20°C, SCOD in the groups of 0, 0.05, 0.10, 0.25, 0.50 and 1.00 g-ZVSI/g-VSS were 178.2 ± 3.5, 178.0 ± 12.3, 181.2 ± 6.4, 179.4 ± 12.9, 184.1 ± 2.4 and

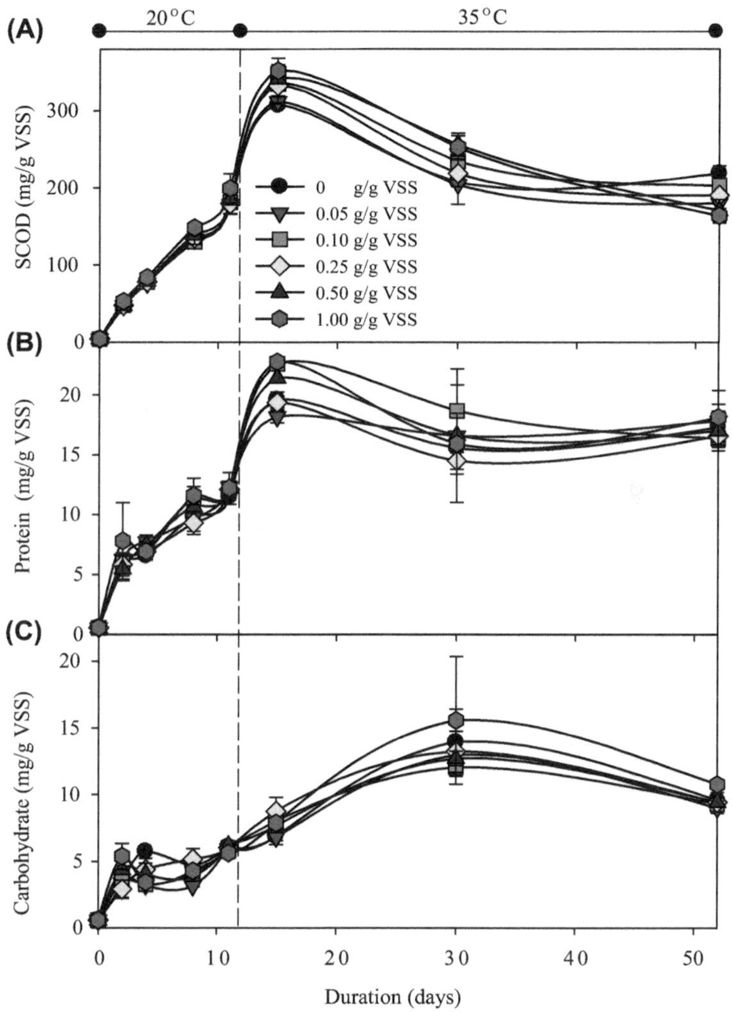

Figure 5.23 Variations of SCOD (A), protein (B), and carbohydrate (C) during the overall digestion process.

199.2 ± 19.1 mg/g-VSS, respectively. The distinct increment of SCOD, particularly as exposed to 1.00 g-ZVSI/g-VSS, in part, hinted the physical disruption of cell membranes and leakage of intracellular substances. The decomposition of bound EPS and glycan cross-linked by peptide chains of the external membranes caused the rupture of partial cells and damage of sludge integrity. In this sense, the solubilization of sludge was advanced.

The disintegration of sludge caused by ZVSI might be effectively upgraded as the operational temperature increased to 35°C. The elevated temperature accelerated ZVSI corrosion and thus greatly promoted sludge solubilization. It can be seen from Fig. 5.23 that the soluble organics in the fermentation liquor increased fleetly with the maximum SCOD and protein in the groups of 0, 0.05, 0.10, 0.25, 0.50, and 1.00 g-ZVSI/g-VSS ranging from 306.9 ± 1.6 to 351.3 ± 16.3 mg/g VSS ($p \approx .05$) and 19.6 ± 0.6 to 22.8 ± 0.4 mg/g-VSS ($p = .03$) only after 3 days of mesophilic digestion. As expected, after a transient rise, these organics, mainly protein (Fig. 5.23B), fast dropped suggesting the rapid start-up of acidogenesis and/or methanogenisis, as indicated by increased methane yield at 35°C, whereas, in comparison with protein, the progressive rise in carbohydrate still persisted until the day 30, then followed by a marked decrease. Besides, temperature influenced not only the ZVSI corrosion but also the composition of microbial community, and even the metabolism pathway of organics during methanogenic degradation. *Coprothermobacter*, having preference for fermentation of protein and amino acids, survive more easily in mesophilic digestion while *Thermotogae*, which excrete hydrolytic enzymes to catalyze a wide range of polysaccharides, are in thermophilic condition. In view of this, the abundant presence of protein-dependent microorganisms in this study might be the core reason for much quicker consumption of protein versus carbohydrate during mesophilic process (12−52 days). The enhanced solubilization of particulate organics and following hydrolysis, particularly in the presence of higher ZVSI, provided the bountiful substrates for subsequent acidogenesis and methanogenisis, and undoubtedly methane yield rose.

5.3.4 Effect of ZVSI on pH and VFAs Production and Composition

Fig. 5.24 shows the variations of pH and VFAs under different dosages of ZVSI. It reveals that the trend of pH highly agreed with the VFAs' accumulation. All pH values dropped below 6.0 throughout the overall low-temperature digestion and the first 3 days of mesophilic process, owing to the accumulation of VFAs. After a distinct drop, a quick pH recovery was detected; moreover, it increased more at higher ZVSI. The pH, for example on day 30, increased from 6.3 ± 0.4 to 6.4 ± 0.0, 6.3 ± 0.1, 6.5 ± 0.4, and 6.7 ± 0.1 with increasing ZVSI from 0 to 0.05, 0.10, 0.25, and 0.50 g-ZVSI/g-VSS, and it further increased to 7.3 ± 0.1 as ZVSI was 1.00 g-ZVSI/g-VSS likely due to the continued dissolution of ZVSI via Eq. (5.7), and consequently enhanced VFA's consumption. These observations evidenced the advantageous buffering capacity of ZVSI, which could

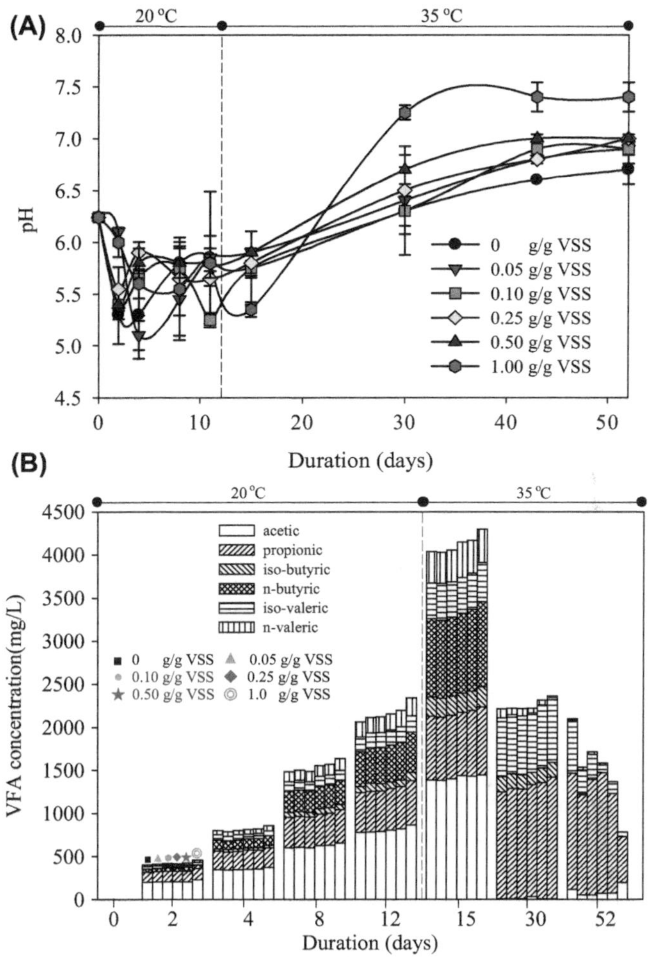

Figure 5.24 Variations of pH (A) and VFAs (B) during the overall digestion process.

prevent pH drop and induce a stable anaerobic system. What is more, during the fermentation of soluble organics, or more exactly protein, some by-products such as CO_2, NH_4^+ and HCO_3^- are generated that can also to some extent alleviate liquor pH.

The results of VFAs components are presented in Fig. 5.24B. Consistent with SCOD, protein, and carbohydrate profiles, total VFAs and acetate showed a gradual accumulation with the operating time and ZVSI dosages. On day 12, for instance, VFAs at 0, 0.05, 0.10, 0.25, 0.50, and 1.00 g-ZVSI/g-VSS were 2059.9 ± 110.8, 2114.7 ± 40.1, 2121.5 ± 35.0, 2156.0 ± 11.2, 2195.9 ± 10.8, and 2340.4 ± 73.0 mg/L;

the corresponding acetate increased from 779.6 ± 18.2 to 783.6 ± 12.2, 790.3 ± 9.4, 807.3 ± 12.0, 820.2 ± 8.0 and then 865.5 ± 24.8 mg/L, respectively. It suggested that ZVSI provoked VFAs, typically acetate production. It is well known that only acetate in VFAs, together with formic acid, hydrogen, methanol, and methylamine can be utilized as the direct substrates for methane production. Therefore, it can be speculated that the enhanced acetate production in the presence of ZVSI would be thermodynamically favorable for methanogenisis.

Furthermore, as operation time elapsed, VFAs and acetate increased drastically and peaked on day 15 (i.e., 3 days after the start of mesophilic digestion). Afterward, a significant decline of VFAs, especially acetate, was observed on day 30 and nearly no sign of acetate was detected, indicating its complete degradation. On the contrary, propionate, at this moment, showed a maximum ranging from 1242.4 ± 47.4—412.4 ± 47.1 mg/L depending on ZVSI dosages, which might derive from conversion of other VFAs. The buildup of propionate is considered to be unfavorable in any anaerobic process unless it is decomposed by β-oxidation into acetate and H_2. However, the conversion is frequently difficult to occur in thermodynamics ($\Delta G^{0'} = +76.1$ kJ/mol) (Eq. 5.12) because of the strict requirements for H_2 partial pressure (3×10^{-4}—10^{-6} atm), acetate concentration, and oxidationreduction potential (ORP) (<−278 mV). Interestingly, these intrinsic defects appeared to be overcome herein in the presence of ZVSI (Fig. 5.24B). Propionate decreased at a considerable level on day 52 while accompanied by the formation of acetate, especially for the sludge treated with 0.50 and 1.00 g-ZVSI/g-VSS, with respect to the continuous buildup in ZVSI-free one.

Based on this finding, it is obvious that ZVSI might have reduced ΔG in the decomposition of propionate through consuming either acetate or H_2, or both together; this provided a favorable free-energy change for propionate β-oxidation, facilitating the final methanogenesis. The reason for this might be principally ascribed to the intrinsic reducing properties of ZVSI, and thus its addition could decrease ORP and create a more reductive atmosphere for AD process. In addition, according to the infinitesimal H_2 content under mesophilic conditions as set forth, we reasonably assumed that the enhanced hydrogen-utilizing methanogenesis and homoacetogenesis by ZVSI depleted H_2 and reduced its partial pressure, making propionate degradation more thermodynamically possible. So, acetogenesis and methanogenisis were intensified.

$$CH_3CH_2COO^- + 3H_2O \rightarrow HCO_3^- + CH_3COO^- + H^+ + 3H_2$$
$$\Delta G^{0\prime} = +76.1 \text{ kJ/mol} \tag{5.12}$$

5.3.5 Variations of Different EPS Fractions in Sludge

EPS (i.e., a three-dimensional, gel-like, and negative-charged matrix) takes a critical part in cells adhesion and cohesion of sludge. Its presence is of great importance in controlling the functional integrity, strength, and dewaterability of sludge as well as biodegradability. Fig. 5.25 compares different

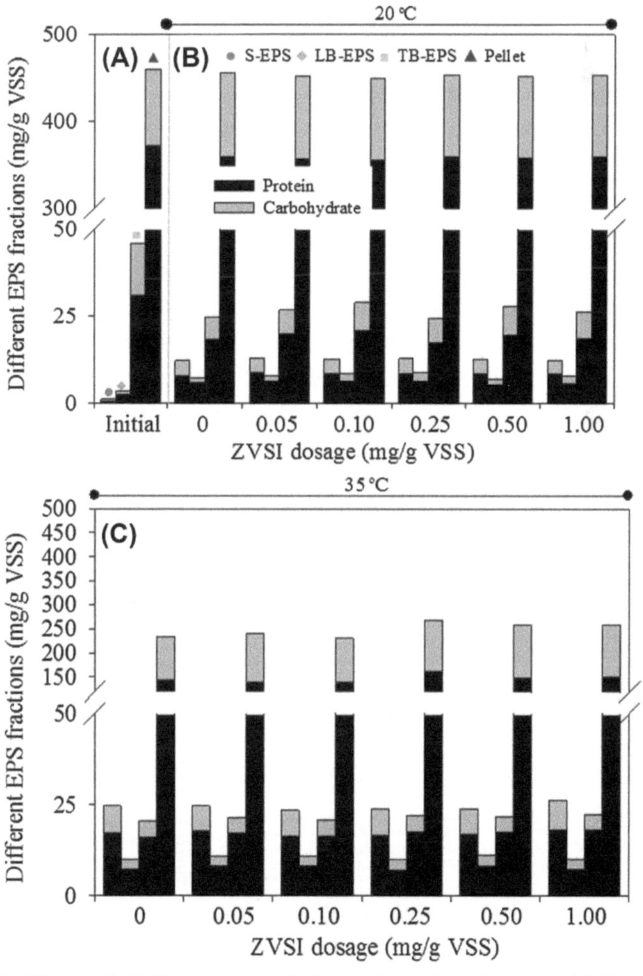

Figure 5.25 Effects of ZVSI on extracellular polymeric substance (EPS): raw (A) and digested sludge on day 12 (B) and on day 52 (C).

EPS fractions during AD processes. It is apparent that EPS in raw sludge were mainly distributed in pellet and tightly bound (TB-EPS) and very few in soluble EPS (S-EPS) and loosely bound (LB-EPS). As well known, S-EPS are well distributed in aqueous phase and LB-EPS, a highly porous slime layer, extend from TB-EPS to the liquid. TB-EPS, in contrast, accumulate both on the outside of cells and in the interior of microbial aggregates (i.e., pellet) and strongly support the mechanical stability of sludge. The sludge with high TB-EPS and pellet contents is thus much stiffer, and has more difficulties to be hydrolyzed. Therefore, the transfer of TB-EPS and pellet into S-EPS and LB-EPS after even 12 days (20°C) was still extremely limited (Fig. 5.25B). Comparatively, the transfer under the help of ZVSI appeared to be enhanced. Of course, owing to the slow dissolution of ZVSI, the promotion looked somewhat slight.

EPS not only act as a protective net-like layer against stressful external disturbance, but also as a carbon/energy reservoir during biological process. Alone with operating time, the solubilization of EPS from TB-EPS and pellet into LB-EPS, and typically into S-EPS increased (Fig. 5.25C), which further were used for methane formation. Therefore, the appropriate rupture of EPS will facilitate the release of more substrates and subsequent methanogenesis. Simultaneously, it is worth recalling that on day 52, the sludge treated by ZVSI displayed a higher level of TB-EPS, and particularly pellet as compared to the ZVSI-free. Microbial cells comprise the sludge pellet, and the increased pellet-indicated ZVSI has a positive effect on the growth of microbes, as a result of which EPS formation was provoked. EPS could provide some protective shielding, helping create a safe environment for bacteria and methanogenic *archaea* to survive. In this case, AD performance was maintained stable, and methane production was reinforced.

5.3.6 Effect of ZVSI on Hydrolysis, Acidification and Methanogenesis of Sludge

To further clarify the exact impacts of ZVSI on methane production, each step (hydrolysis, %H; acidogenesis, %A; and methanogenesis, %M) involved in the anaerobic process as well as biodegradability (%B) was calculated after 52 days of digestion. As can be seen from Fig. 5.26A, the influence of ZVSI on hydrolysis of solubilized products (i.e., soluble protein and carbohydrate) was dosage-dependent. At 0.05—0.25 g/g-VSS, the influence of ZVSI was inappreciable ($p > .05$), but the degradation efficiency of solubilized products at 0.50 and 1.00 g/g-VSS significantly increased in comparison with the ZVSI-free ($p = .019$ and .003, respectively), and reached up to

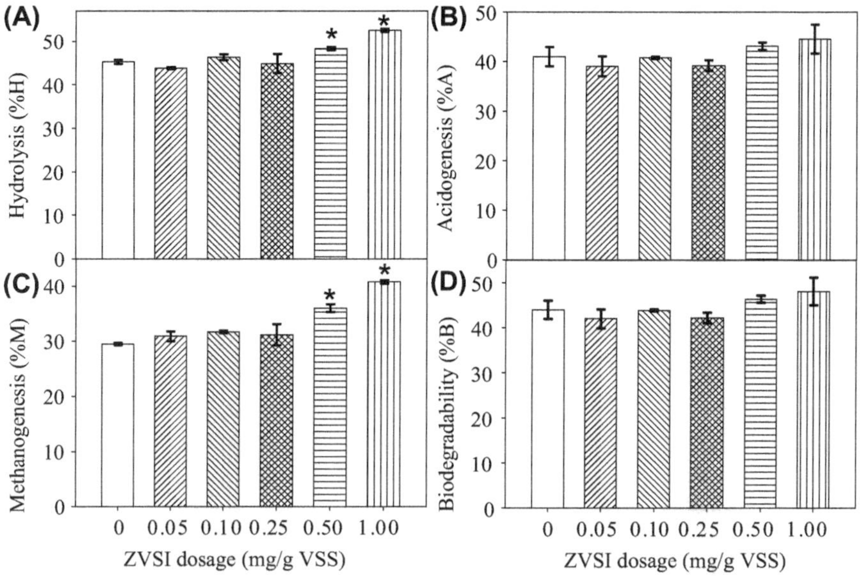

Figure 5.26 Effects of ZVSI on hydrolysis (A), acidogenesis (B), methanogenesis (C), and biodegradability (D) of sludge at 52 days. Asterisks indicate statistical significance ($p < .05$) from the ZVSI-free.

48.3 ± 0.3% and 52.6 ± 0.4% (versus 45.2 ± 0.5% in the ZVSI-free) at the end of test, respectively. Evidently, ZVSI could considerably accelerate the hydrolysis involved in sludge digestion.

The data in Fig. 5.26B relation to acidogenesis suggested that no significant difference in VFAs production existed among the ZVSI-free and others. Despite this, with increasing ZVSI the acidogenesis was promoted. Acidogenesis rate was 40.9 ± 2.0%, 39.1 ± 2.0%, 40.8 ± 0.2%, and 39.2 ± 1.1% for the reactors of ZVSI-free, 0.05, 0.10 and 0.25 g/g-VSS, respectively, which was further raised into 43.1 ± 0.7% and 44.5 ± 2.9% as ZVSI was 0.50 and 1.00 g/g-VSS, respectively. The aforementioned observation manifested the previous finding that ZVSI could effectively accelerate the hydrolysis-acidification of sludge. This might be one of the pivotal reasons for the increased methane yield treated by higher ZVSI.

For the effects of ZVSI on methanation step, the results from Fig. 5.26C hinted that ZVSI contributed more to the methanogenesis. Methanogenesis for those exposed to 0.05, 0.10, and 0.25 g/g-VSS increased to 30.9 ± 0.9%, 31.7 ± 0.3%, and 31.2 ± 1.9%, respectively, even not very significantly versus 29.5 ± 0.2% of the ZVSI-free ($p > .05$). The remarkable progress was obtained at 0.50 and 1.00 g-ZVSI/g-VSS, in which methanogenesis ran up to 36.1 ± 0.7% ($p = .006$) and 40.8 ± 0.3% ($p = .001$), respectively, suggesting that ZVSI was beneficial to the growth of methanogens and enhanced the bioconversion of acetate to methane, corresponding well with methane profiles (Fig. 5.20) and COD balance (Fig. 5.27). In contrast, the effect of ZVSI on biodegradability was insignificant ($p > .05$) (Fig. 5.26D). The responsible reasons are still not well known and more research is required. Whatever, by comparing the data in

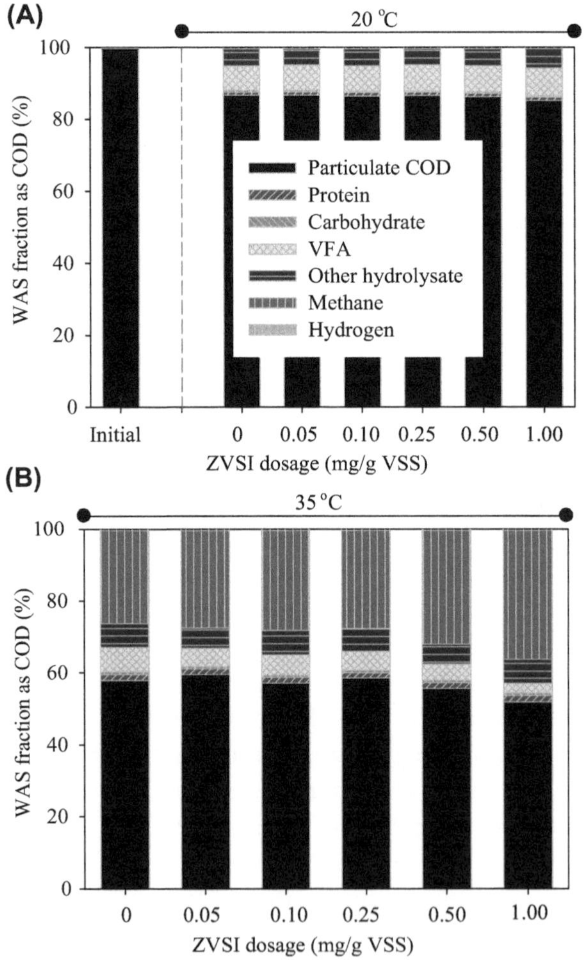

Figure 5.27 Chemical oxygen demand (COD) balance calculation of sludge: (A) after 12 days and (B) 52 days. Data reported here were average values from duplicate experiments.

Fig. 5.26A–D, it can be concluded that the accelerated hydrolysis and methanation steps might be the fundamental mechanisms behind the enhanced methane formation by ZVSI.

5.3.7 Evidence for Dissolution of ZVSI and Its Stimulation on Anaerobic Process

Scanning electron microscopy (SEM) with energy-dispersive X-ray spectroscopy (SEM-EDX) analysis was conducted to support our findings that ZVSI could stimulate the methane production of sludge. SEM images in Fig. 5.28 revealed the surface of unused ZVSI was flat with high-purity iron detected by EDX examination, while it became scraggy and rough at the end of digestion. A higher magnification of SEM revealed that many "flower-shaped" crystals composed of prismatic iron (hydr)oxides clusters (e.g., $FeCO_3$, Fe_3O_4, Fe_2O_3, $Fe(OH)_3$, and FeOOH) prevailed on the surface of used-ZVSI, showcasing the microbe-driven dissolution of ZVSI throughout the digestion process.

To comprehensively verify the underlying enhancement mechanisms, the morphology of digested sludge and distribution of iron in sludge were

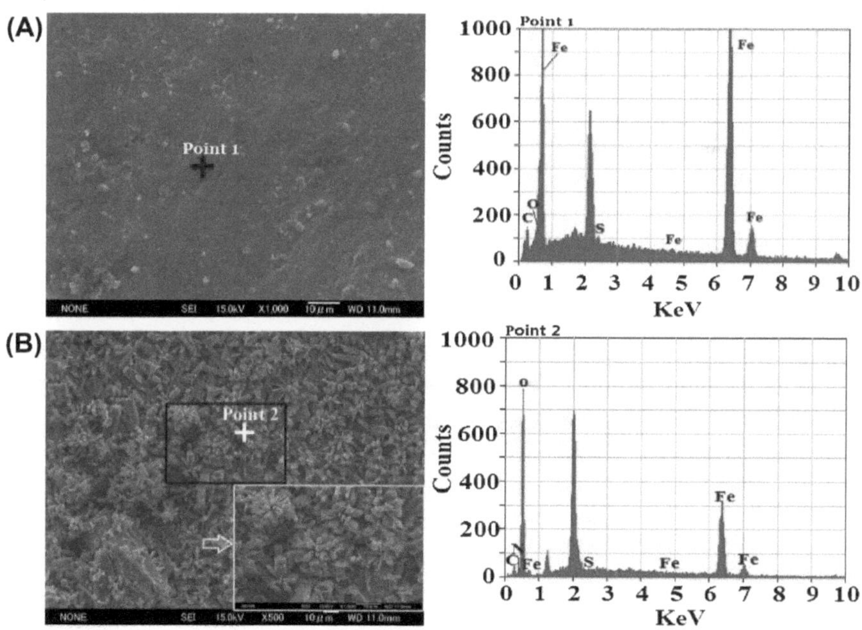

Figure 5.28 Scanning electron microscopy with energy-dispersive X-ray spectroscopy (SEM-EDX) analysis (+) of ZVSI before (A) and after (B, 1.00 g-ZVSI/g-VSS) sludge AD.

characterized (Fig. 5.29). The group of ZVSI-free had relatively low background iron and sludge flocs displayed a compact and rigid microstructure (Fig. 5.29A), whereas, with the increase of ZVSI, the level of iron in sludge elevated, agreeing well with the results of iron EDX mapping. Correspondingly, the sludge became well-regulated, distinct, and stable, and the amount of microbial cells increased as well, in accordance with the increasing trends of methane (Fig. 5.20A), EPS (Fig. 5.25), and methanogenesis rate (Fig. 5.26C). From this point of view, the microbes observed in sludge matrix were presumably related to methanogens and hydrogen-utilizing microbes. Based on a higher magnification of SEM for the sludge treated with 1.00 g-ZVSI/g-VSS, some crystalline dots on the surface of cells were identified, possibly attributed to iron-bearing compounds according to EDX (Fig. 5.29C).

Unlike the commonly observed inhibition from ZVI due to the formation of iron precipitates (e.g., iron—phosphorous complexes) or physical disruption to cell membrane, the bactericidal effects did not occur in our study. By contrast, the released iron ions could penetrate into the inner part of cells and stimulate the synthesis of key enzymes and growth of microbes, particularly methanogens. It is still unclear whether iron ions binding the specific DNA may directly upgrade the enzymatic functions of methanogens, and speed up cell replication. Nevertheless, based on the above results, several possible pathways of ZVSI prompting methane production could still be proposed (Fig. 5.30). These biotic/abiotic factors all eventually led to the observed increase in the methane production. Our work here builds solid support for the potential applicability of ZVSI for sludge digestion and provides an additional route for enhancing the methane production accompanied with waste iron scrap reuse.

5.4 MESOPHILIC ANAEROBIC CODIGESTION OF WASTE-ACTIVATED SLUDGE (WAS) AND *EGERIA DENSA*: PERFORMANCE ASSESSMENT AND KINETIC ANALYSIS

Anaerobic treatment using sludge as a sole substrate is, however, not always satisfactory due to the poor hydrolysis caused by rigid cell walls and substantially secreted extracellular biopolymers, as well as the possible VFAs and ammonia inhibition. Codigestion has unique benefits over the traditional AD including: balances carbon to nitrogen (C/N) ratio and

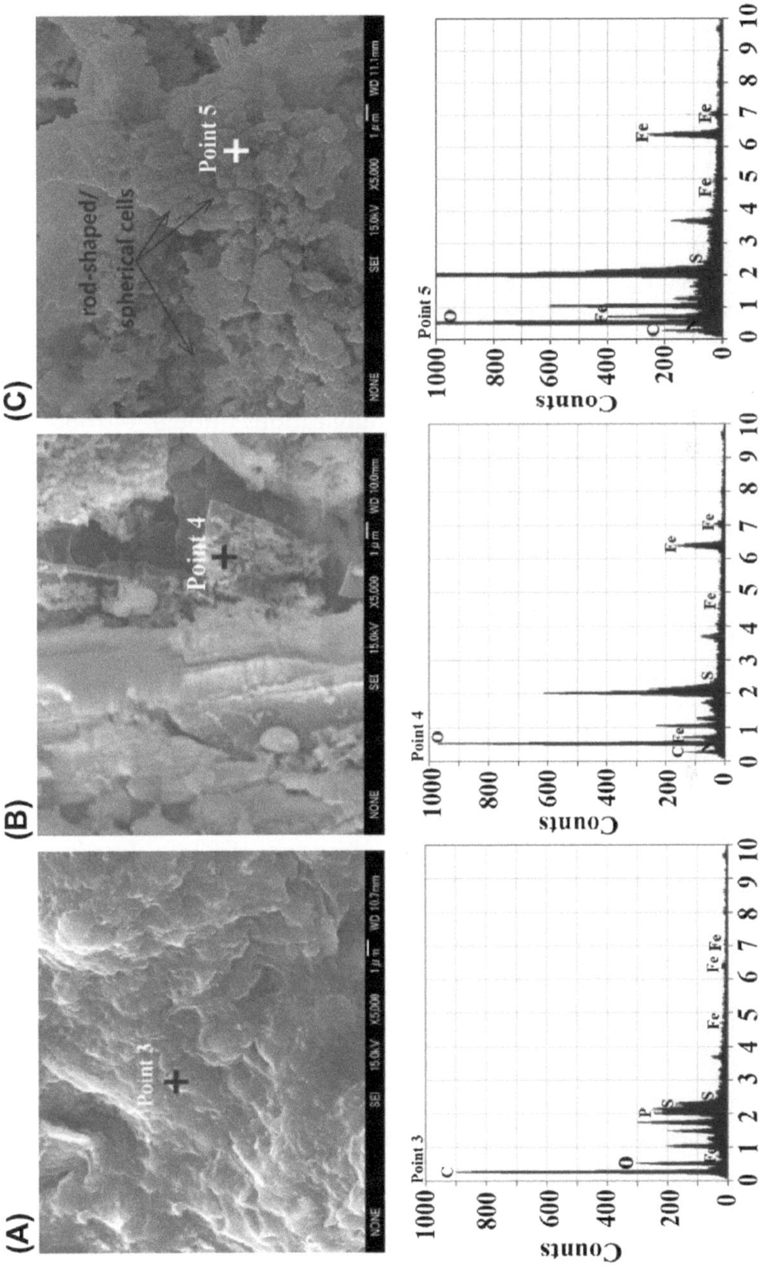

Figure 5.29 SEM-EDX analysis (+) of sludge exposed to 0 (A), 0.25 (B), and 1.00 g-ZVSI/g-VSS (C).

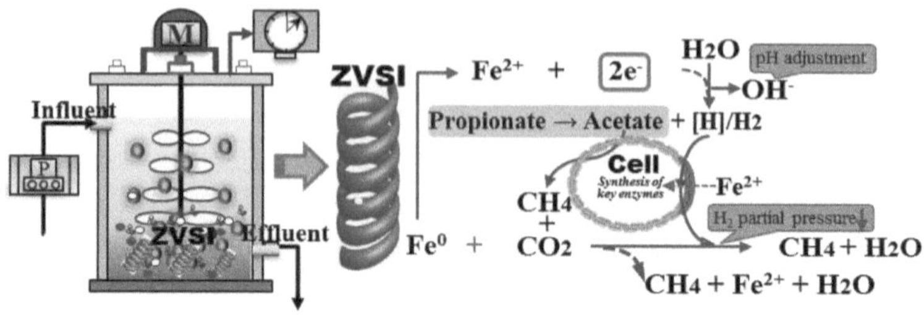

Figure 5.30 Proposed enhancement mechanisms for methane production from sludge in the presence of ZVSI.

nutrient, increases pH buffering capacity, decreases ammonia toxicity and accumulation of VFAs, and upgrades biochemical conditions for microbial growth. The potential increases in biogas production from sludge have been indeed documented in the previous publications through introducing different organic wastes like fruit-juice industrial waste, fatty waste water, pig/poultry manure, sugar beet pulp lixiviation, *Quercus serrata* chips, used disposable diapers, food waste, fat, oil, and grease.

One type of potential source for sludge codigestion is grass biomass. The use of grass for recovering bioenergy is promising not only because of its high C/N ratio (38.5—50) compared with typical SL (8.3) but also the plentiful yields. If sludge is codigested with grass as cosubstrate, the C/N ratio in anaerobic systems will be well balanced, which could eliminate the accumulation of VFAs and ammonia occurring with high or low C/N ratio, and increase methane production. Simultaneously, this codigestion may contribute to a more efficient use of grass and diminish the improper disposal such as in situ burning while avoiding the emission of GHG. In addition, codigestion is a complex microbiological process involving the multiple biochemical reactions. Understanding the principal kinetic of methane evolution involved in the digestion is thus definitely fundamental for designing the anaerobic digesters, predicting the behavior of anaerobic systems and optimizing the process.

This section evaluated the feasibility of mesophilic anaerobic codigestion of SL and grass *Egeria densa*, that is, one kind of aquatic plants harvested from a testing flume of water purification in Japan. The methane yield and the changes of different parameters (e.g., COD, protein, carbohydrate, TS, VS, pH, VFAs, TAN [total ammonia nitrogen], and [NH_3][free ammonia] under different VS ratios of WAS: *E. densa*) were all determined to assess the overall codigestion performance. Besides, three types of classical kinetic models (e.g., first-order kinetic, modified Gompertz, and Cone models) were chosen and compared to more accurately reveal the principal kinetics of methane production from the anaerobic codigestion.

5.4.1 Biochemical Methane Potential Tests and Kinetic Models

BMP assays were batch conducted in series of 120-mL sealed glass serum bottles with a working volume of 90 mL. Different combinations of WAS/*E. densa* cosubstrates were performed based on VS ratio: 1.0: 0, 0.9: 0.1, 0.8: 0.2, and 0.7: 0.3. The total amount of substrates added in each vial was all 0.75 g-VS. The biogas volume was measured periodically. First-order kinetic, modified Gompertz, and Cone models were used to describe the kinetics of methane production from the codigestion of WAS and *E. densa* (Table 5.6). The parameters f_d, k_{hyd}, R_m, λ, and n were estimated by fitting the BMP data into the models via Sigmaplot 12.5. Root mean-square prediction error (rMSPE) and Akaike's information criterion (AIC) were calculated to evaluate the accuracy of the used models.

5.4.2 Codigestion Effect on Methane Production

Methane yields with different mixing ratios of WAS and *E. densa* are shown in Fig. 5.31. Methane yield was strongly WAS: *E. densa* ratio-dependent. Final methane yield for sludge alone (i.e., 1.0: 0) was only about 176.36 ± 0.00 mL/g-VS$_{added}$ after 70 days of monodigestion because of the presence of stiff bacteria cells and EPS. Digesting sludge alone for methane production has been proven to be less judicious in many cases.

Comparatively, in the case of codigestion, although the content of methane in the biogas remained constant ranging from 60% to 65% irrespective of WAS: *E. densa* ratios applied, the addition of *E. densa* greatly upgraded the WAS methane production. A marked rise in methane yield to 181.69 ± 0.97 mL/g-VS$_{added}$ ($p = .0572$) and then to 187.63 ± 8.75 mL/g-VS$_{added}$ ($p = .1455$), respectively, were observed when WAS: *E. densa* ratios of 0.9: 0.1 and 0.8: 0.2 were employed. A further rise in *E. densa* dosage contributed more significantly to methane production. The highest methane yield averaged 198.32 ± 2.61 mL/g-VS$_{added}$ ($F_{observed} = 38{,}181.16$, $p = 2.62 \times 10^{-5} < .01$, $F_{crit} = 18.51$) for WAS: *E. densa* ratio of 0.7: 0.3, up to 18.72 ± 0.14% higher respective to that of WAS alone. The enhanced methane production achieved herein might be attributable to the rich trace nutrients, the well-balanced C/N ratio, as well as the alleviated toxicity of VFAs and ammonia caused by the coexistence of sludge and grass, which provided a stable environment for

Table 5.6 Models Used to Describe the Kinetics of Methane Production From the Codigestion of WAS and *E. densa*

Model	Equation	Parameters	N
First-order kinetic	$Y_{(t)} = f_d(1 - \exp(-k_{hyd}t)),\ t \geq 0$	f_d, k_{hyd}	2
Modified Gompertz	$Y_{(t)} = f_d \bullet \exp\left\{ - \exp\left[\frac{R_m e}{f_d}(\lambda - t) + 1 \right] \right\},\ t \geq 0$	f_d, R_m, λ	3
Cone	$Y_{(t)} = \frac{f_d}{1 + (k_{hyd}t)^{-n}},\ t > 0$	f_d, k_{hyd}, n	3

e, exponential 1, which is 2.71,828; f_d, maximum methane potential (mL/g-VS_{added}); k_{hyd}, hydrolysis rate constant (1/day); n, shape factor; N, number of model parameters; R_m, maximum methane production rate (mL/g-VS_{added} day); *t*, digestion time since the setup of BMP tests (days); $Y_{(t)}$, specific methane yield at a given time (mL/g-VS_{added}); and λ, lag-phase (days).

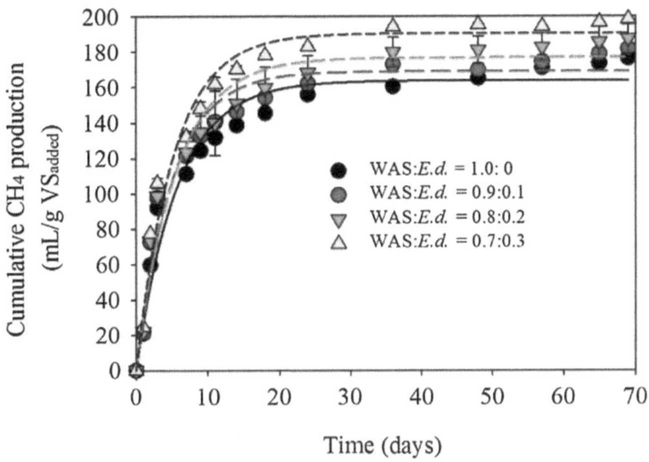

Figure 5.31 Cumulative methane production for codigestion of waste-activated sludge (WAS) with *E. densa* at mixing ratios of 1.0: 0, 0.9: 0.1, 0.8: 0.2, and 0.7: 0.3.

anaerobes to survive and work. Thus, addition of grass *E. densa* to sludge could profoundly benefit methanogenesis of sludge digester. Based on these results, grass *E. densa* can be considered an interesting substrate for codigestion with SL.

5.4.3 Codigestion Effect on TS and VS Removal

Fig. 5.32 illustrates the final TS and VS contents for the digestion under different WAS: *E. densa* ratios. Adding *E. densa* into sludge notably promoted fermentation performance with respect to the degradation of organic

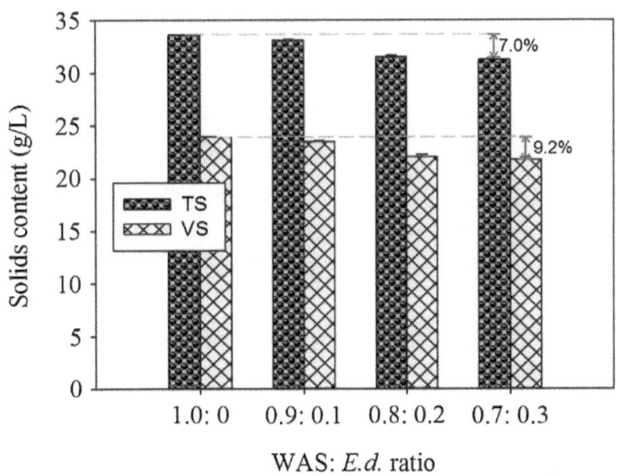

Figure 5.32 Total solid (TS) and volatile solid (VS) contents on day 70 for codigestion of WAS with *E. densa* at mixing ratios of 1.0: 0, 0.9: 0.1, 0.8: 0.2, and 0.7: 0.3.

solids, corresponding well with the results measured for methane production. The average TS and VS values after 70 days of anaerobic operation for sludge as a sole substrate were about 33.6 g/L and 23.97 g/L, respectively. However, both of them dropped sharply in accordance with the increasing *E. densa* dosage, particularly at WAS: *E. densa* of 0.7: 0.3. The TS and VS in this situation averaged 31.27 ± 0.19 g/L and 21.75 ± 0.22 g/L, reducing by up to 7.0% and 9.2% more compared with sludge alone (TS: $F_{observed} = 292.76$, $p = .0034 < .01$, $F_{crit} = 18.51$; and VS: $F_{observed} = 197.76$, $p = .0050 < .01$, $F_{crit} = 18.51$), respectively. It is clear that despite the presence of lignin, cellulose, and hemicellulose in grass as commonly reported in the literature; the grass *E. densa* used here likely depicted a better biodegradability compared to sludge.

Low biodegradability of the sludge is related to its complex physicochemical structure. Sludge flocs is a multimedium, composed of bacteria cells, organic (i.e., EPS) and inorganic matters. EPS, located on the outer layer of flocs, aside from enhancing cells' resistance to environmental stress, also control flocs' strength, dewaterability, and even biodegradability. The sludge with high content of EPS is usually difficult to dissolve or hydrolyze. Many different methods such as electricalalkali, onsite photocatalytic, high-frequency ultrasounds processes have been developed in recent years to promote the sludge methanogenesis. Despite the increased solubilization, the use of pretreatments does not always induce a considerable improvement in the methane production while requiring extra energy input. High energy consumption and low benefits consequently hinder the wide application of the pretreatment techniques in practice. In contrast, the codigestion with other types of biomass such as *E. densa* used in this study with negligible energy demand, besides adjusting C/N ratio, could also increase the portion of hydrolyzable organic matters in sludge fermentation systems, simultaneously compromising the unbeneficial impact exerted by sludge through a diluting effect. As a result of this, the mixture with *E. densa* was degraded more easily, eventually giving a more satisfying solids removal and output of methane-rich biofuel. In this sense, it is apparent that the codigestion of sludge with grass *E. densa* is a certainly feasible and cost-efficient approach for the organic wastes stabilization and renewable energy recovery and thus is more promising.

5.4.4 Release and Biodegradation of SCOD, Protein and Carbohydrate

The solubilization of substrate is commonly regarded as a key step before hydrolysis. SCOD, protein, and carbohydrate evolution in the course of operational time were analyzed to further verify the above noted findings, and the corresponding results are presented in Fig. 5.33. As expected, of course, the initial organic concentrations increased in accordance with the increasing *E. densa* dosage in the feed. The higher the *E. densa* was applied,

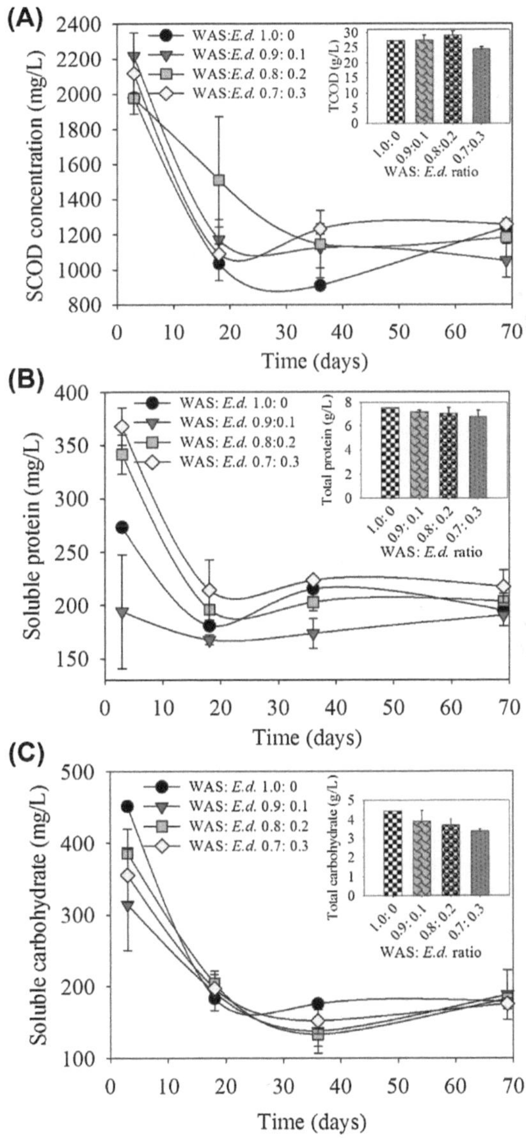

Figure 5.33 SCOD (A), protein (B) and carbohydrate (C) evolution throughout the codigestion operation of WAS with *E. densa*; the inserts of the figures gave the final total chemical oxygen demand (TCOD), total protein, and total carbohydrate concentrations (i.e., on day 70) for codigestion of WAS with *E. densa* at mixing ratios of 1.0: 0, 0.9: 0.1, 0.8: 0.2, and 0.7: 0.3.

the more the biodegradable organics were dissolved, again confirming the solubilization properties of *E. densa*. The initial SCOD for digesting sludge alone, for instance, was estimated to be around 1981.02 mg/L, and the corresponding value for protein was measured to be 273.44 mg/L approx., while it increased up to 2119.20 ± 229.76 mg/L when WAS: *E. densa* ratio was 0.7: 0.3, and the corresponding protein also climbed up to 367.73 ± 17.62 mg/L, making the cosubstrates more accessible to hydrolysis.

Initial carbohydrate, oppositely, exhibited a slight decline with the addition of *E. densa* presumably owing to the strong binding force of lignin and carbohydrates (e.g., cellulose and hemicellulose). Intense cross-linking with cellulose and hemicellulose appears to influence or limit the solubilization and effective degradation of carbohydrate. In spite of this, the overall benefits of *E. densa* in the decomposition and subsequent methanogenesis of cosubstrates were not compromised at all. SCOD, protein, and carbohydrate, as shown in Fig. 5.33, were rapidly consumed during the first 20 days of operation in conjunction with methane production, and then stabilized at different levels with negligible variations until the end of the run, while the final total easily biodegradable organics in the systems after 70 days of operation exhibited an evident decline with the addition of *E. densa* (inserts in Fig. 5.33). The final concentrations for TCOD, total protein, and total carbohydrate dropped significantly from 27.33 g/L to 24.50 ± 0.91 g/L ($p = .0480$), from 7.55 g/L to 6.78 ± 0.55 g/L ($p = .1881$), and from 4.45 g/L to 3.37 ± 0.12 g/L ($p = .0060$), respectively, with the reducing WAS: *E. densa* ratio from 1.0: 0 to 0.7: 0.3, in line with methane profiles (Fig. 5.31). This confirmed that there was a synergy between the input streams that accelerated the hydrolysis/conversion rate of cosubstrates and the subsequent fermentation performance.

5.4.5 Characteristics of pH and VFAs During Codigestion Process

The pH and VFAs (i.e., the important intermediate products of AD) are the two critical parameters controlling the overall stability of AD. Fig. 5.34 illustrates the pH and VFAs variations throughout the codigestion operation of sludge with *E. densa*. The pH decreased gradually during the first 20 days of digestion, particularly when in the presence of *E. densa*. The average pH on day 20 for sludge alone was 7.0 whereas those fed with *E. densa* averaged 6.7–6.8, possibly caused by the low pH of *E. densa* itself (i.e., 5.45 ± 0.07). Soon later, the pH in sludge alone kept decreasing while

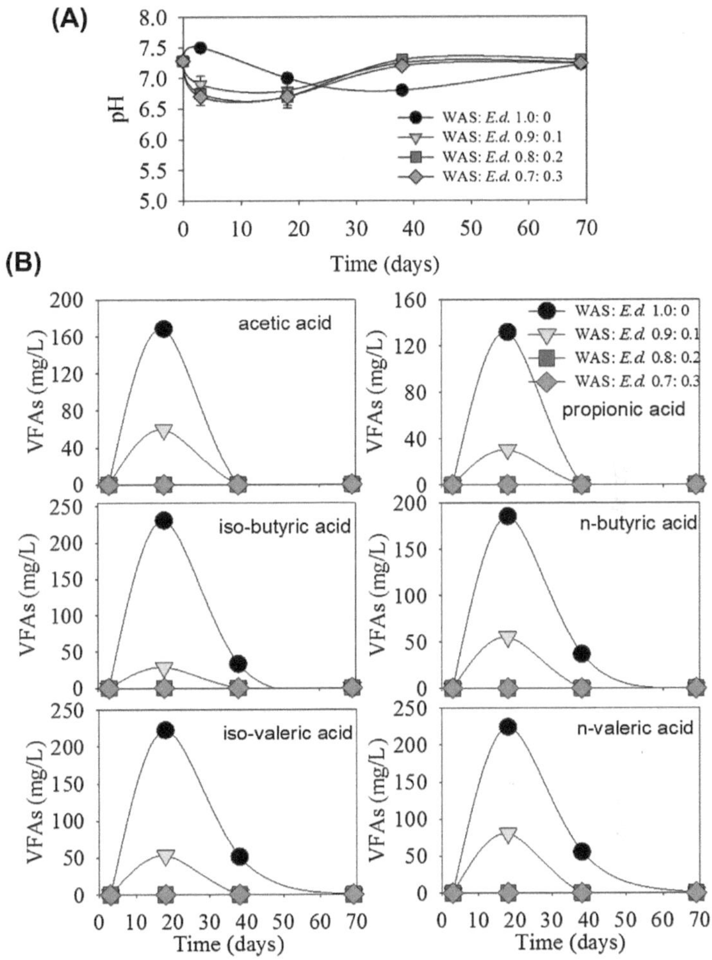

Figure 5.34 pH (A) and volatile fatty acids (VFAs) (B) variations throughout the codigestion operation of WAS with *E. densa*.

that with addition of *E. densa* started to recover, subsequently exceeding sludge alone and remaining at 7.2–7.3 until the termination of tests. Even though the codigestion resulted in a slight pH drop initially, this level was still within the optimal range (6.5–8.0) for the growth of methanogens and biogas production, hence no triggering any imbalance or instability of process.

Unlike the general relationship between pH and VFAs as commonly documented in the literature, VFAs monitoring along experimental time in the current study followed a pattern completely different from that of pH possibly due to the excellent buffering capacity of cosubstrates. As described in the figure, the accumulation of VFAs decreased in conjunction with the ascending WAS: *E. densa* ratio. The lowest VFAs values were measured in WAS: *E. densa* ratio of 0.7: 0.3 and the highest in WAS: *E. densa* ratio of 1.0: 0 (i.e., sludge alone). Acetic, propionic, iso-butyric, n-butyric, iso-valeric, and n-valeric acids in the monodigestion reached a maximum concentration of 168.51, 132.03, 230.91, 185.80, 222.85, and 224.50 mg/L, respectively, within the first 20 days, and then kept falling. Comparatively, the VFAs in all codigesters consistently stayed below the detection limit over time, indicating a stable digestion process. Notwithstanding the level of VFAs measured herein for digesting sludge alone was far lower than VFAs inhibition thresholds (4000−5000 mg/L), the considerably accumulated VFAs still revealed the inadequate methanogenesis in this condition as compared with the codigestion. The codigestion of sludge with grass *E. densa*, in contrast, not only provided the high pH buffering capacity, but also effectively abated the sharp accumulation of VFAs. VFAs produced during the codigestion process could be transformed into methane promptly and completely by methanogens with the help of *E. densa*. for avoiding acidification, inducing a stable and efficient performance and thus concomitant increase in methane potential. The above experimental data for the first time confirmed the beneficial role of *E. densa* in eliminating VFAs' accumulation during WAS AD, and evidenced the sustainability and attractiveness of codigesting sludge with grass *E. densa* as the cosubstrate.

5.4.6 Release of TNA and [NH_3] During Codigestion Process

TNA and [NH_3] under different WAS: *E. densa* ratios were analyzed in the end of experiments as important descriptors for process stability of the codigestion, and the results were graphically reported in Fig. 5.35. Similar to that on VFAs, WAS: *E. densa* ratio exerted a crucial influence on TNA and [NH_3]. Both TNA and [NH_3] concentrations decreased substantially with exception of WAS: *E. densa* 0.9: 0.1 when the *E. densa* ratio in the feedstock increased due to the low nitrogen content of the grass. TNA was 242.51 mg NH_4^+−N/L or so for the monodigestion on day 70, and it decreased down to 188.43 ± 33.73 mg NH_4^+−N/L with WAS: *E. densa* ratio of 0.8: 0.2, then further to 150.16 ± 68.91 mg NH_4^+/L at WAS: *E. densa* 0.7: 0.3. The estimated free [NH_3] concentration also declined from approx. 4.44 mg NH_3/L to 4.01 ± 0.21 mg NH_3/L, and then to 2.80 ± 1.08 mg NH_3/L.

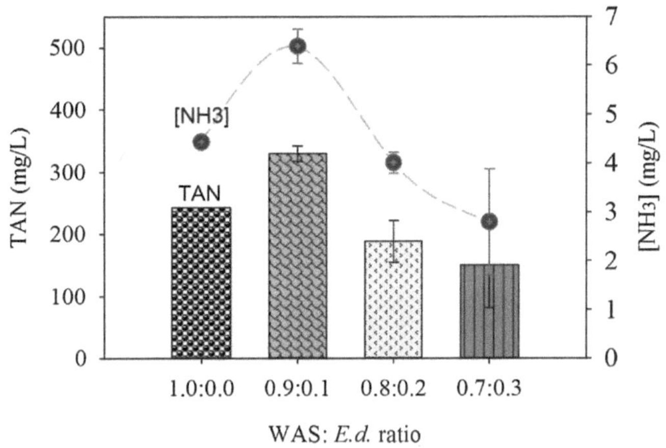

Figure 5.35 TNA and [NH_3] on day 70 for codigestion of WAS with *E. densa* at mixing ratios of 1.0: 0, 0.9: 0.1, 0.8: 0.2, and 0.7: 0.3.

It is well known that ammonia plays a critical role in the overall digestion efficiency. Low ammonia concentration, for example, can reduce the acetoclastic methanogenic activity and subsequent methane production; oppositely, the elevated ammonia accumulation liberates the overload free [NH_3], which can cause the intercellular pH variation, proton imbalance, and potassium deficiency, and even cease the growth of methanogens by diffusing into the cells. Most evidences have confirmed the high sensitivity of methanogens to free [NH_3]. Although there is a considerable discrepancy in the scientific literature with respect to the levels of free [NH_3] inhibition, the [NH_3] level obtained herein particularly in the presence of grass *E. densa* were extremely lower than the inhibitory range reported before, indicating that the inhibitory effect exerted by free [NH_3] did not occur in this study, and more importantly, the build-up of ammonia could be partially counteracted by means of adding *E. densa*. The codigestion of sludge and grass *E. densa* reduced the concentrations of potential inhibitors such as VFAs and ammonia, thus upgrading the system stability and the conversion of substrate to methane.

5.4.7 Kinetic Analysis of Codigestion Processes With Various Sludges to *Egeria densa* Ratios

Three types of models including first-order kinetic, modified Gompertz, and Cone models were subsequently employed to simulate the principal kinetic patterns of biogas production obtained from BMP tests. The kinetic parameters such as f_d, k_{hyd}, R_m, λ, and n were estimated based on the best fit of the studied models, and the results are summarized in Table 5.7.

Table 5.7 Estimated Parameters of the First-Order Kinetic, Modified Gompertz and Cone Models

Model	Parameters	$VS_{\text{waste-activated sludge}}$: $VS_{\textit{Egeria densa}}$			
		1.0: 0	**0.9: 0.1**	**0.8: 0.2**	**0.7: 0.3**
	$Y_{(70)\text{-measured}}$ (mL/g-VS_{added})	176.36 ± 0.00	181.69 ± 0.97	187.63 ± 8.75	198.32 ± 2.61
First-order kinetic	f_d (mL/g-VS_{added})	163.80	168.94	176.82	190.60
	$Y_{(70)\text{-predicted}}$ (mL/g-VS_{added})	163.80	168.94	176.82	190.60
	Diff. (%)	7.1	7.0	5.8	4.0
	k_{hyd} (/day)	0.175	0.200	0.182	0.194
	R^2	0.9656	0.9641	0.9639	0.9774
	Adjusted R^2	0.9599	0.9581	0.9579	0.9736
	S.E.E.	11.13	11.67	12.25	10.47
	rMSPE	9.95	10.49	10.95	9.43
	AIC	73.34	74.92	76.21	71.70
Modified Gompertz	f_d (mL/g-VS_{added})	163.45	168.24	176.88	189.63
	$Y_{(70)\text{-predicted}}$ (mL/g-VS_{added})	163.45	168.24	176.88	189.63
	Diff. (%)	7.3	7.4	5.7	4.4
	R_m (mL/g-VS_{added} day)	13.44	16.04	14.60	18.02
	λ (day)	0	0	0	0
	R^2	0.9410	0.9363	0.9383	0.9529
	Adjusted R^2	0.9312	0.9257	0.9281	0.9450
	S.E.E.	14.58	15.55	16.01	15.11
	rMSPE	17.70	18.15	19.86	18.06
	AIC	91.35	92.10	94.80	91.96

Cone	f_d (mL/g-VS_{added})	181.34	183.55	195.07	203.34
	$Y_{(70)\text{-predicted}}$ (mL/g-VS_{added})	171.70	176.56	185.24	196.41
	Diff. (%)	2.6	2.8	1.3	1.0
	k_{hyd} (/day)	0.244	0.288	0.257	0.290
	n	1.02	1.08	1.02	1.16
	R^2	0.9887	0.9864	0.9881	0.9907
	Adjusted R^2	0.9868	0.9841	0.9861	0.9892
	S.E.E.	6.38	7.20	7.04	6.70
	rMSPE	5.70	6.37	6.29	6.14
	AIC	57.37	60.69	60.33	59.60

Diff. (%), difference between $Y_{(70)\text{-predicted}}$ and $Y_{(70)\text{-measured}}$; R^2, correlation coefficient; *S.E.E.*, standard error of estimate; $Y_{(70)\text{-measured}}$, measured methane yield on day 70; $Y_{(70)\text{-predicted}}$, predicated methane yield on day 70.

For all the studied models, the predicted maximum methane potential (f_d) increased with increased portions of *E. densa* in the feed. The introduction of *E. densa* into the codigestion mixture resulted in not only higher f_d but also the elevated hydrolysis and methanogenesis rates. The hydrolysis rate constant (k_{hyd}) estimated for first-order kinetic and Cone models increased from 0.175 to 0.194/day and 0.244 to 0.290/day; meanwhile, the maximum methane production rate (R_m) estimated for modified Gompertz model increased from 13.44 to 18.02 mL/g-VS_{added} day, apparently indicating that the codigestion with *E. densa* can indeed facilitate the methane production from WAS.

It is worthwhile to note that although these classical models have been used for predicating the methane production in numerous lab- and full-scale trials, the suitability and precision of models always vary considerably depending on the experimental conditions, operating parameters as well as origin of inoculum and type of substrates employed. Moreover, for the same operational conditions, the reliability of different models might differ as evidenced hereafter. According to the parameters achieved from the model simulation, it can be observed that the highest difference (i.e., diff., %) between the predicated methane yield ($Y_{(70)\text{-predicted}}$) and measured methane yield ($Y_{(70)\text{-measured}}$) on day 70 was observed in modified Gompertz model (4.4−7.3%), followed by first-order kinetic model (4.0−7.1%). The lowest values originated from Cone model ranging from 1.0% to 2.6% for all studied substrates. Clearly, among the proposed models, Cone model best fitted the actual evolution of methane production, which was also strongly supported by its high correlation coefficient (R^2 and adjusted R^2, ranging 0.9864−0.9907, and 0.9841−0.9892, respectively), and the low standard error of estimate (standard error of estimate., i.e., 6.38−7.20). The rMSPE and AIC were calculated as well. Similar to the results of diff. (%), Cone model gave the lowest rMSPE and AIC values as shown in Table 5.7, proving the high suitability and adequate precision of this model. To further verify the above observations, the predicted values of methane yield from the Cone model were also plotted versus the actual values, as presented in Fig. 5.36 (the plot of predicted and actual values for first-order kinetic, modified Gompertz models were given in Fig. 5.37). Very high Pearson's correlation coefficients ($R_p = 0.9935-0.9955$, $p << .01$) under all studied substrates again suggested a reasonably adequate agreement between the actual and predicted values. The finding drawn from this study seems a little surprising since the first-order kinetic and modified Gompertz models are always

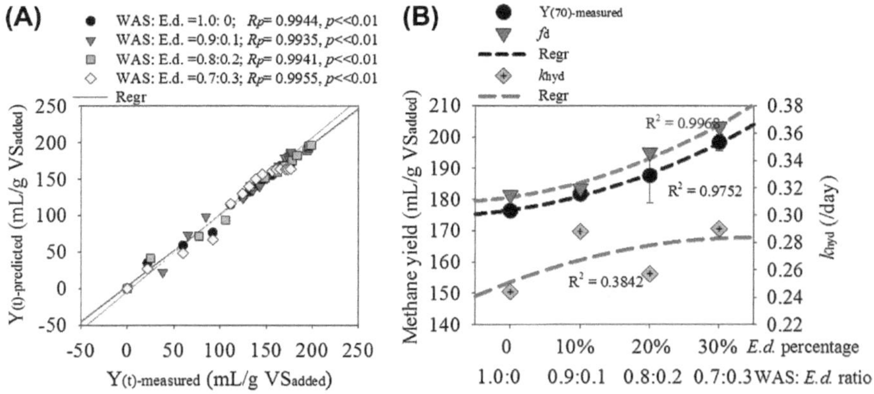

Figure 5.36 Cone model: (A) Pearson's correlation of predicated methane yields vs. actual values; and (B) quadric correlations between k_{hyd}, f_d, $Y_{(70)\text{-measured}}$ and *E. densa* percentage (or WAS: *E. densa* ratio).

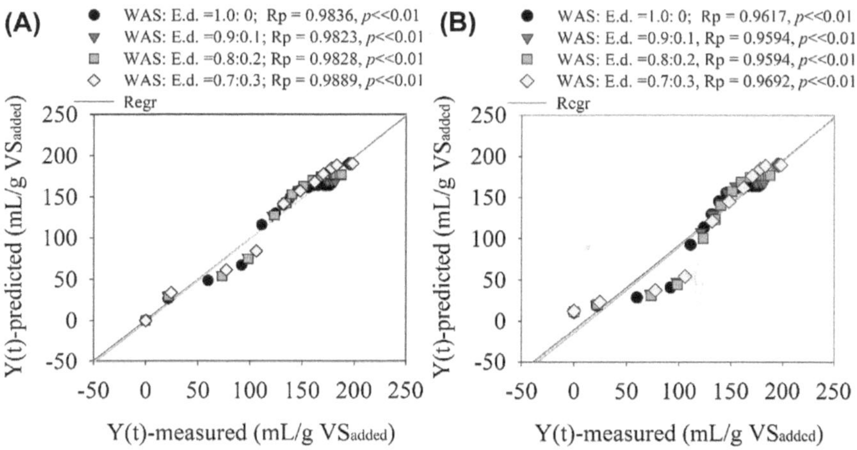

Figure 5.37 Pearson's correlation of predicated methane yields vs. actual values for first-order kinetic (A), and modified Gompertz (B) models.

regarded as the most suitable methods for the predication of biogas evolution. Most of the works undertaken in the past have often chosen the first-order kinetic or modified Gompertz models or both of them to model the experimental data, whereas placing very limited emphasis on Cone model possibly owing to the low familiarity with this model. For that reason, even until now the Cone model is still not widely used in spite of its high precision and credibility. Hence, it can be expected that this work will arouse much more interest from the scientific community in the potential development of Cone model in revealing the principal kinetics of methane production from the fermentation.

Besides, in order to provide an in-depth insight into the precise role of grass *E. densa* in biodegradation kinetics and methane production of sludge, the quadric correlation analysis between *E. densa* percentage (χ, %) and k_{hyd}, f_d as well as $Y_{(70)\text{-measured}}$, which were achieved from Cone model, were subsequently carried out. The results given in Fig. 5.36 show that there were fairly high quadric correlations existing between k_{hyd}, f_d, $Y_{(70)\text{-measured}}$, and *E. densa* percentage ($k_{hyd}(\chi) = 0.25 + 0.19\chi - 0.285\chi^2$, $R^2 = 0.3842$; $f_d(\chi) = 180.7 + 32.1\chi + 151.5\chi^2$, $R^2 = 0.9752$; and $Y_{(70)\text{-measured}}(\chi) = 176.6 + 31.6\chi + 134.0\chi^2$, $R^2 = 0.9968$, respectively). The increase of *E. densa* in the codigestion mixture clearly increased the hydrolysis rate (k_{hyd}) of particulate components and methane potential (f_d), which fully explained the upgraded solids removal and final output of methane ($Y_{(70)\text{-measured}}$). In the view of this, it, therefore, can be said that the choice of *E. densa* as a cosubstrate for the sludge fermentation is technically feasible. Overall, this study would offer a valuable basis for practical applications of biomass stabilization and simultaneous renewable biofuel recovery via codigestion.

5.5 APPLICATION OF AGED REFUSE TO BOOST BIO-HYDROGEN PRODUCTION FROM FOOD WASTE AND SEWAGE SLUDGE

Hydrogen is a promising energy source due to its clean product and high-energy yield (122 kJ/g). Electrolysis of water, steam reforming of hydrocarbons, and autothermal processes are well-known methods for hydrogen gas production, but cost-intensive due to high-energy requirements. Environmental friendly anaerobic bio-hydrogen fermentation from organic wastes has gained increasing concern due to energy generation while reducing the wastes.

The quantity of food wastes amounts to 65,000 t/d in China, accounting for 15.3% of municipal solid wastes, which are the main source of decay, odor, and leachate in collection and transportation processes due to its high VS (85—95% of TS) and moisture content (75—85%). Anaerobic acidification of food waste produces VFAs, hydrogen, and other intermediates. Not only hydrogen gas itself is a beneficial energy source, but also VFAs can be further digested for methane production. Food wastes are more suitable for bio-hydrogen production, especially its cofermentation with SL because the latter contains rich easily hydrolyzable carbohydrate. The chief challenge to bio-hydrogen production is the prompt acidification caused by the accumulation of VFAs. The shift of predominant metabolic flow from hydrogen- and acid-forming pathway to solvent-forming pathway may be delayed by control of dilution rates.

The refuse in landfills becomes aged or stabilized after 8−10 years of placement, and the resultant partly or fully stabilized refuse thus obtained in this work is referred to as aged refuse (AR). The AR contains a wide spectrum and large quantity of microorganisms which have been proved to have a strong decomposition capability for both biodegradable and refractory organic matter present in some wastewaters. In this section, the role of the AR in avoiding prompt acidification and enhancing hydrogen production from anaerobic cofermentation of food waste and SL was investigated.

5.5.1 Test Materials and Experimental Procedures

Food wastes, collected from a dining hall at Tongji University, in Shanghai, China, were smashed after manual separation of plastic scoops, cups, and wooden chopsticks, etc., before use as the substrate for bio-hydrogen production. The basic properties were as follows: moisture content 85.3%, TS 14.7%, VS 88.8%-TS. SL was taken from an anaerobic digester in a WWTP, in Shanghai. The moisture content, VS, pH, alkalinity, and VSS were 82%, 36%, 7.18, 687 mg/L $CaCO_3$, 15.7 g/L, respectively.

AR was excavated from a 10-year-old closed landfill compartment in the *Shanghai Refuse Landfill*. The larger inorganic substances such as stones, glass, plastic, etc., were manually separated. The air-dried AR was screened by a conventional mechanical screener and a granule size of less than 2 mm was employed in this work (Table 5.8).

The exploratory and pasteurization tests were conducted in 13 serum bottles (250 mL), seven for the former and six for the latter, respectively (Table 5.9). Moisture content was adjusted to 90% and the proportion of sludge to food waste was 30: 100 (dry basis), while in the pasteurization test, Bottle 5# had no sludge. Fresh food wastes of 20 g were carefully put into each serum bottle except Bottle 6#.

In exploratory test, granule-activated carbon (GAC) and powder-activated carbon (PAC) were added individually as porous materials for comparison. In order to analyze the biological influence of domestic microorganisms in AR on the codigestion of food waste and sludge, a condition of

Table 5.8 Characteristics of the Aged Refuse (AR)

Parameter	Value
Moisture content (%)	32.0
Organic content (g/kg)	102.5
pH	7.81
TN (%)	0.41
TP (%)	1.02
Cationic exchange capacity (CEC) (meq/100 g)	65.4
Specific surface area (m^2/g)	5.46
Total bacteria (cfu/g)	$9-10^6$
Degree of porosity (%)	44
Specific gravity (g/cm^3)	1.89

160°C and 2 h was selected in pasteurization test to eliminate the spore of microorganisms. The experimental installation was equipped with a gas-gathering bottle for the detection of biogas volume (Fig. 5.38). The serum bottles were placed in an incubator at 36°C. Biogas was sampled from the silica gel tube and collected by the drainage of saturated brine. The modified Gompertz equation was adopted to model the hydrogen production course as

$$H = H_p \exp\left\{-\exp\left[\frac{R_{max^e}}{H_p}(\lambda - t) + 1\right]\right\} \quad (5.13)$$

where H was accumulative hydrogen production (mL), H_p the ultimate hydrogen production potential (mL), R_{max} the maximal hydrogen production rate (mL/g-VS·h), λ the lag-phase time (h), and e = 2.718281828.

5.5.2 Effects of Porous Materials on Hydrogen Fermentation From Food Wastes

5.5.2.1 Effects of Porous Materials on Hydrogen Production

The first set of batch experiments was performed in the presence of AR, GAC, and PAC, and the accumulative hydrogen concentration curves are shown in Fig. 5.39. It was found that AR (50%) had a remarkable effect on bio-hydrogen fermentation. The cumulative hydrogen concentration had reached to 26.6% (50% AR, the fifth day) and 8.4% (25% AR, the seventh day), in comparison with 6.37%, 3.4%, and 4.8% in the bottles with

Table 5.9 Feed Schedule of the Two Groups of Tests for Aged Refuse-Boosted Bio-Hydrogen Production From Food Waste and Sewage Sludge

Item	Bottle	Aged Refuse (g)	Food Waste (g)	Sewage Sludge (g)
Exploratory test	5% AR + SL	1	20	4.9
	15% AR + SL	3	20	4.9
	25% AR + SL	5	20	4.9
	50% AR + SL	10	20	4.9
	Sludge	—	—	4.9
	GAC + SL	3	20	4.9
	PAC + SL	3	20	4.9
Pasteurization test	1#	10 (×)	20 (√)	4.9 (√)
	2#	10 (×)	20 (√)	4.9 (×)
	3#	10 (×)	20 (×)	4.9 (√)
	4#	10 (√)	20 (√)	4.9 (√)
	5#	10 (×)	20 (√)	—
	6#	10 (×)	—	4.9 (√)

'√', high-temperature sterilization treatment (HTS); '×', not HTS treated; '—', not added; *AR*, aged refuse; *GAC*, granule-activated carbon; *PAC*, powder-activated carbon; and *SL*, sludge.

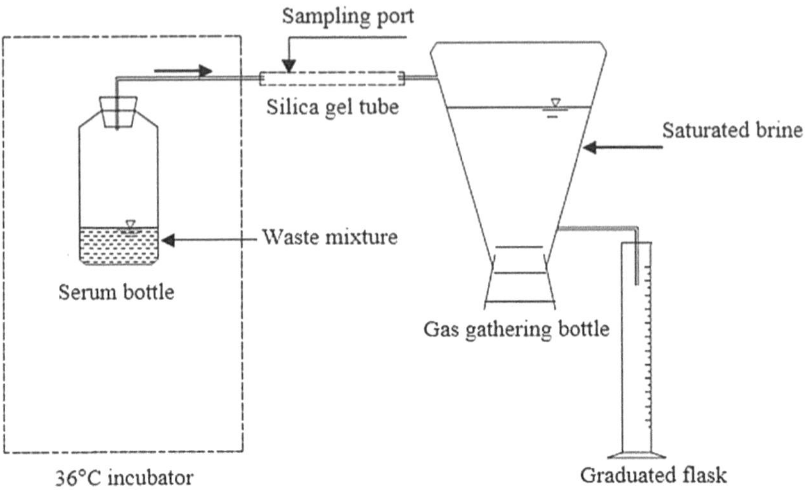

Figure 5.38 Experimental installation for aged refuse-boosted bio-hydrogen production from food waste and sewage sludge.

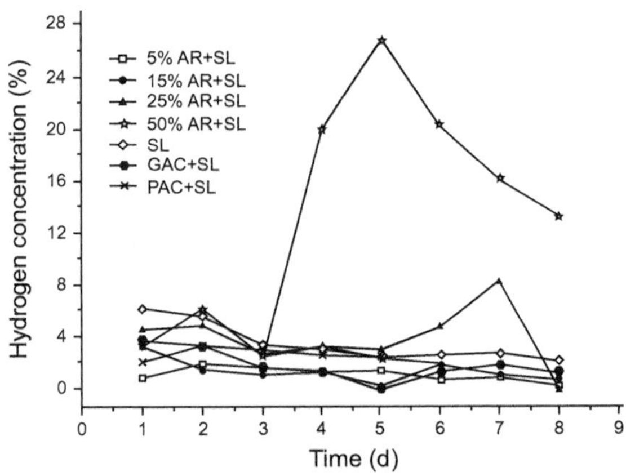

Figure 5.39 Hydrogen strengths of food waste and sludge with different porous materials. *AR*, aged refuse; *GAC*, granule-activated carbon; *PAC*, powder-activated carbon; and *SL*, sludge.

heat-treated sewage, PAC, and GAC, respectively. On the other hand, hydrogen pressure increased quickly if the gas outlet was not installed, which gradually restrained the activity of hydrogen-producing bacteria. It was found that the acetic acid decreased with slight fluctuation in the liquid products, while the propionic acid increased remarkably from third to eighth day (Table 5.10). Hence, it was inferred that a branch reaction occurred

Table 5.10 Individual VFAs on Different Porous Materials for Hydrogen Production From Food Waste and Sludge

Additive	Time (d)	Hac (mg/L)	HPr (mg/L)	iHBu (mg/L)	nHBu (mg/L)	HVa (mg/L)	B/A
5% AR + SL	1	—	—	—	—	—	—
	2	2,583.50	—	—	—	—	—
	3	7,299.00	—	—	—	—	—
	4	5,840.50	—	—	—	—	—
	5	2,814.50	99.50	287.00	—	—	0.10
	6	2,508.00	809.50	—	—	—	—
	7	2,889.50	246.50	—	—	—	—
	8	3,611.50	323.00	—	385.50	—	0.11
15% AR + SL	1	10,647.00	—	—	—	—	—
	2	6,854.50	—	—	—	—	—
	3	2,718.50	304.50	—	—	—	—
	4	20,370.50	—	—	—	—	—
	5	3,381.50	154.50	—	120.50	—	0.04
	6	4,921.50	343.50	—	256.00	—	0.05
	7	5,329.50	364.50	—	242.00	—	0.05
	8	2,992.50	—	—	1,570.50	—	0.52
25% AR + SL	1	10,490.50	—	—	—	—	—
	2	14,542.00	—	—	—	—	—
	3	2,183.00	303.00	—	—	—	—
	4	10,446.00	7,697.00	—	21,111.50	706.50	2.02
	5	4,108.00	214.00	—	125.50	—	0.03
	6	4,431.50	—	—	9,098.00	—	2.05
	7	1,263.50	411.50	—	12,037.50	—	9.53
	8	473.00	4,294.50	—	13,199.50	—	2.79
50% AR + SL	1	12,416.50	—	—	—	—	—
	2	44,207.50	18,592.00	—	—	—	—
	3	364.00	606.50	—	218.50	—	0.06
	4	28,597.50	1,243.50	—	1,998.50	—	0.07
	5	1,329.00	3,416.00	—	28,115.00	316.00	21.16
	6	4,381.00	3,597.00	—	6,653.50	—	1.52
	7	4,407.00	4,319.00	271.50	9,902.00	320.50	2.31
	8	11,986.50	380.50	—	4,996.00	—	0.42
Sludge	1	7,348.00	—	—	—	—	—
	2	2,443.00	—	—	—	—	—
	3	1,623.00	258.00	—	190.50	—	0.12
	4	4,581.50	—	—	—	—	—

(Continued)

Table 5.10 Individual VFAs on Different Porous Materials for Hydrogen Production From Food Waste and Sludge—cont'd

Additive	Time (d)	Hac (mg/L)	HPr (mg/L)	iHBu (mg/L)	nHBu (mg/L)	HVa (mg/L)	B/A
	5	1,242.00	—	—	—	—	—
	6	4,312.50	439.50	—	737.00	—	0.17
	7	2,751.50	242.00	—	—	—	—
	8	670.50	155.50	—	—	—	—
GAC + SL	1	—	—	—	—	—	—
	2	—	—	—	—	—	—
	3	356.50	—	—	—	—	—
	4	11,398.50	—	—	—	—	—
	5	1,395.00	—	—	—	—	—
	6	2,002.00	—	—	—	—	—
	7	2,152.00	201.00	—	—	—	—
	8	214.00	527.00	—	951.50	—	0.44
PAC + SL	1	752.50	—	—	—	—	—
	2	—	—	—	—	—	—
	3	1,296.50	—	—	—	—	—
	4	3,562.00	—	—	—	—	—
	5	2,087.50	22.00	—	—	—	—
	6	2,404.00	—	—	—	—	—
	7	3,598.50	247.00	—	—	—	—
	8	6,415.50	863.00	—	158.00	—	0.25

B/A, (nHBu + iHBu)/Hac; *Hac*, acetic acid; *HPr*, propionic acid; *Hva*, valeric acid; *iHBu*, iso butyric acid; *nHBu*, normal butyric acid; and '—', not detected.

(Eq. 5.14) for the adaptation to the increased hydrogen press, which was a main feedback inhibition mechanism in bio-hydrogen production process, and the chief microorganisms involved in this bypass reaction were *Propionibacteria*. This could cause the decrease of hydrogen production, especially in the bottles with 50% and 25% of AR:

$$2H_2 + \text{glucose} \rightarrow 2\text{propionate} + 2H_2O \tag{5.14}$$

5.5.2.2 Effects of Porous Materials on pH and Metabolites of Food Waste Bio-Hydrogen Fermentation

The pH of the cofermentation bottle with AR (50%) increased from 4.5 to 6 due to its alkalescent characteristic. During the 7-day fermentation, the butyric acid/acetic acid (B/A) increased from 2.02 to 2.79 in the food waste added with 25% AR while a mass of propionic acid was produced causing the termination of biohydrogen production. When a 50% dose of AR

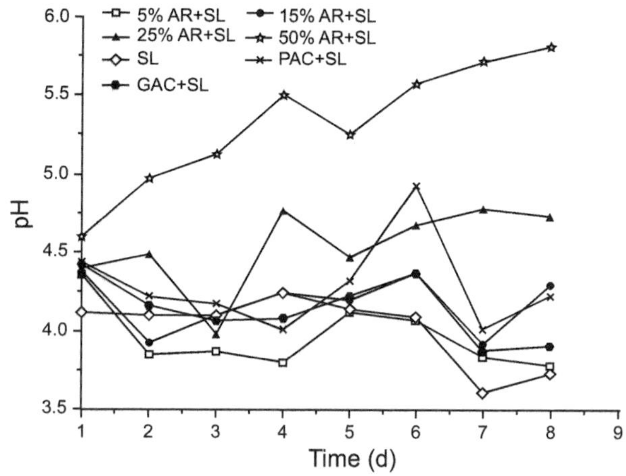

Figure 5.40 pH changes at different porous materials for hydrogen production from food waste and sludge.

was added into the food waste, B/A ratio rose to 21.16 and then decreased to 0.42. The distribution of metabolites generated during hydrogen fermentation is a crucial signal in the assessment of the efficiency of hydrogen-production course. In the bottles added with AR (>15%), the propionic acid increased significantly in the third day while hydrogen concentration started to decrease (Table 5.10). Hence, it was inferred that the propionic acid was an important indicator of the variation of bio-hydrogen process. In fact, in the bottle with 25% AR, the butyric acid became the main ingredient in VFA. However, in the system with 50% AR, the butyric acid decreased drastically while high-hydrogen pressure caused the increase of propionic acid. In contrast, hydrogen production of food waste without AR was terminated for the accumulation of acetic acid as shown in Fig. 5.40.

5.5.3 Biological Effects of AR on Hydrogen Production From Food Waste

5.5.3.1 Bio-Hydrogen Production

As shown in Fig. 5.41A, hydrogen concentrations ascended distinctly from the first day, and reached the climax (53.3%) on the second day in Bottle 1#. The accumulative hydrogen production reached to 176.3 mL/g-TS (Fig. 5.41B) (i.e., 195.6 mL/g-VS) (Fig. 5.42, Table 5.11) on the fourth day. But hydrogen production decreased gradually since the fifth day, due to the metabolism of *Propionibacterium*. Bottle 4# witnessed a low hydrogen

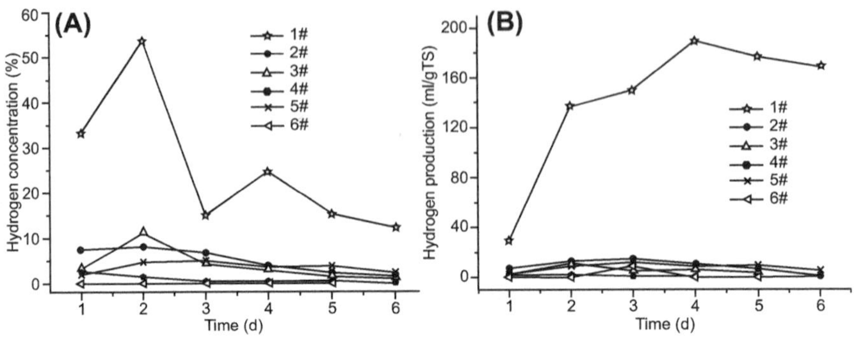

Figure 5.41 Hydrogen concentration (A) and accumulative yield (B) for hydrogen production from food waste and sludge.

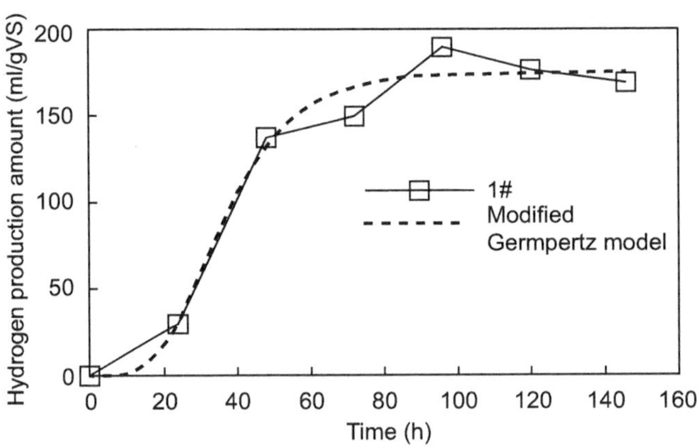

Figure 5.42 Modified Germpertz model of hydrogen production from food waste and sludge.

Table 5.11 Modeling Results of Bottle 1# for Hydrogen Production From Food Waste and Sludge

H (mL/g-VS)	H_p (mL/g-VS)	R_{max} (mL/g-VS·h)	*l* (h)	R_2
187.46	193.85	94.35	15.28	0.9821

production, indicating that the hydrogen-producing bacteria derived mainly from the microorganisms in the AR. However, only 15.6 mL/g-VS and 11.1 mL/g-VS were obtained in Bottles 2# and 3#, respectively. Hence, the activity of the bacteria in AR might be inactivated due to potential competition with the bacteria in the feedstock, which exceeded in quantity and exerted depressed matters. Moreover, since the hydrogen yield of Bottle

2# surpassed that of Bottle 3#, it can be inferred that residual microorganism in SL was not only a rival to native bacteria in AR, but also an underlying hydrogen consumer if not heat treated. Besides, low hydrogen productions were also obtained in Bottle 5# (without sludge) and Bottle 6# (without food waste), indicating the significance of the unitive fermentation of food waste, sludge, and AR.

5.5.3.2 Change of pH and VFA

Butyric acid type fermentation is considered to be one of the most efficient ways in bio-hydrogen production, especially for *Clostridium* with endogenesis spore. The quick increase of propionic acid usually implies the low hydrogen yield. In the four unitive fermentation bottles containing food waste, sludge, and AR, similar quantities of total VFA were found in Bottles 2# (3.49×10^4 mg/L) and 3# (3.47×10^4 mg/L) (Fig. 5.43). However, slightly different valeric acid was produced in Bottles 2# (1.02×10^4 mg/L) and 3# (9.60×10^3 mg/L) while the remaining four bottles did not produce, indicating the difference of biological property of fermentation bottles. The lowest total VFA was obtained in Bottle 5# (1.90×10^4 mg/L) probably due to the deficiency or unbalance of the microbial development factors.

Bottles 1# and 6# experienced typically butyric acid-type fermentation, with total acetic and butyric acid reaching 78%, 75%, and pH value 4.70, 4.77 (Fig. 5.43), respectively. However, as sewage SL was not a proper

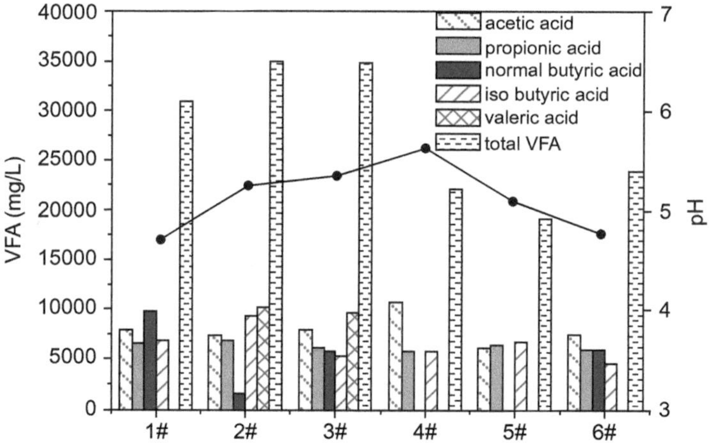

Figure 5.43 Individual VFAs and pH for hydrogen production from food waste and sludge.

substrate for hydrogen fermentation, lower hydrogen production was obtained in Bottle 6#. The variation of VFA in Bottles 2# and 3# indicated that both bottles experienced mixing acid type fermentation, with pH values reaching 5.24 and 5.35, respectively, which also indicated the specific competition between communities in substrates and those in AR. The largest acetic acid yield (1.02×10^4 mg/L) was found in Bottle 4#, and hydrogen production course ceased at pH = 5.63 (Fig. 5.41B), as AR acted only as porous material after sterilization treatment. Propionic acid-type fermentation was performed in Bottle 5# with pH = 5.09, indicating that the microorganisms in AR cannot be effectively activated if food waste was taken as a single feedstock for bio-hydrogen production.

5.5.4 Fundamentals of the Addition of Aged Refuse in Promoting Hydrogen Recovery

Organic waste is a plentiful source for fermentative hydrogen production as shown in the following equations:

$$\text{Glucose } 2H_2O \rightarrow \text{Acetate} + 2CO_2 + 4H_2, \quad \Delta G = -184.2 \text{ kJ} \tag{5.15}$$

$$\text{Glucose} \rightarrow \text{Butyrate} + 2CO_2 + 2H_2, \quad \Delta G = -184.2 \text{ kJ} \tag{5.16}$$

Carbohydrates are the favorite substrates for hydrogen-producing bacteria while hydrogen was hardly produced from protein and lipid. In recent years, various organic matters including food waste, activated sludge, wastewaters, and municipal solid wastes as anaerobic substrates for hydrogen production have been examined. The maximum hydrogen-production potential and hydrogen-production rate reported are in the range of 25.9–180 mL/g-VS and 10.14–111.2 mL-H_2/g-VSS·h, respectively. In this work, the AR excavated from a 10-year-old closed landfill compartment was employed as a biological inoculum for the enhancement of bio-hydrogen production from food wastes, and a good result was obtained (i.e., 94.35 mL/g-VS·h).

The AR experiences a series of aerobic, facultative, and anaerobic decomposition courses in landfill, and consequently, it possesses various kinds and large quantities of effective microorganisms. Hydrogen-producing bacteria participate in the initial anaerobic decomposition of refuse and predominant community gradually evolves with the variation of organic materials in refuse landfill. The mining of AR causes the change of moisture content and anaerobic microenvironment since the AR is exposed to the air and air-dried ones. Hence, most of the anaerobic microorganisms die while those microorganisms such as *Fusobacterium* survive due to the formation of spores. Hence, the enrichment of hydrogen-producing bacteria and the elimination of hydrogen consumers such as methanogen and *Thiobacillus* are accomplished simultaneously. Gradually, the interactions between marinating and washing by leaching, biological degradation, and mineral soakage make the landfilled refuse transform into alkalescent humic substance attached with rich effective microorganisms. It contains mineral and organic solid materials, water, and gas, becoming a good porous microbial carrier with high-mass transfer capacity finally. In general, the addition of AR can highly promote hydrogen fermentation efficiency in comparison with sludge, as it functions chiefly as an inoculum due to its rich effective microorganisms, instead of a simple porous additive due to its alkalescent carrier with high-mass transfer capacity.

A two-stage sequential fermentation process with a food waste disposal capacity of 100 kg/d had been successfully adopted for simultaneous generation of both bio-hydrogen and methane. In the acidification phase, AR was employed for the enhancement of the bio-hydrogen production from food waste, and then the hydrogen fermentation residue was further digested for methane production. The potential of food waste for hydrogen and methane production arrived at 150 mL/g-VS and 630 mL/g-VS, respectively.

5.6 TOXIC EFFECT OF ANTIBIOTIC CEFALEXIN ON METHANE PRODUCTION FROM SEWAGE SLUDGE

A high proportion of antibiotics can enter aquatic ecosystems directly, through the discharge of wastewater from the pharmaceutical industries, households, and hospitals, or indirectly, by leaching and runoff of agricultural soils amended with manure from livestock. The release of antibiotics is becoming of considerable concern since it can cause negative effects not only on the environment but also on the human health. Cefalexin (CLX), as one of the most prescribed antibiotics, belongs to the first-generation cephalosporin-type used in human medicine to treat aspiratory path infection, urine pathway infection, and skin tissue infection, and also in veterinary medicine due to its enhanced oral activity. The widespread and overuse of CLX has led to its detection in the aquatic environment. For example, the concentration of CLX present in municipal wastewater was found to be 339.4–375 ng/L, suggesting that the potential effect of antibiotics in the environment requires further attention.

Owing to heavy use of pharmaceutical drugs, high concentrations in the slurry influent to WWTPs will inevitably result in the effects on the mixed population of anaerobic bacteria during the anaerobic process due to these molecules' biological residual activities. It is therefore of considerable importance to make great efforts for better understanding the long-term effect and exact role of CLX on/in digestion efficiency and stability of AD process. The inhibitory effect of CLX was assessed by monitoring methane production, variations of EPS, and accumulation of individual VFAs. Removal of CLX was evaluated based on measured bulk phase concentrations of CLX.

5.6.1 Test Materials and Experimental Procedures

Antibiotic cefalexin (CLX, $C_{16}H_{17}N_3O_4S \cdot H_2O$) used was purchased from Wako (99% purity), Japan. BMP assay was conducted at $35 \pm 1°C$ to evaluate the potential effect of CLX on WAS AD. The modified Gompertz model was employed for analysis of the BMP data. The maximum cumulative biogas production (H_m, mL), the maximum methane production rate (R_m, mL/d), and the lag-phase (λ, day) were estimated by fitting the data into the Gompertz model.

5.6.2 Effect of Cefalexin (CLX) on Methane Production

The methane production during AD is commonly used as an indicator of anaerobic digester performance. Cumulative methane production (CMP) during AD trials for each dose of CLX tested (50, 200, 400, 600, 1000, and 2000 mg/L) was recorded and compared with no-CLX control. Fig. 5.44 shows the variation of the methane yield (mL) obtained against digestion time (day) and corresponding methane content. The modified

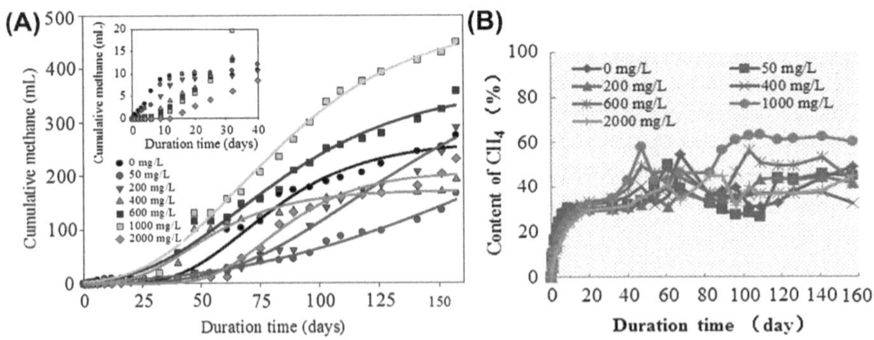

Figure 5.44 Net cumulative methane production (A) and methane content (B) at different cefalexin (CLX) concentrations in the course of duration time.

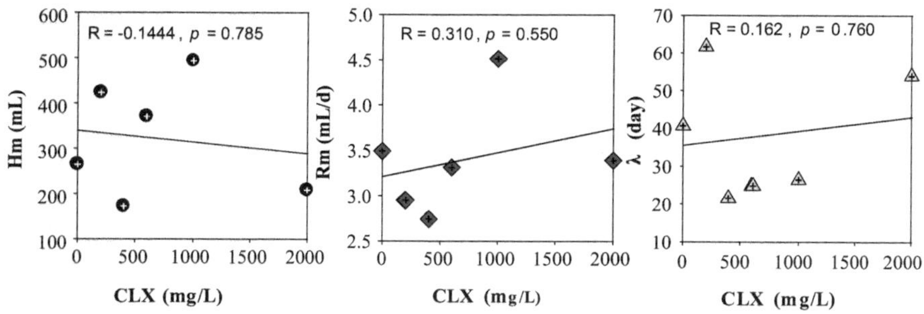

Figure 5.45 Pearson's correlation between H_m, R_m, λ, and CLX dosages used.

Gompertz model was used to fit the data of each CMP curve. Model lines presented in the figure are based on the best fit of the maximum cumulative gas production (H_m, mL), maximum methane production rate (R_m, mL/d), and the lag-phase time (λ, day) (Fig. 5.45).

As illustrated in Fig. 5.44, a dose effect was found; with the rising concentration of CLX the net methane yield decreased substantially at the beginning (0−25 days), suggesting that addition of CLX has exerted some pressure on methanogens, which reduced the activity of microbes. As a result of this, a temporary adverse impact on methane generation occurred and a lag-phase, thus, could be observed at this stage. As the AD process advanced, the acclimation to CLX increased the tolerance of anaerobic inoculum to CLX toxicity. In this respect, the activities of methanogens were effectively revived after a short period of acclimation depending on the CLX concentrations applied; the adverse impact induced by CLX ceased gradually and the generation of methane began to increase for the duration of experiment (i.e., 26−157 days). For instance, the methane yield with 200 and 600 mg/L of CLX at the end of fermentation (157 days) was 5.7% and 30.3% higher than the CLX-free control, respectively. Moreover, methane yield could be further increased and reached the peak of 450.84 mL as 1000 mg/L of CLX was added, increasing by around 63.8% with respect to the control. But a sharp drop of methane yield was detected when increasing CLX into 2000 mg/L.

Evidently the effect of CLX was no strict dose-dependent, in conformity with the results obtained from Pearson's correlation ($p > .05$) (Fig. 5.45). From the aforementioned results, it can be noted that the increased CLX concentration does not always cause the elevated inhibition to methane production; the use of an appropriate dose of CLX might be, sometimes, helpful to enhance methane production. Unlike the commonly believed adverse effect related to antibiotics before, the results demonstrated the beneficial role of CLX in methane production, providing new insights into its true environmental impacts.

5.6.3 Effect of Cefalexin (CLX) on Specific Methanogenic Activity

Sludge AD usually undergoes solubilization of particulate organic-carbon, hydrolysis, acidification, and methanation. Methanogenesis, as the central step of digestion, is considered the rate-limiting for fermentation of soluble substrates and methane production. It can be anticipated that different activity of methane-producing microbes, relying on their sensitivity toward CLX, would differ in methane yield during anaerobic process. To further evidence this, the effect of CLX on the methanation was assessed through methane production activity (MPA) tests using synthetic wastewaters with sodium acetate as substrate (2000 mg-COD/L). CLX dose applied during MPA tests ranged from 0 to 2000 mg/L and digestion period was 16 days. The MPA was then calculated according to the slope of methanogensis in the test and was expressed in terms of g-COD per gram VSS per day (Fig. 5.46).

As given in Fig. 5.46A, the addition of CLX yielded an evident but limited influence on the activity of methanogens. An abrupt drop in the MPA from 0.010 to 0.007 g-COD/g-VSS-day could be found as CLX concentration increased from 0 to 200 mg/L. When further elevating CLX dose to 2000 mg/L, however, the inhibitory effect was not provoked too much and the MPA exhibited a slight but gradual reduction. The results agreed well with the trend of methane yield during the beginning of BMP tests in the inset of Fig. 5.44 but apparently conflicted with the ultimate methane yield at day 157. Considering these observations in the context of activity of methanogens as well as ultimate methane yield, it can be speculated that the impact of CLX is time-dependent, showing that short-term tests (e.g., MPA procedure used herein and previously) might not be suitable for disclosing the true role of antibiotics in the sludge AD. Incipient inhibition observed in both MPA tests (16 days) and the first 25 days of BMP might be due to the inhibition of CLX to protein synthesis involved in methanogens metabolisms because of a sudden change in external environment, which temporarily deteriorated their MPA and reduced methane production.

As the incubation progressed (i.e., 26–157 days), the methanogens and precursor fermentative populations were well established, and the

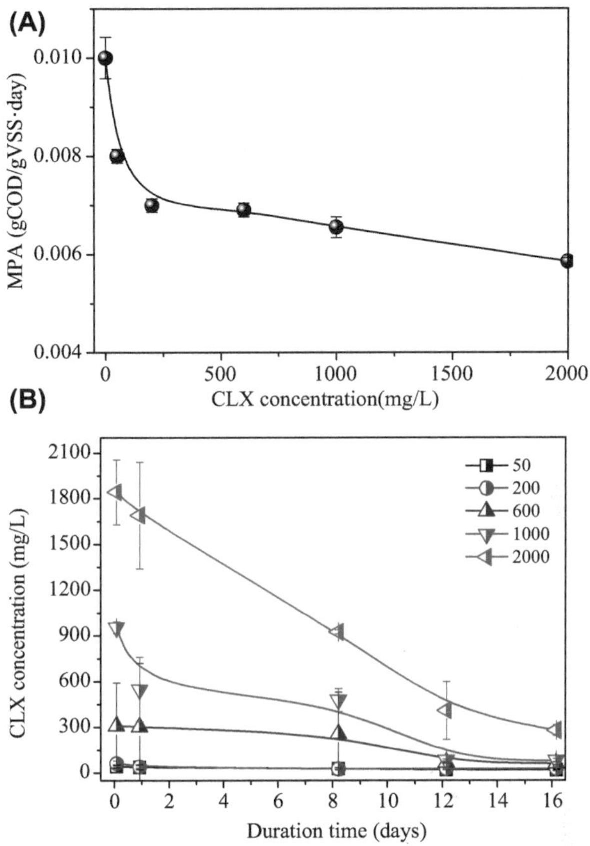

Figure 5.46 (A) Methane production activity (MPA) values at different CLX concentrations, and (B) removal of CLX during 16 days of MPA tests.

production of methane therefore increased. Mechanisms of microbes' resistance to antibiotics might be ascribed to the development of resistance in existing bacteria and/or an increase in abundance of resistance genes, etc. Although several antibiotic resistance genes in anaerobic bacteria have been identified in the literature, knowledge regarding CLX resistance genes is very rare. Nevertheless, the enhanced methane yield after 157 days of digestion could still be best explained by the acclimation of inoculum to CLX or the presence of CLX resistance genes. This is because that daily use of CLX to treat aspiratory path infection, urine pathway infection, and skin tissue infection will inevitably bring about its liberation into environmental receptors (e.g., WWTPs), hence benefiting the growth of CLX resistant microorganisms in WAS.

5.6.4 Cefalexin (CLX) Biodegradation and Removal During Anaerobic Digestion

The levels of CLX in MPA tests throughout the entire digestion were detected to ascertain its fate and possible removal in the anaerobic digester, and the resulting data are depicted in Fig. 5.46B. It can be found that the CLX was effectively removed after 16 days of digestion even the considerable inhibition of MPA was achieved as noted in Fig. 5.46A, and the corresponding removal efficiencies of CLX in the different assays ranged from approximately 65−92% by the end of experiments depending on the concentrations used. The transformation and substantial removal of CLX obtained might result from two reasons as follows: (1) CLX, containing one NH_2 and one COOH active groups that provide grafting sites and have a broad antimicrobial spectrum, is biodegradable and could be metabolized by the one or more of active bacterial groups during biological process; and (2) on the other hand, part of CLX remained entrapped/adsorbed by sludge flocs or solid matters associated with their strong affinity, similar to other types of cations, particles, and organics. Owing to the rapid CLX mineralization or bioadsorption (e.g., via EPS), the performance of CLX-bearing anaerobic systems can recover after a short-term inhibition, which, therefore, would allow methane production to resume during the latter phase of exposure experiments.

5.6.5 Effect of Cefalexin (CLX) on TSS and VSS Reduction

Fig. 5.47 presents the changes of TSS and VSS along with rising cefalexin dose after 157 days of fermentation. The initial TSS and VSS were

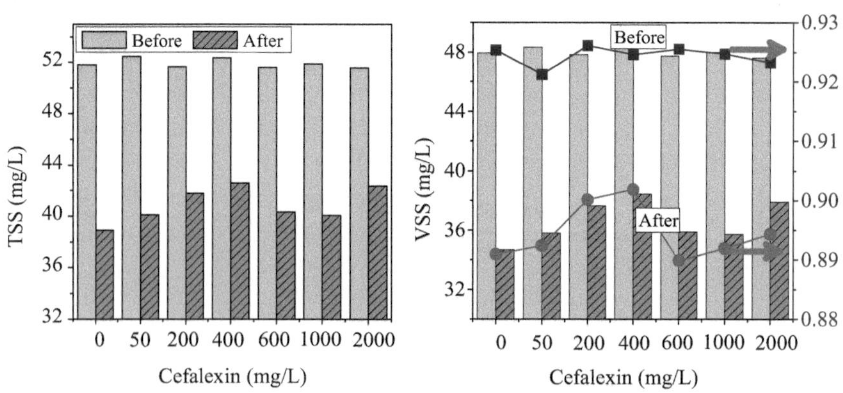

Figure 5.47 TSS and VSS removals after 157 days of digestion.

determined to be 51.9 ± 0.4 mg/L and 48.0 ± 0.3 mg/L on average, respectively. Regardless of the concentrations of CLX, the marked reduction in both TSS and VSS took place over the duration of digestion. By the termination of tests, TSS and VSS decreased to 38.9–42.6 mg/L and 34.6–38.4 mg/L, with corresponding removal efficiencies ranging from 17.8% to 24.8%, and 20.4–25.9%, respectively. The aggressive and deleterious effects that CLX exerted on digestion of sludge, as expected, were not considerable. Comparatively, higher TSS and VSS removals occurred when 1000 mg/L CLX was dosed with a lower VSS/TSS ratio of 0.89, in accordance to the data regarding methane yields observed before. This might be mainly attributable to the striking differences in biodegradability of antibiotics investigated and doses, as well as the types of inoculums used and their resistance capabilities to toxins. Further evidences to support the resistance mechanisms of anaerobic microbes to CLX are thus required.

5.6.6 Role of Extracellular Polymeric Substances (EPS)

5.6.6.1 Variations of Different EPS Fractions During Anaerobic Digestion

It is commonly believed that EPS have a crucial influence on sludge dewatering and sedimentation. Over the past decade, the essential role played by EPS in the toxicity response of microbial aggregates to chemical toxins has drawn much more attention in scientific communities. Information about EPS changes during AD thus is very important for better understanding of the toxicity response of microbes to antibiotics. Fig. 5.48 compares the overall variations of different EPS fractions including slime EPS, LB-EPS,

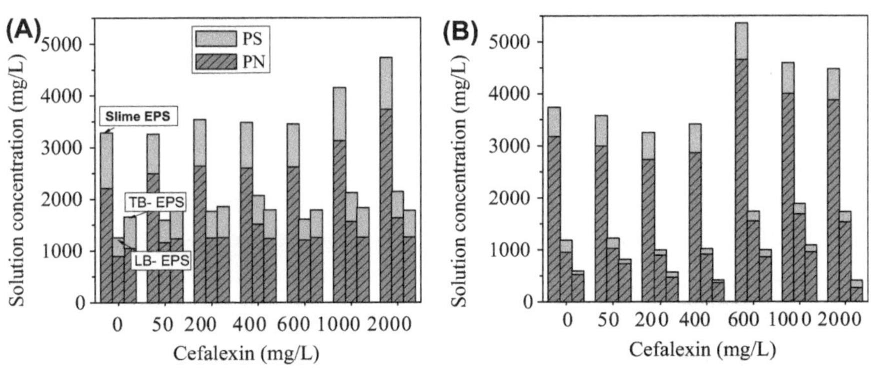

Figure 5.48 Variations of different EPS fractions before (A) and after (B) 157 days of digestion.

and TB-EPS before and after 157 days of digestion. The addition of CLX contributed to a stepwise rise in EPS as presented in Fig. 5.48A, in particular in slime EPS at the beginning of digestion. It might be associated with the acute toxicity of CLX, which resulted in the lysis of some planktonic cells in bulk liquor or absorbed on the surface of sludge flocs and subsequent release of biopolymers. Of course, possibly because of the toxic impact, a short but clear lag-phase in methane production depending on the doses of CLX added occurred (inset of Fig. 5.44).

Furthermore, it is generally believed that the sludge with high EPS, which can embed microcolonies through both electrostatic and hydrophobic forces that contribute to the strength and cohesiveness of a floc, would have a stiff texture and thereby be more difficult to be hydrolyzed during anaerobic process. The effective rupture of EPS is thereby regarded as the prerequisite for sludge solubilization and succeeding AD. As expected, an appreciable decrease in EPS, especially in both LB-EPS and TB-EPS was identified irrespective of the concentrations of CLX (Fig. 5.48B), implying that sludge had undergone normal decomposition and fermentation processes. However, it does not mean that higher the rupture of EPS, the better the performance of digestion is. EPS, which accumulate on the bacterial cell surface, beyond influencing sludge dewatering can also serve as a protective layer against external stress and as carbon and energy source during starvation. Hence, if EPS matrix are subjected to severe destruction or are removed markedly during anaerobic process, more microbial cells trapped within sludge flocs as well as cell membranes would be liberated and become more exposed to toxins. In this situation, CLX would become more permeable and may diffuse into the cells faster, thus causing the awful lysis and death of microbes. The hypothesis is strongly supported by the findings herein (Fig. 5.48B) that the sludge flocs, at moderate doses of CLX ranging from 600 to 1000 mg/L, had an abundant but not superfluous level of EPS (i.e., LB-EPS and TB-EPS), which corresponded to the increased methane outputs (Fig. 5.44). This could be due to the fact that the presence of EPS can adsorb/bind CLX and retard its diffusion into the cells to some extent; on the other hand, anaerobic microbes, after experiencing an adaption process to CLX, can retain viability at the high concentration of CLX and utilize CLX as carbon sources for self-growth and metabolisms while, in turn, stimulating the excretion of EPS and subsequent fermentation of WAS. Because of the complementary roles played by EPS and CLX, the performance of WAS AD thus maintained stable, and methane production was promoted greatly.

Obviously, the results obtained here confirm the idea that the floc EPS matrix plays a vital role in protecting microbial cells from the chemical toxins through impeding the access of toxins into cells. Thanks to the key functions of EPS demonstrated herein and in the literature, the microbes involved in sludge flocs survive well even under environmental shocks, which is the core reason for enhanced methane yield observed here as well.

5.6.6.2 UV–Vis Spectroscopic Properties of Different EPS Fractions

To further approach to the protective mechanisms of EPS, UV–Vis spectroscopy was used to characterize EPS derived from the sludge before and after 157 days of digestion. A Typical UV–Vis spectra of different EPS fractions is given in Fig. 5.49. UV–Vis absorbance values decreased steadily according the wavelength. The spectra usually include two main absorption bonds: the first one located at λ of 200–223 nm and the second present at λ between 250 and 270 nm. The first bond belongs to the presence of carboxylic bonds while the second bond might be attributable to n→π* and

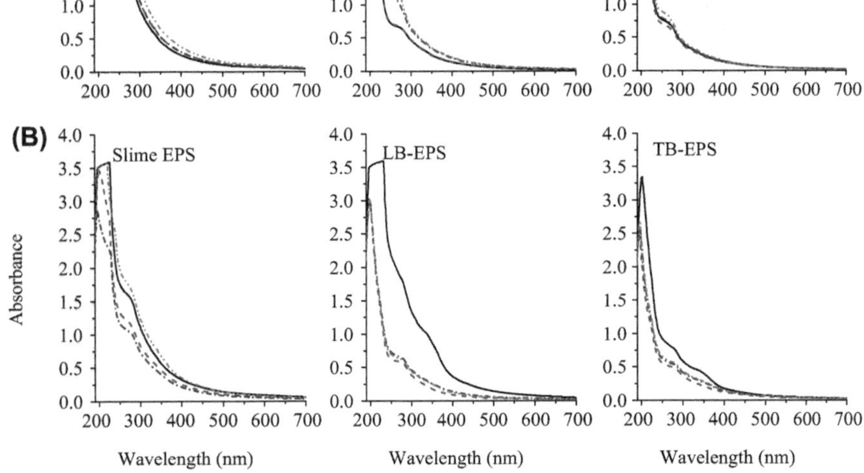

Figure 5.49 Relative quantity of EPS derived from the sludge before (A) and after (B) 157 days of anaerobic digestion.

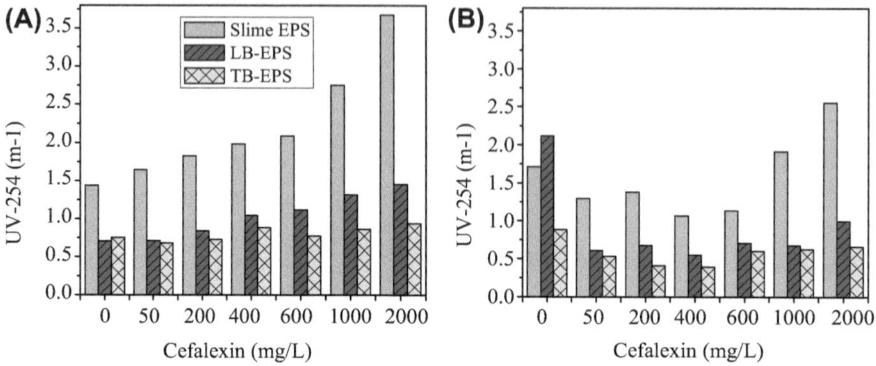

Figure 5.50 UV—Vis absorbance of the dissolved organic matters in the EPS of sludge before (A) and after (B) 157 days of anaerobic digestion.

$\pi \rightarrow \pi^{\star}$ electron transition of C=O and C=C structure of substituted benzenes or polyphenols.

To better understand the potential differences in UV—Vis spectroscopic properties of sludge EPS before and after anaerobic process, the absorption of dissolved organic matters at 254 nm (i.e., a rough indicator of the density of conjugated aromatic molecules [UV-254]) is thus given in detail (Fig. 5.50). The change of UV-254 reflects the possible degradation, synthesis, or transformation of key organics (e.g., humic and aromatic compounds) involved in slime EPS, LB-EPS, and TB-EPS. As exhibited in Fig. 5.50A, UV-254 values of different EPS fractions before AD exhibited a similar trend to EPS (i.e., PN and PS) with increasing CLX doses (Fig. 5.48A), demonstrating the deleterious influences of CLX on damage of partial planktonic cells. Only discrepancy between UV-254 values and measured EPS seems that the former increased comparatively fast with increased CLX. One possible explanation might be that the absorbance in 254 nm achieved here not only resulted from PN and PS, but also from CLX. Furthermore, Fig. 5.50B shows the variation of absorbance of sludge flocs' EPS after AD. As expected, the absorbance of different EPS fractions dropped first, followed by a considerable increase as the applied CLX elevated from 600 to 1000 mg/L. High absorbance observed herein, agreeing well with increased EPS content (Fig. 5.48B), strongly proved the central role of EPS in keeping stable performance of sludge fermentation and stimulating methane production (see Fig. 5.51). Based on the aforementioned observations, it can be noted that unlike the commonly accepted adverse effect, this study demonstrated the beneficial role of CLX in restoration of EPS matrix,

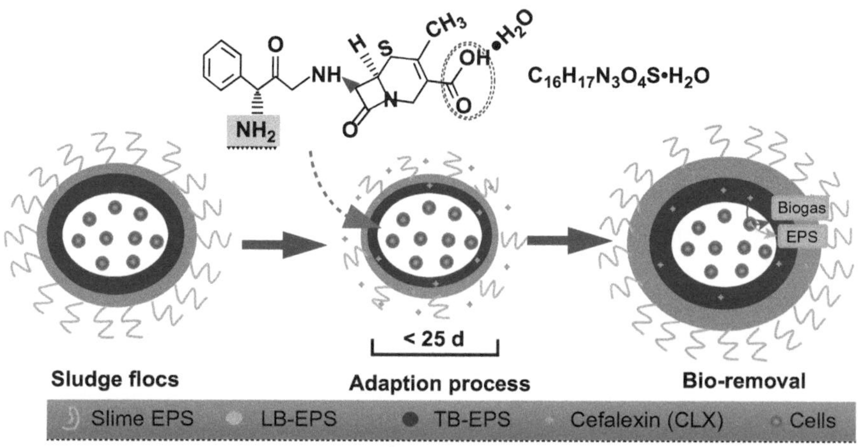

Figure 5.51 Role of EPS in the CLX-treated sludge anaerobic digestion.

stability of microbial consortia, as well as methane production, providing innovative insights into its true environmental impacts.

5.6.7 Volatile Fatty Acids (VFAs) Assessment

VFAs are considered as the key parameter influencing the performance of AD. VFAs, including acetic, propionic, butyric, iso-butyric, valeric, and iso-valeric acids, before and after AD were determined. Fig. 5.52 gives the

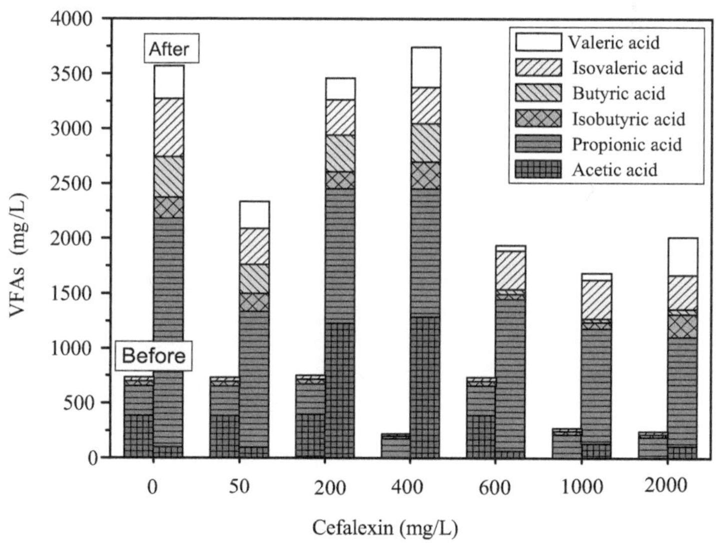

Figure 5.52 Variations of VFAs under different dosages of CLX before and after 157 days of digestion.

initial and final VFAs concentrations in this test. At the beginning of the test, the total VFAs level (sum of acetic, propionic, butyric, iso-butyric, valeric, and iso-valeric acids) were usually below 750 mg/L independent of the CLX concentration with acetic and propionic acids being the dominant components and even in some cases were less than 250 mg/L. After 157 days of digestion, the VFAs accumulation occurred, particularly when low concentrations of CLX were used. Total VFAs reached up to 3569, 2337, 3457, and 3746 mg/L with CLX concentrations of 0, 50, 200, and 400 mg/L, respectively. This elevated VFAs' level suggested that antibiotic CLX reduced methane output at low concentrations mainly via hindering the utilization of VFAs by homoacetogenic bacteria or acetoclastic methanogens instead of the conversion of amino acids, sugars, and fatty acids into VFAs through fermentative or acid-producing bacteria. Presumably because of the significant inhibition of methanogenesis process induced by antibiotics, methane production thus dropped.

Comparatively, the VFAs accumulation seemed to be effectively alleviated, as CLX was further increased to 600 mg/L and above. Efficient conversion of VFAs, in particular acetic acids, by acetoclastic methanogens such as *Methanosaeta* spp. into methane and carbon dioxide was observed, which corresponded to the pH variation (Fig. 5.53) and increased methane yield. Whereas, even though rapid use of acetic acids took place throughout the duration of experiment, the degradation of other VFAs, especially propionic acids, was not actively progressing. Propionic acids consistently kept a high level by the end of test regardless of applied CLX dose, possibly owing to

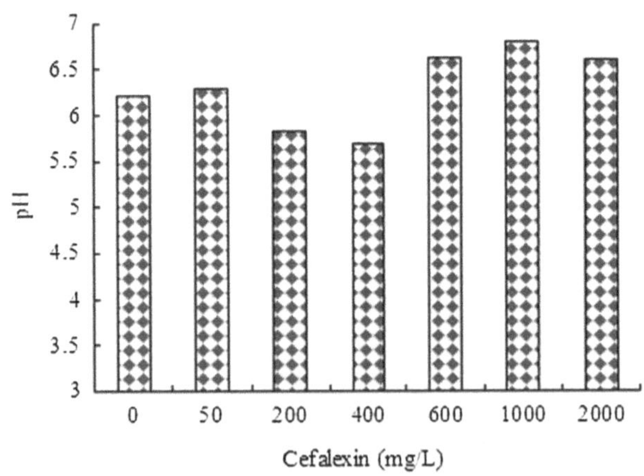

Figure 5.53 pH under different dosages of CLX after 157 days of digestion.

high Gibbs free energy value ($\Delta G^{0'}$) of +76.1 kJ/mol as well as the relatively low activity of propionate degraders such as Gram-negative bacteria (e.g., *Syntrophobacter* species). Nevertheless, the process stability of the digesters in this situation has been maintained without obvious upset or failure except the lag-phase observed before, providing new information about the true roles of CLX on methane generation from sludge.

CHAPTER SIX

Pollution Control and Recycling of Sludge in Sanitary Landfill

6.1 DESIGN AND CONSTRUCTION OF A SLUDGE LANDFILL

Dumping of sludge must be prohibited. Sanitary landfill is a cost-effective and fast disposal technique for dewatered sludge, especially for those with a lower organic matter (OM) and higher inorganic substance. The sanitary landfill can be constructed and operated as that of municipal solid wastes, but the moisture of sludge to be filled should be below 60%. Leachate and landfill gas should be properly treated as done in the municipal solid wastes' landfill. Qualified design of a landfill can prevent, or reduce as far as possible, negative effects on the ecological environment as well as the resultant risks to human health arising from the landfilling of sludge. There are four critical components in a standard landfill: lining systems, leachate collection systems, gas collection systems as well as capping and closure systems. This chapter first outlined the design and operation of standard landfill elements. Details on the stabilization process and risk assessment of sludge landfill were discussed.

6.1.1 Lining Systems

The principal functions of a lining system are to prevent the leachate from polluting the underground water and to limit landfill gas migration.

1. Bottom Lining Systems
   This lining system from the bottom-up are composed of the following components (Fig. 6.1):
   a. Compacted layer with a minimum fall of 2% toward the main leachate collection sump and compacted density of equal to or higher than 93% (clay or aged refuse);
   b. The lower protective layer (staple fiber geotextile with mass per unit area of 400 g/m^2);
   c. Primary liner (a minimum 1.5 mm-thick high-density polyethylene (HDPE) geomembrane or equivalent flexible membrane liner);

Pollution Control and Resource Recovery: Sewage Sludge
ISBN: 978-0-12-811639-5
http://dx.doi.org/10.1016/B978-0-12-811639-5.00006-1

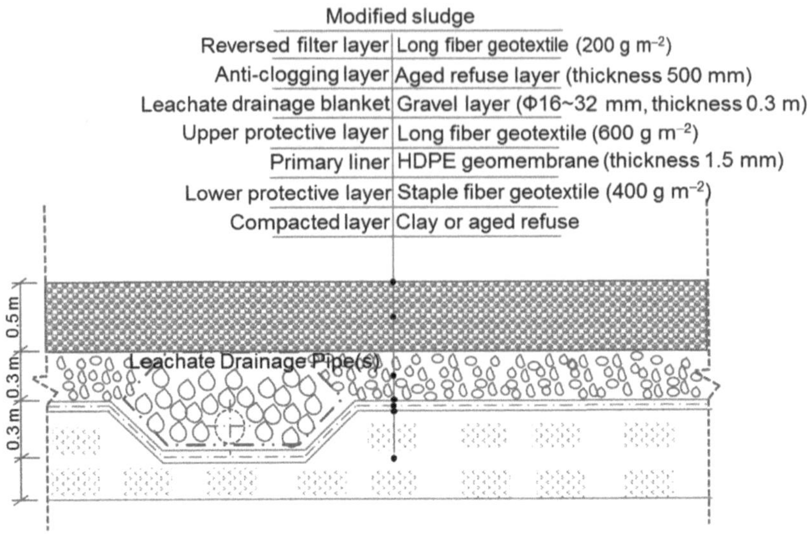

Figure 6.1 Basal lining systems for sludge sanitary landfills.

Figure 6.2 Construction of horizontal leachate drainage lines in a sludge landfill demonstration project in Shanghai, China.

d. The upper protective layer (long-fiber geotextile with mass per unit area of 600 g/m^2);

e. Leachate drainage blanket (a minimum 0.3 m-thick layer of gravel with a diameter of 16–32 mm) (Fig. 6.2);

f. Anticlogging layer (an aged refuse layer of minimum 0.5 m thickness);

   g. Reversed filter layer (long-fiber geotextile with mass per unit area of 200 g/m^2).
2. Slope Lining Systems
   The liner system should at minimum consist of the following components from the bottom up (Fig. 6.3):
   a. Compacted layer with the slope of 1:2 and compacted density of equal to or higher than 93% (clay or aged refuse);
   b. The lower protective layer (staple fiber geotextile with mass per unit area of 400 g/m^2);
   c. Primary liner (a minimum 1.5 mm-thick HDPE geomembrane or equivalent flexible membrane liner);
   d. The upper protective and drainage layer (5 mm-thick composite drainage geotextile);

The seaming must be performed with the slope geomembrane as a top sheet and the bottom one as a bottom sheet in order for the lining systems to perform satisfactorily. Fusion wedge welding is the commonly used method in the field for jointing two adjacent overlapped geomembranes. It creates a fusion weld by heating the facing overlapped surfaces and then pressing both together while in a molten state (Fig. 6.4). The seaming is done across slopes and the minimum distance from the toe of slope for geomembrane seaming is 1.5 m. In addition, to prevent slippage, the geomembrane should be bedded in an anchor trench. The trench should be 1.2 m back from the top edge of the slope, 0.8 m deep, and 0.8 wide. Fig. 6.5 gives the typical anchor trench.

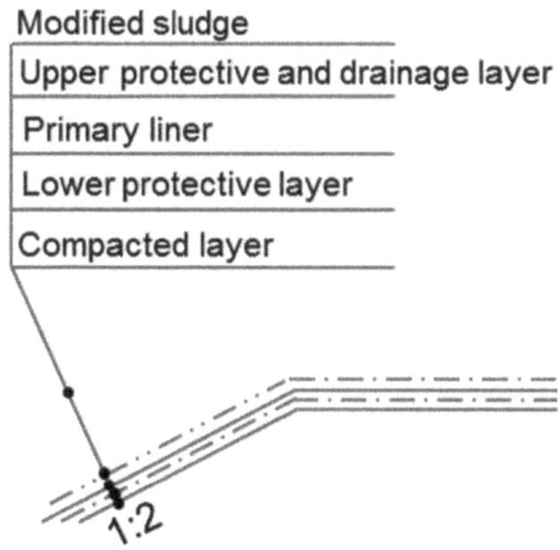

Figure 6.3 Slope lining systems for sludge sanitary landfill.

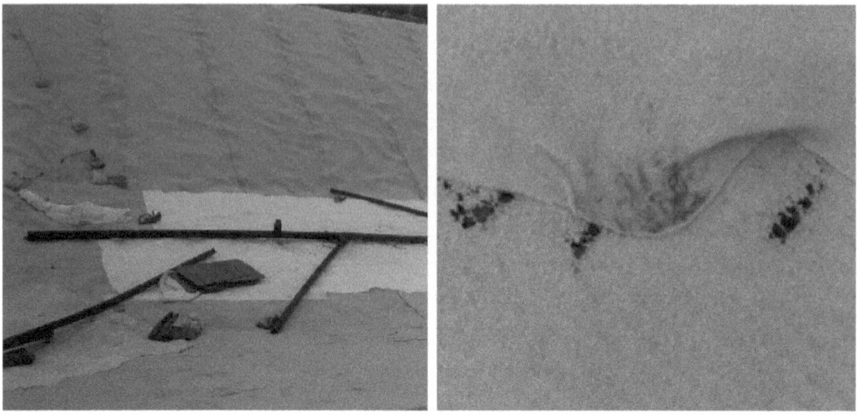

Figure 6.4 Typical fusion weld details for sludge sanitary landfill.

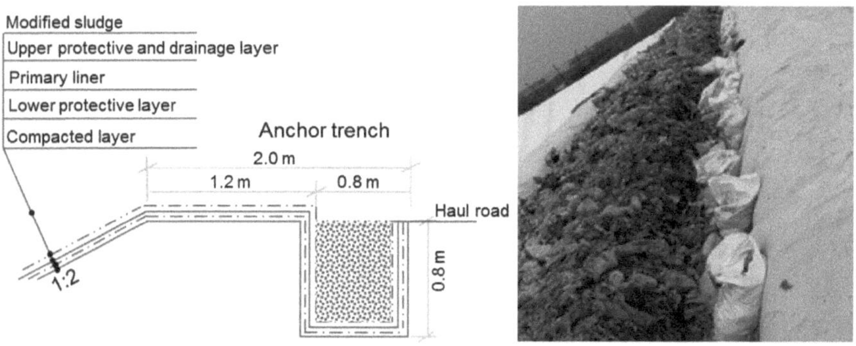

Figure 6.5 Typical anchor trench for sludge sanitary landfill.

6.1.2 Leachate Collection and Treatment Systems

An effective leachate collection and treatment system is necessarily required for a sludge landfill. The system should include the following components: drainage blanket, drainage ditch, perforated collection pipes, leachate collection sumps, collection inclined shaft, lift pump, regulating reservoir, leachate treatment facilities, etc. Among them, a leachate drainage blanket is constructed of gravel (Φ 32–100 mm) (Fig. 6.6), with perforated collection pipes (Φ 225 mm HDPE) laid within the drainage blanket to collect leachate. If required, a protective filter layer (woven geotextile with mass per unit area of 200 g/m^2) over the drainage blanket is adopted to prevent physical clogging of drainage blanket by fine-grained materials. The leachate

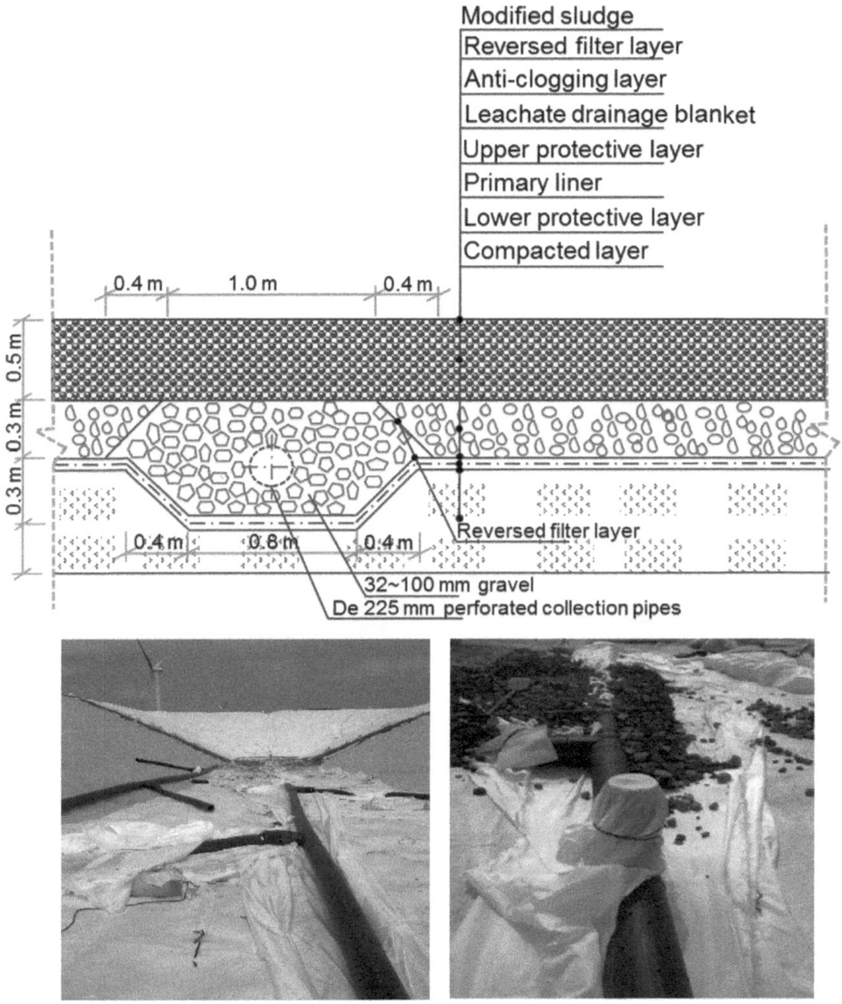

Figure 6.6 Typical leachate collection systems for sludge sanitary landfill.

is collected and then removed by submersible pump from the inclined shaft (Fig. 6.7).

6.1.3 Landfill Gas Collection and Recovery

Landfill gas results from the decomposition of organic material in landfills. The major components of landfill gas are methane and carbon dioxide with a member of minor constitutes. In order to control greenhouse gas

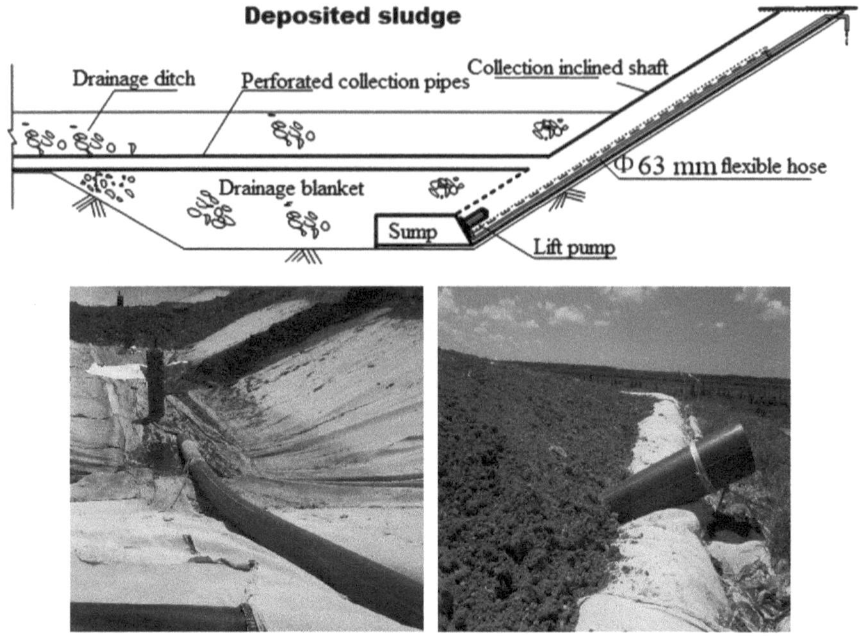

Figure 6.7 Leachate collection inclined shaft for sludge sanitary landfill.

emissions and minimize the risk of migration of landfill gas, effective extraction systems must be developed. An example of a typical vertical gas collection system is given in Fig. 6.8. The collection system (e.g., vertical extraction well) comprises vertical gas well (Φ 160 mm HDPE), wellheads, and collection pipe. The vertical well has perforations over the lower part of pipe length, and is surrounded by a suitably sized aggregate fill (0.64-m thickness) and a composite protective layer (including steel wire gauze, 200 g/m^2 woven geotextile, 0.3 m-thickness aged refuse or construction waste, and 200 g/m^2 woven geotextile). The welding between HDPE geomembrane and perforated pipe should be performed to prevent air admission (Fig. 6.9). Vertical gas wells are normally spaced between 20 and 25 m apart (Fig. 6.10); the collected landfill gas is then conveyed via the collection pipe network to the point of thermal destruction (e.g., flare unit) or energy production (Fig. 6.11). The recovery of landfill gas as an energy resource can compensate the part of the costs of sludge disposal, but the potential depends upon the methane content of the gas. Normally, the purification of gas should be taken into account prior to entering generator set.

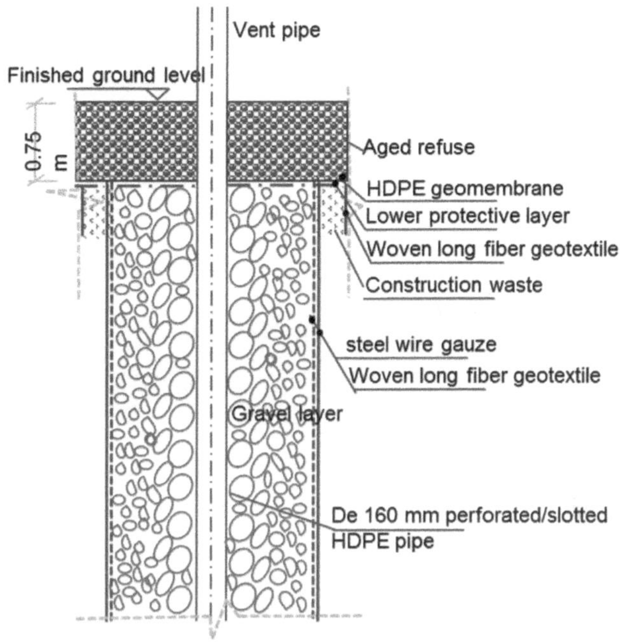

Figure 6.8 Cross-sectional view of vertical extraction well for sludge sanitary landfill.

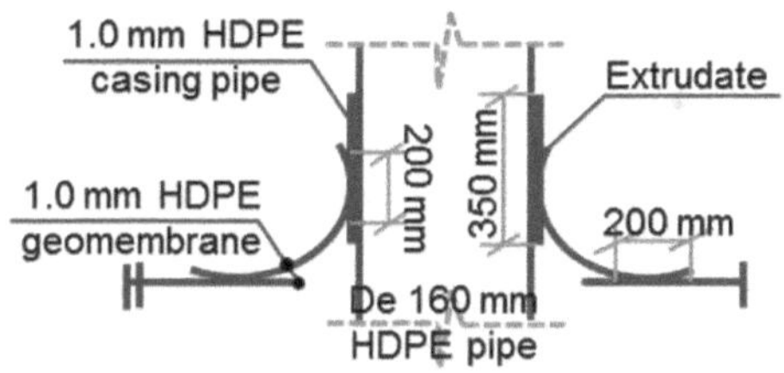

Figure 6.9 Typical extrusion weld details between HDPE geomembrane and perforated pipe for sludge sanitary landfill.

6.1.4 Capping and Closure Systems

The design objectives of a capping are to: (1) separate the deposited sludge from the environment; (2) limit water infiltration to minimize leachate production; (3) reduce the migration of gas, carbon loss, and odor release; and (4) provide a sustainable land form fit for its intended future purpose. The major components of a landfill capping system usually comprise: top soil of at least 0.75 m thickness, 1.0 mm HDPE geomembrane, lower protective layer (200 g/m^2 woven long fiber geotextile), and 0.3-thickness clay layer

Figure 6.10 Landfilling operation around vertical extraction well and landfill gas collection for sludge sanitary landfill.

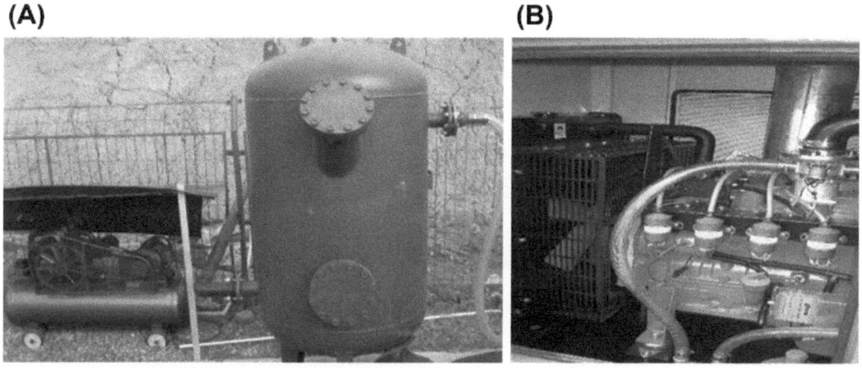

Figure 6.11 Purification units of landfill gas (A) and power generator set (B) installed in a sludge landfill demonstration project in Shanghai.

(or aged refuse layer) within which horizontal gas wells can be laid horizontally (Fig. 6.12). Fig. 6.13 is an example of a landfill capping system in a sludge landfill demonstration project in Shanghai, China.

6.2 SANITARY LANDFILL OF DEWATERED SLUDGE AND CHARACTERIZATION OF STABILIZATION PROCESS BY PARTICLE SIZE DISTRIBUTION AND HUMIC SUBSTANCES CONTENT AS WELL AS FT-IR

The stabilization process in the landfill should be explored to facilitate the landfill management. Sludge could be said to be stable when the original solid residues have been converted into nonputrefying odorless humic matter with very slight and very slow degradation. Then, and bearing in mind the basis of stabilization, a parameter for the characterization of stability should be related to the principal stabilization process taking place. Many parameters have been proposed for the evaluation of wastes' stability, with varying degrees of reliability and technical complication, including chemical tests, microbiological assays, and determination of humified OM.

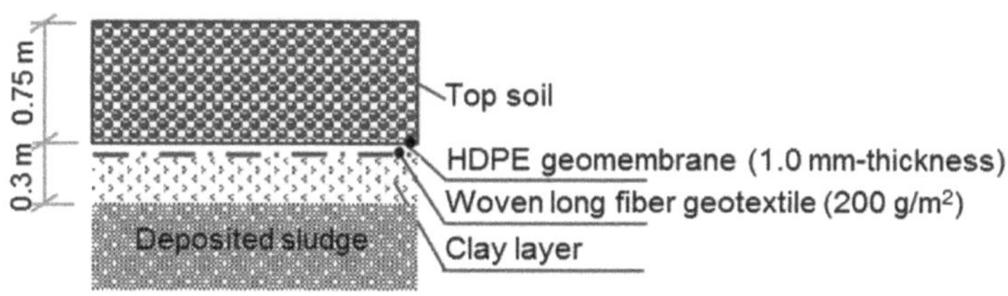

Figure 6.12 Typical landfill capping and closure systems for sludge sanitary landfill.

Figure 6.13 Photos of capping and closure systems in a sludge landfill demonstration project in Shanghai, China.

Particle size analysis has been used for a long time and is widely used in the process of wastewater treatment, but its application to follow the degradation process of organic wastes has been limited to date. FT-IR spectroscopy has served as a valuable tool to characterize qualitatively and nondestructively the principal classes of chemical groups. FT-IR spectroscopy has contributed substantially to environmental analysis including humic substances, organic wastes-soil OM, and their compost. Sludge undergoes a series of physical and chemical changes in the landfill. Some materials will be stabilized into humic substances. It has been proposed that the amount of humic substances and their fractions [i.e., fulvic acids (FAs), HAs, and humins] can be used as indexes for maturity of compost.

The aim of this section is to assess the feasibility of particle size distribution as a tool for evaluating the evolutionary process of sludge, assessing stability of sewage sludge with different stabilization degrees by FT-IR spectroscopic characterizations and humic substances analysis.

6.2.1 Landfill Unit Construction and Operation

A pilot-scale sanitary landfill unit with lining system was constructed at the Shanghai Refuse Landfill in Shanghai, China. A total weight of 1800 tons of sewage sludge was placed in the landfill unit without any pretreatment (Table 6.1). The top and bottom areas of the unit were 1024 m^2 (32 m × 32 m) and 100 m^2 (10 m × 10 m) respectively. The total height of the unit was 6 m and it was divided into three layers on average by the ladder shape of the landfill slope. During the positioning of every layer, triplaner geocomposites were laid to accelerate the drainage and concretion. Samples were taken from the same sampling points for an interval from weeks to months. The sludge was air dried at room temperature for about 24 h, ground, and screened through 0.63 mm.

6.2.2 Variation of Particle Size Distribution of Sludge with Landfill Time

The particle size distribution of sludge at the different landfill stages is illustrated in Fig. 6.14a, which shows the percentage of particles, by volume, between 0.37 and 2000 μm. The size range of 0-d sludge was 0.37−449 μm. As the landfill proceeded, the size range increased to 0.37−1821 μm of 300-d sludge. There existed a main distinct peak and a small shoulder peak in the graph of every particle size distribution. As for 0-d sludge, the main peak was near 15 μm and the shoulder peak was near 257 μm. There was no evident change in the position of the main peak in the former 90-d sludge except it became less pronounced. However, there was a shift in particle size distribution toward larger particles as a new peak developed at about 84 μm and was more pronounced in the later landfill periods. As the landfill progressed, the particle size distribution of the shoulder peak shifted to larger particles and reached about 948 μm for 300-d sludge. The intensity of this shoulder

Table 6.1 Characteristics of Sewage Sludge Tested

Parameter	pH	VS (%/Dry Basis)	TOC (%/Dry Basis)	TN (%)	Water Content (%)
Value	7.34	44.7	23.5	3.60	83.0

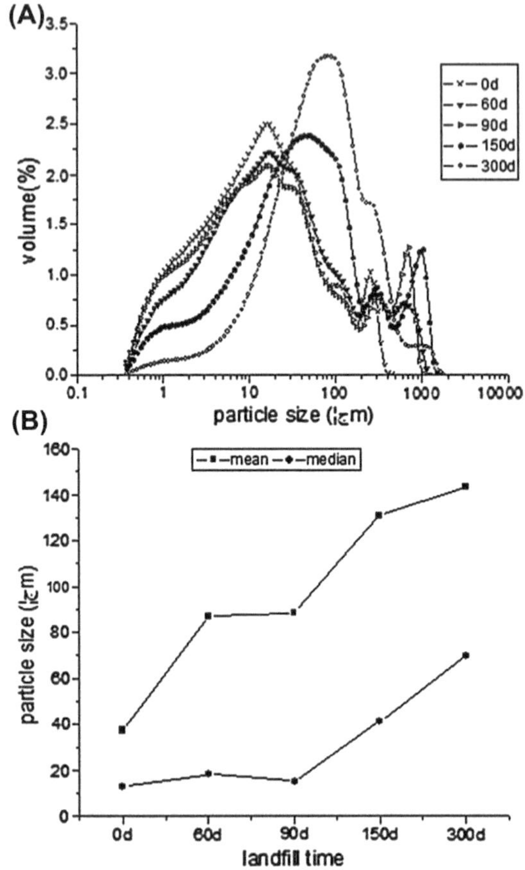

Figure 6.14 Particle size distribution in sludge (A) and mean and median particle size distribution (B) in sludge landfill.

peak first showed an increasing and then a decreasing tendency. The particle size distribution of 300-d sludge showed only one broad peak and fitted better to the log-normal distribution function than that of the former sludge.

A comparison of the mean and median particle sizes of sludge at different landfill times is shown in Fig. 6.14B. The result indicated that the mean particle size increased from 37 μm of 0-d sludge to 143 μm of 300-d sludge and the corresponding median particle size from 13 to 70 μm. The phenomena described above indicated that there were much smaller particle-size materials in the sludge of earlier periods than those of the latter periods. A larger amount of fine materials in sludge is related to the poorer dewaterability. On the contrary, materials of large and dense particle size are always desirable for good settling and dewatering. This can be tested by the slow settlement in the earlier landfill period and quick settlement in the latter landfill period. As the landfill progressed, many soluble and fine substances were first degraded, which led to the decrease of the volume percentage at the main peak. At the same time, the shift toward lager particle size can be attributed to the decrease of water content (WC), degradation of OM, flocculation–deflocculation, and increase of polymers content, as well as all kinds of physical and chemical changes. Different temperatures are also likely to be the cause of the shift in particle size distribution.

The stability of particle size distribution can be assessed by the mean or median particle size, the shift and intensity of peak, knowledge of the degree of mineralization, characterization of penetration and flocculation, to evaluate the maturity of sludge and provide a basis for the resource use of landfill sludge.

6.2.3 Variation of Humic Substances in Sludge With Landfill Time

Landfill sludge of 0, 60, 150, 300, 400-days were used to extract HAs and FA. The variation of HA and FA contents for all samples with landfill time are summarized in Fig. 6.15A. The contents of HA and FA increased from 4.2% and 2.7% of 0-day to 5.6% and 3.1% of 400-days, respectively, which indicated that the stabilization process of sludge in landfill was also a humification process. Fig. 6.15B shows that the contents of volatile materials in dry sludge generally decreased with landfill time. The decrease of volatile solids (VS) during the stabilization process indicated degradation of the total OM. The proportion of HA and FA contents to volatile materials' contents increased with landfill time, which showed the transformation of degradable organic materials to humic materials in the landfill process.

6.2.4 Element Composition of Humic Substances in Landfill

The elemental analysis results for HA and FA of all the samples are summarized in Table 6.2. In general, the nitrogen contents of HA and FA in sludge were higher than that in the soil. This may be attributed to the combination of humic molecules with decomposition products of protein materials. Both C and H contents in HA and FA showed an increasing tendency whereas O and N showed a decreasing tendency. This indicated that the degree of maturity and humification generally improved with landfill time. The HA samples contained more carbon and nitrogen, and less hydrogen and oxygen than the FA samples, indicating a high degree of maturity and humification of these HA samples.

Atomic ratios of H/C, O/C, and N/C are important indicators of the degree of aromatic condensation and maturity of humic substances. To further investigate the characteristics of the HA and FA, atomic ratios of H/C, O/C, and N/C were calculated from the elemental analysis results. The H/C ratio values of HA and FA are between 1.6 and 1.9, higher than 1 for humic materials in soils. The H/C ratios suggest that the degree of maturity and aromatic condensation of HA and FA increases with landfill age. The O/C ratio is an indicator of the carbohydrate and carboxylic acid content of humic substances. Compared with that of HA, the high O/C

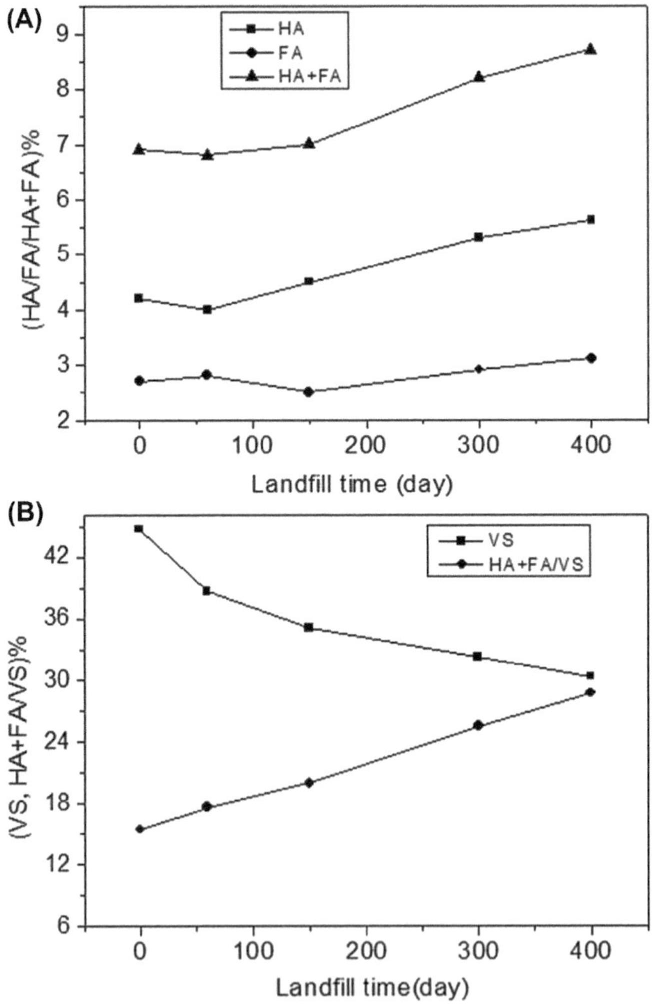

Figure 6.15 Variations of HA, FA (A), and VS (B) with landfill time for sludge.

Table 6.2 Elemental Composition of HA and FA in Sludge ($n = 3$, $p = .05$)

Sample	Mass Ratio (%)				Atom Ratio		
	N	C	H	O	N/C	O/C	H/C
HA (0 day)	10.33	51.94	7.33	30.40	0.17	0.44	1.69
HA (60 days)	10.22	52.05	7.29	30.44	0.17	0.44	1.68
HA (150 days)	10.14	53.35	7.36	29.15	0.16	0.41	1.66
HA (300 days)	8.97	55.05	7.70	28.28	0.14	0.39	1.68
HA (400 days)	9.13	54.80	7.97	28.10	0.14	0.38	1.75
Variation tendency	↓	↑	↑	↓	↓	↓	↑
FA (0 day)	4.59	46.82	7.18	41.41	0.08	0.66	1.84
FA (60 days)	4.18	46.26	7.01	42.55	0.08	0.69	1.82
FA (150 days)	4.18	47.65	7.38	40.79	0.08	0.64	1.86
FA (300 days)	4.08	48.03	7.60	40.29	0.07	0.63	1.90
FA (400 days)	3.96	50.74	8.07	37.23	0.07	0.55	1.91
Variation tendency	↓	↑	↑	↓	↓	↓	↑

↓, decrease; ↑, increase.

ratio of FA suggests that the FA contains a high proportion of O-alkyl and carboxylic acid functional groups. The O/C ratio trend implies that the number of O-containing functional groups in FA decreases with landfill age. The N/C ratio in HA was higher than in FA probably because HA retained N, while FA lost N. The N in both HA and FA may have been formed from the biodegradation of protein in the sludge.

6.2.5 Infrared Spectra (FT-IR) of Sludge in Landfill

For FT-IR analyses, five samples (0, 60, 90, 150, and 300 days) were selected during the landfill process. The FT-IR spectra of the entire sample (Fig. 6.16) show the changes of functional groups of compounds that characterize different stages of decomposition. Table 6.3 gives a summary of relevant infrared absorption bands and their development with time during the landfill process. The infrared spectra of the sludge in different landfill stages were very similar and showed the same peak pattern, the differences lying in the intensity of certain bands.

The intense broad band at about 3400 cm^{-1} is usually attributed to the O—H vibrations of water. A band at about 2920 and 2850 cm^{-1} is due to aliphatic C—H group stretching and assigned to fats and lipids, whose relative intensity tends to decrease during the landfill process. A small shoulder at about 1740 cm^{-1}, indicating C=O stretching of COOH and other carbonyl groups disappeared within the first 60 d. So, it can be inferred that this band was primarily assigned to early decomposed or volatile compounds.

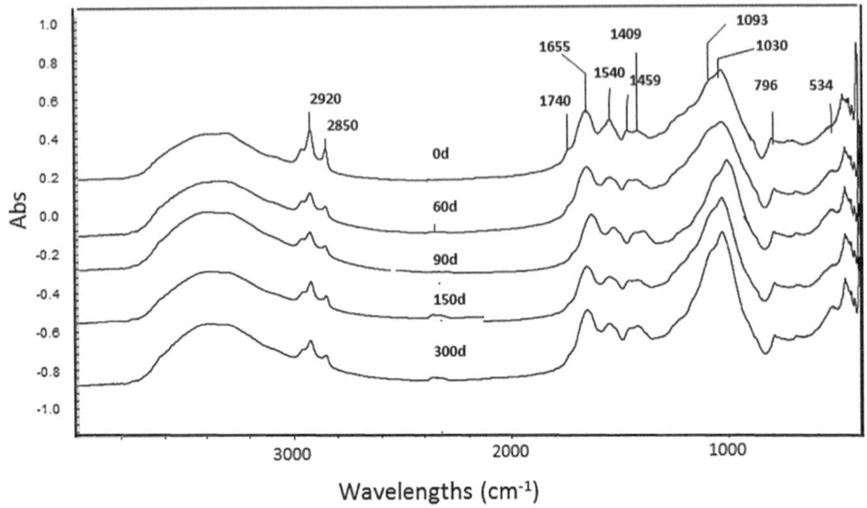

Figure 6.16 Indicator bands and IR-spectroscopic characteristics of different landfill stages of landfill.

Table 6.3 Summary of Infrared Absorption Bands and Their Developments With Time During the Landfill Process of Sludge

Wavelengths (cm^{-1})	Assignments	Evolution
2920	Aliphatic C—H group stretching	↓
2850	Aliphatic C—H group stretching	↓
1740	C=O stretching of COOH and other carbonyl groups	↓
1655	Aromatic C=C bonds and the C=O group of amides or carboxylates	↑
1540	N—H deformation	↓
1460—1400	O—H in-plane bend of carboxylic acids, the CO_2 stretching of carboxylates and the aliphatic CH_2 group of alkanes	↑
1093, 1030	C—O stretching of polysaccharides or polysaccharide-like substances, Si—O asymmetric stretching of silicate impurities	↑
796, 696, 534	Inorganic materials	↑

↓, decrease; ↑, increase.

The broad band at 1655 cm^{-1} can be assigned to the absorption of aromatic C=C bonds and of the C=O group that is part of amides or carboxylates. The relative intensity of this band tends to increase in the landfill process. The increase of this band is caused by enriched chemical compounds that can be attributed to the aromatic C=C vibration, such as lignin moieties and other aromatic compounds. Formation of carboxylates due to the release of carboxylic acids from decomposed lipids contributes to this rise as well. A peak at about 1540 cm^{-1} is preferentially ascribed to N—H deformation, whose relative intensity appears to decrease during landfill.

The broad band at 1460—1400 cm^{-1} is caused by the absorbance of O—H in-plane bend of carboxylic acids, the CO_2 stretch of carboxylates, and the aliphatic CH_2 group of alkanes. An absorption at about 1093 and 1030 cm^{-1}, is generally attributed to C—O stretching of polysaccharides or polysaccharide like substances and Si—O of silicate impurities, whose relative intensity tended to increase with landfill time. The absorbance of sludge at this band was the strongest absorbance of the infrared spectra, which showed the high contents of polysaccharides and silicate impurities in sludge. Bands at 796, 696, 534 cm^{-1} are due to inorganic materials, such as clay and quartz minerals. Bands representing strictly inorganic functional groups tend to rise due to the increasing concentration of inorganic compounds as the OM decomposes.

In general, easily degradable OM components such as aliphatic chains and protein were distinctly decomposed during landfill. At the same time, there was an increase in the relative intensity of the aromatic polycondensed structures of higher stability, silicate impurities, and inorganic materials. According to the changes of the relative band intensity, the degree of degradation and maturity can be predicted.

6.3 COMBINATION OF COMBUSTION WITH PYROLYSIS FOR THE STABILIZATION PROCESS OF SLUDGE IN LANDFILL

Degradation and stabilization of sludge in landfill are manifested by the changing energy content. This fact leads to consider thermal methods as promising techniques through the control of the energy content of the sludge and the loss of OM by oxidation/ignition. To date, many thermal analysis techniques have been utilized such as differential scanning calorimetry (DSC), thermogravimetry (TG) and differential thermogravimetry (DTG) to study the thermal degradation kinetics of biowastes. They are based on a programmed heating of the samples in controlled atmosphere, which finally provides qualitative and quantitative informations regarding the OM. In addition, the entire sample can be analyzed avoiding chemical extraction steps that often affect the nature and the ambience of OM. For complex samples, this approach proves advantageous and provides additional information. This section mainly introduced the applicability of thermal analysis to assess stability of sewage sludge by analyzing TG-DTG, DSC of combustion, and corresponding pyrolysis of sludge with different stabilization degree.

6.3.1 Procedures of Thermal Analysis

Thermal analyses were performed with an SDT Q600 (TA) (Fig. 6.17) that enables the recording of thermograms and DSC curves, and all samples were combusted with synthetic air gas flow: 100 mL/min composition 21 $\pm$ 1% O_2 and 79 $\pm$ 1% N_2; (purity $\geq$ 99.9994%) and pyrolyzed with N_2 (purity $\geq$ 99.9994%). The sample (10 mg) was subjected to a heating rate of 10°C/min from 50 to 900°C. A total calibration with indium/zinc, followed by a check DSC exo-indium, was performed before the analysis.

6.3.2 Physical, Chemical, and Biological Parameters

During different periods, several parameters of the sludge were analyzed to follow the stabilization process. Table 6.4 shows the variations of these parameters. It can be observed that pH decreased during the first 60 d and then increased. This indicated that the anaerobic process entered the stage of methanogenesis after finishing the acetogenesis stage. The decrease of volatile matter (VM) and total organic carbon (TOC) along the stabilization

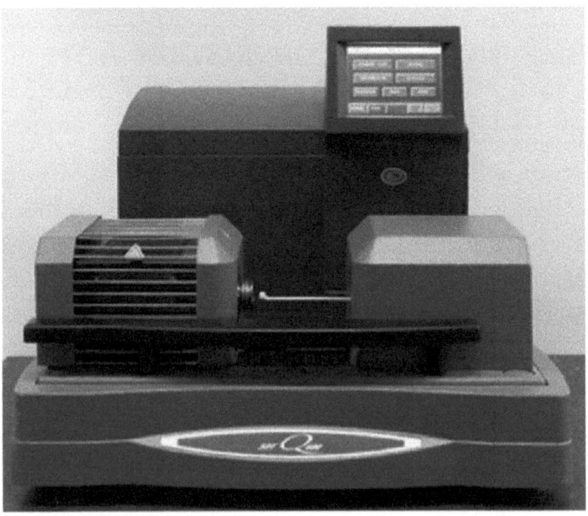

Figure 6.17 Thermogravimetric analyzer SDT Q600 (TA).

Table 6.4 Changes of Parameters During Landfill Process

Parameters	0 d	60 d	90 d	150 d	300 d
pH	7.34	7.04	7.25	7.44	7.93
VM (%/Dry Basis)	44.7	38.6	36.3	35.1	32.2
TOC (%/Dry Basis)	23.5	19.8	18.6	18.0	17.1
DHA (mg TF/g 6 h)	6.72	13.6	13.8	12.0	11.4
STOC (mg/L)	469	747	758	673	651

process indicated degradation of total OMs. Moreover, the degradation rate of the first 60 d was faster than that of the later stages.

Microbial activity can change during biodegradation of organic compounds under anaerobic conditions. Dehydrogenase activity (DHA) concentration was used to indicate microbial activity in this study. There was relatively high DHA content in 60 and 90 d. This was related to the acetogenesis stage of anaerobic process. In these two stages, the microorganism activity is higher than that in the later methane production stage. Soluble total organic carbon (STOC) can reflect the content of soluble OMs already directly degraded. There was a good correlation relationship between STOC and DHA content, and the degradation ability of microorganisms could be determined by measuring the DHA.

6.3.3 Thermogravimetry (TG) and Comparison of Combustion and Pyrolysis

Fig. 6.18 shows the TG combustion curves of the sludge in different landfill periods. It can be seen that there was a general tendency of decomposition. Owing to the mineralization process taking place during stabilization treatment, sludge residues from temperature-programmed combustion increased as stabilization progressed. The residual weight fraction at 900°C varied between 55.6% and 67.8%.

Considering the undistinguishable characteristics of so many TG combustion curves and the corresponding pyrolysis curves, only combustion curves and corresponding pyrolysis curves of two landfill periods are listed in Fig. 6.19 for comparison. C and P are representatives of the combustion and pyrolysis processes, respectively. From Fig. 6.19, it can be found that 0-d and 150-d sludges had combustion curves that were coincident or close to the corresponding pyrolysis curve in the 50–400 and 50–260°C temperature range, respectively (the first part) and then showed decay till 900°C (the second part). This phenomenon indicated that the weight loss of the first part was mainly caused by the thermal separation of volatiles. The decrease

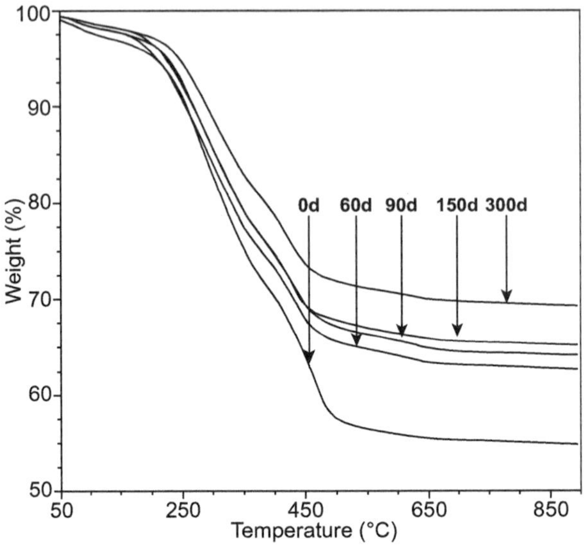

Figure 6.18 TG combustion curves of sludges at different stages in the landfill process.

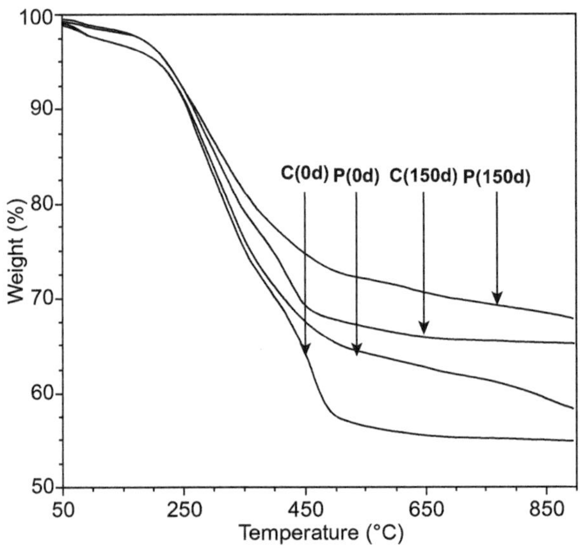

Figure 6.19 Comparison of TG combustion and pyrolysis curves of sludges at different stages in the landfill process.

of volatiles due to degradation in landfill made the first part of C150-d curve end earlier than that of 0-d curve. With the temperature rise, in the second part, the oxygen accelerated the decomposition rate in comparison to the pyrolysis, OMs started to combust, and as a consequence of the char formation, there was a final sharp decay of the curve due to the combustion of the char. So, combustion process was faster than pyrolysis process. There was almost equilibrium for combustion at the time when the temperature of the furnace reached 900°C, but this did not seem to be the case for pyrolysis. In this part, the C150-d curve was closer to the corresponding pyrolysis curve than that of 0-d. This can be attributed to the degradation of 150-d sludge, and the char formation was also less than that of 0-d sludge. Here, it should be noted that the degradation of organic materials in the second part might increase the amount of volatiles at the same time, because VM was usually the easiest to be digested by microorganisms and it has been the first to disappear along the landfill process. The amount of volatiles produced in the second part was less than that of the degraded in the first part, so the temperature at which combustion starts for most of materials in 150-d sludge was still lower than that of 0-d sludge.

Hence, it can be concluded that the TG combustion curve and corresponding pyrolysis curve of landfill sludge in different degradation stages can not only present some information about the weight variation of the sludge in certain temperature, but also let us to know about the degradation of simple and complex OMs generally.

6.3.4 Differential Thermogravimetry (DTG) Analysis

6.3.4.1 DTG Comparison of Combustion and Pyrolysis

Fig. 6.20 shows changes undergone by the DTG combustion and pyrolysis profiles corresponding to sludges along the stabilization process. In accordance with the shape of the curves, the combustion sludge curves can be divided into four obvious phases, while only three obvious phases can be detected in the pyrolysis curves. For both of these kinds of curves, a slight weight loss within the temperature range of 50—130°C was mainly caused by the dehydration reactions. The WC of the DTG combustion curve of this phase was almost coincident with the corresponding pyrolysis curve. The WCs of air-dried samples were determined to be 1.7%, 1.5%, 1.4%, 1.3%, and 1.1% for 0-d, 60-d, 90-d, 150-d, and 300-d sludges, respectively. This further confirmed the dehydration reactions at this temperature.

Two obvious peaks (between 130°C and 375°C, and 375°C and 590°C, respectively) were observed in the DTG combustion curve. While for the

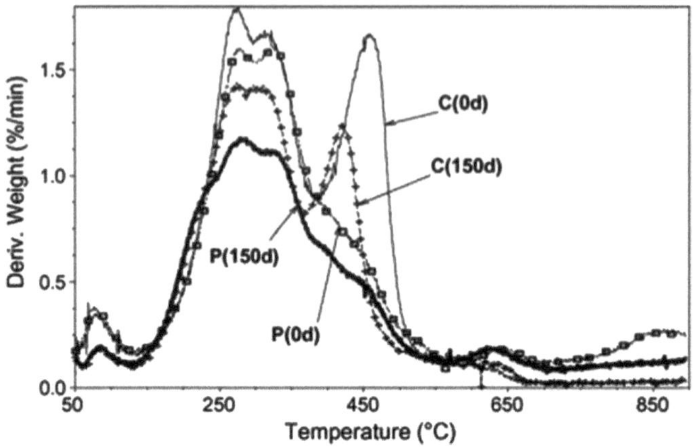

Figure 6.20 Comparison of DTG combustion and pyrolysis curves of sludges at different stages in the landfill process.

pyrolysis curve, only the first peak could be differentiated, as the second one did not take place. It indicated that the first peak was caused mainly by the thermal decomposition and loss of volatiles. The first peak had the maximum mass loss compared with the other stages. By comparison, it can be seen that the combustion curve was a little higher than the corresponding pyrolysis curve, which could be attributed to the combustion of a small amount of organic materials with shorter hydrocarbon chains. According to the disappearance of the second peak of the pyrolysis curve, the second peak of the combustion curve was caused mainly by the combustion of OMs with longer hydrocarbon chains.

From 590°C to 900°C, the rate of mass loss became small, and there was a small peak between 590°C and 670°C. The rate of mass loss for the pyrolysis curve was higher than that of the combustion curve in this phase. Several factors, including thermal decomposition of residual organic materials and the mineral materials, the char combustion and oxidation of metals in the presence of oxygen, etc., might contribute to the results mentioned above.

6.3.4.2 DTG Analysis of Sludge Combustion

DTG combustion curves of different landfill periods are presented in Fig. 6.21. Owing to the dehydration and solidification process of sludge, the WC decreased and the intensity of the first peak showed a little decrease along the stabilization process. In the 130−375°C range, corresponding mainly to the loss of volatiles and a little to the combustion of

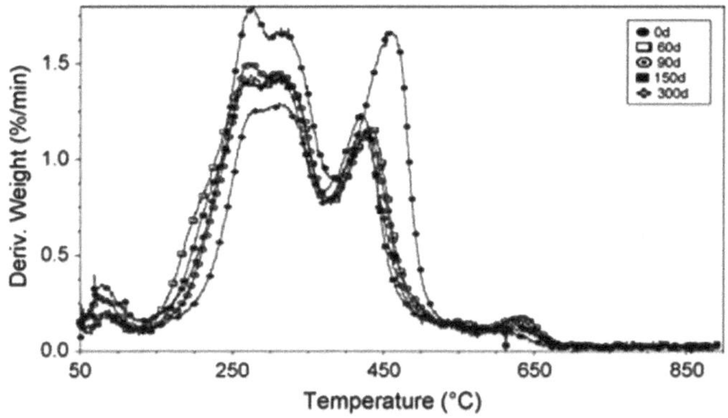

Figure 6.21 DTG combustion curves of progressive stages in the landfill process.

OMs, the weight-loss rate decreased as the landfill process proceeded. This indicates the degradation of OM s during landfill. Intensities of the third peak were also affected by the degradation process. Moreover, more metabolic products, which had different chemical structure and were preferentially combusted, were produced with the stability of OMs. This affected the peak positions and the shift toward lower temperatures. From the decrease tendency of this peak, it can be seen that the stabilization process was still ongoing. In the anaerobic process, the third peak of DTG combustion curve was mainly caused by the combustion of OMs. The peak intensities were mainly affected by the degradation process. On the contrary, in the aerobic process, the peak at around 475°C could be attributed to scarcely degradable molecules enriched during stabilization process. It is easy to see that the peak intensities at this temperature by aerobic stabilization process were less affected by the degradation process than by anaerobic stabilization process. Between 590°C and 900°C, the weight loss rate of different stabilization periods had no visible variation. Consequently, DTG analysis of sludge combustion can be used to know about the variation of different kinds of organic matters and the degree of the anaerobic stabilization process.

From the results in Table 6.4 and especially from TOC and VM reduction, it can be observed that the most evident progress toward stabilization took place in the first 60 days of anaerobic process. This was in agreement with the weight-loss results and DTG profiles. Therefore, we can acquire the general information about the degradation degree according to these characterization parameters.

6.3.5 Differential Scanning Calorimetry (DSC) Analysis

6.3.5.1 DSC Comparison of Combustion and Pyrolysis

Fig. 6.22 presents the heat flow values of combustion and pyrolysis for 10-mg sludge samples. DSC combustion curves showed two distinct exothermic peaks while the pyrolysis curves did not show evident exothermic behavior. In general, the sludge combustion has more energy release than the corresponding pyrolysis. Both DSC combustion curves and corresponding pyrolysis curves presented a little endothermic peak at low temperature, which was caused by the dehydration reactions. By comparing two distinct exothermic peaks, it was found that the area of the first peak was bigger than that of the second peak for DTG combustion curves.

On the contrary, the area of the first peak was smaller than or almost the same as that of the second peak for DSC combustion curves. The reason was that devolatilization as well as combustion contributes to the first peaks whereas only combustion to the second peaks. Endothermic reaction for devolatilization led to the reduction of the first-peak's intensities for DSC combustion curves. The energy for DSC pyrolysis curves were almost in a balanced state at 100−650°C. This confirmed that the energy release for the products of OMs thermal decomposition reaction with some components of sludge was equal with the energy needed for pyrolysis. At 650−900°C, DSC pyrolysis curves presented endothermic tendency, which was the result of pyrolysis of some in OMs.

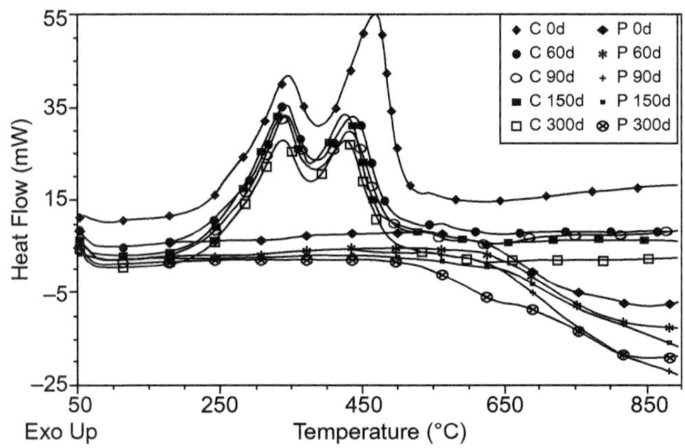

Figure 6.22 DSC combustion and pyrolysis curves of progressive stages in the landfill process.

It is noteworthy that the temperature of the two distinct DTG peaks did not correspond exactly with the two peaks' maximum of the heat flow in the aged samples, especially the first peak. This was mainly because of the thermal inertia of the heat flux sensor (delay on signal).

6.3.5.2 DSC Analysis of Sludge Combustion

According to the declining energy content, the energy release was higher in the undecomposed material than in the degraded one. The decrease of organic compounds resulted in a reduction of DSC peak intensities. As the landfill process advanced, because of the preference-combustion of the degraded OMs during landfill, energy release of the second peak began at lower temperature. The declining tendency of heat flow indicated that stabilization of sludge was still ongoing. Therefore, the DSC curve's analysis could be used to evaluate the stability of the sludge.

6.4 VARIATION OF PAHs', PCBs', AND OCPs' CONTENTS AND INFLUENCING FACTORS IN SLUDGE LANDFILL PROCESS

Polycyclic aromatic hydrocarbons (PAHs), polychlorinated biphenyls (PCBs), and organochlorine pesticides (OCPs) are persistent organic pollutants (POPs) which are usually found in the sludge. Many kinds of these hazardous organic compounds are defined as the "priority control pollutants" by the United States Environmental Protection Agency (USEPA). Just like the heavy metals, the organic pollutants in the sewage sludge used for agriculture can be accumulated in the soil and then pollute the crops.

The aim of this section is to measure the contents of PAHs, PCBs, and OCPs in sewage sludge buried in the landfill of different periods in order to find out the changes of the above two kinds of POPs as well as to investigate the main factors that affect the PAHs', PCBs', and OCPs' contents so as to provide scientific proofs for the agricultural recycling of aged sludge.

6.4.1 Contents and Transformations of PAHs in Sludge Landfilled in Different Stages

The contents and the regularities of distribution of PAHs in varieties of landfilling duration are shown in Table 6.5 and Fig. 6.23, respectively. It can be seen according to the single PAHs compound that fluorine, phenanthrene, anthracene, fluoranthene, and pyrene were all not detected in every stage of landfilling, and neither were naphthaline, acenaphthylene, benzo [*a*]

Table 6.5 Contents of PAHs in Sludge Landfilled in Different Stages (mg/kg)

PAHs	0 d	60 d	90 d	150 d	300 d	400 d
Naphthalene (2)	ND	0.093	ND	ND	0.872	0.113
Acenaphthylene (2)	0.051	1.421	0.096	0.011	ND	0.874
Acenaphthene (2)	1.07	2.347	0.034	1.028	0.08	0.539
Fluorine (2)	ND	ND	ND	ND	ND	ND
Phenanthrene (3)	ND	ND	ND	ND	ND	ND
Anthracene (3)	ND	ND	ND	ND	ND	ND
Fluoranthene (3)	ND	ND	ND	ND	ND	ND
Pyrene (4)	ND	ND	ND	ND	ND	ND
Benzo [*a*] Anthracene★ (4)	0.022	ND	3.811	0.97	0.23	0.479
Chrysene★ (4)	0.009	0.008	ND	ND	ND	ND
Benzo [*b*] Fluoranthene★ (4)	0.002	ND	0.021	ND	ND	0.023
Fluoranthene Benzo [*k*]★ (4)	0.014	0.612	0.907	0.052	0.851	0.497
Benzo [*a*] pyrene★ (5)	0.033	0.978	0.225	1.319	0.32	0.509
Diphenyl and [*a*, *h*] anthracene★ (5)	4.212	1.693	1.971	1.961	2.87	1.591
Indene and [1, 2, 3-*cd*] pyrene★ (5)	0.045	0.206	0.045	0.053	0.278	0.035
Benzo [*g*, *h*, *I*]★ (6)	4.003	2.65	1.974	1.576	2.12	1.985
$\sum$noncarcinogenic compounds	1.121	3.861	0.13	1.039	0.952	1.526
$\sum$carcinogenic compounds	8.34	6.147	8.954	5.931	6.669	5.119
$\sum$16PAHs	9.461	10.008	9.084	6.970	7.621	6.645

ND stands for not detected, $\sum$16PAHs represents the content of the total 16 kinds of PAHs compounds, and the asterisk "★" was used to mark a carcinogenic compound.

anthracene, chrysene and benzo [*b*] fluoranthene in some stages of landfilling. Except for acenaphthylene, acenaphthene, benzo [*a*] pyrene, diphenyl, and [*a*, *h*] anthracene and benzene [*g*, *h*, *i*], which were all with a content of more than 1 mg/kg, other compounds' contents were all very low. Generally speaking, the PAHs' compounds with more than three phenyls were dominant in every landfilling stage, while those organics with less than three phenyls occupied only a small proportion. It is also aware that the content of PAHs compounds, which can be carcinogenic, is higher than the noncarcinogenic ones.

In the initial sewage sludge landfilled, the content of PAHs was 9.461 mg/kg, and the content of the total PAHs compounds (represented

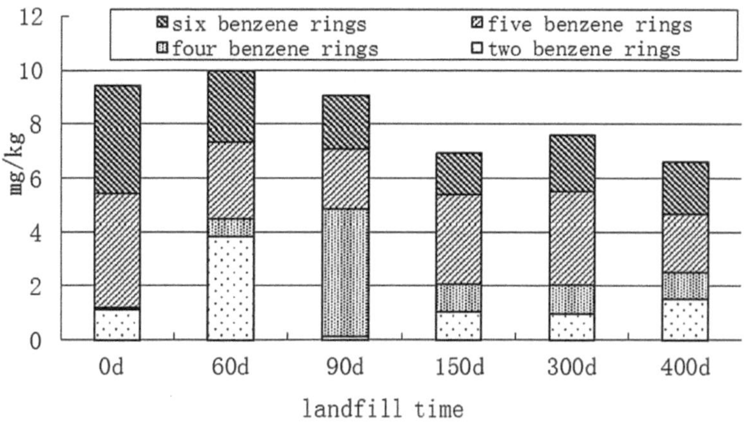

Figure 6.23 Regularities of distribution of PAHs in the sludge landfilled in different stages.

by $\sum 16$ PAHs) ranged from 6.645 to 10.008 mg/kg. The distribution characteristics of PAHs are different in sewage sludge obtained from various cities, and the most common PAHs are those with three or four phenyls, while the ones with two, five, and six phenyls are rarely to be found. The contents of total PAHs contained in sludge of other countries range from 1 to 50 mg/kg, and the common values are between 1 and 10 mg/kg. The compounds with low values of phenyls and low molecular weights (compounds such as naphthaline, acenaphthylene, and acenaphthene) showed inconspicuous changing rules in the stage of landfill. Contents of PAHs with high molecular weight and more than three phenyls showed a decreasing trend, especially the kinds of organics with five or six phenyls.

After the processed sludge is landfilled, the content of PAHs would be influenced by the following main factors: evaporation, interception (namely, the formation of some residual substances that cannot be extracted) and conversion. But generally speaking, it's very difficult to distinguish the three processes, and the superimposed effect on the content of PAHs would also be generated in different processes.

The effect of evaporation can lead to a reduction of every PAH compound, especially in summer when the temperature can reach up from 50°C to 60°C. But in this case, evaporation may not be the dominating influencing factor for the content of PAHs as the PAHs compounds with less phenyls and low molecular weight were not observed an apparent decreasing trend. It must be stressed that the PAHs with more phenyls and high molecular weight can be more easily influenced by the effect of interception than the ones with less phenyls and low molecular weight because the octanol–water partition coefficient of the PAHs with more phenyls and high molecular weight is much larger, and thus they also show strong hydrophobicity. According to the decreasing trend of PAHs with more phenyls and high molecular weight landfilled in different stages, the interception performance could be the factor that caused the changes in content of PAHs, but the phenomenon that the contents of PAHs compounds would become higher as the time went by could not be explained. The conversion process includes bioconversion and abiotic conversion, and the conversion of PAHs with more phenyls and high molecular weight is the reason why the content of PAHs is reduced; moreover, they can be degraded to some compounds with less phenyls and low molecular weight and thus lead to the incensement of them, but they can also be degraded meanwhile. The double effect in which a compound can be generated and degraded leads to inconspicuous change rule on the contents of PAHs with less phenyls.

Except for the above three main influencing factors, other situations can also influence the contents of the PAHs compounds, such as the possible pollution in the pretreatment process. Through the above analysis, it can be concluded that the degradation of conversion is the main factor that influences the changing contents of PAHs.

6.4.2 Contents of PCBs and Its Change in the Landfill Sludge of Different Periods

Table 6.6 shows the contents of PCBs contained in sewage sludge landfilled in different periods, and the contents of PCBs, which were not listed, indicated that the PCBs were not detected in every landfill period. Seen from the content of a single PCBs' compounds, the one contained less chlorine such as PCB6 and PCB24/27 had a higher content than the one with more chlorine. Many of the PCBs' compounds can only be detected in a certain landfill period, and their contents could be ignored in other durations. The total contents of the PCBs were always in a low level in all the landfill periods, with concentration ranging from 15.655 to 25.569 μg/kg. The highest content of PCBs, which was 25.569 μg/kg, was obtained in the first period of landfill, and it was much lower than the national standard value of 0.2 mg/kg.

The distribution figure of PCBs in sludge landfilled in different periods indicates that the total contents of PCBs are decreasing as the landfill duration lasts (Fig. 6.24). In the first stage, the compounds mainly contained 3-Cl

Table 6.6 Contents of PCBs Contained in Sludge of Different Landfill Periods (μg/kg)

PCBs	0 d	60 d	90 d	150 d	300 d	400 d
PCB6	ND	1.78	13.245	19.434	13.897	11.675
PCB24/27	9.821	8.482	6.331	ND	ND	ND
PCB66	0.598	ND	ND	ND	ND	1.788
PCB97	0.348	1.368	0.122	ND	ND	ND
PCB82	ND	0.342	ND	ND	ND	ND
PCB91/121	ND	0.837	ND	ND	0.212	ND
PCB90/101	ND	ND	ND	ND	1.511	ND
PCB99/113	ND	3.917	ND	ND	ND	0.436
PCB83/112	11.561	3.781	ND	ND	0.376	ND
PCB97	ND	0.791	ND	ND	ND	ND
PCB151/124/135/144	ND	ND	ND	ND	0.746	ND
PCB107/109	ND	ND	ND	ND	0.357	ND
PCB134/143	0.746	ND	0.873	ND	1.711	0.953
PCB146	0.221	ND	ND	1.275	ND	0.295
PCB132/161	ND	ND	ND	0.204	ND	ND
PCB153	ND	ND	ND	ND	0.25	ND
PCB138/163/164	ND	ND	ND	ND	ND	0.22
PCB158	ND	0.363	ND	ND	ND	ND
PCB174	ND	ND	0.317	ND	0.117	ND
PCB171	1.017	ND	ND	0.603	ND	0.288
PCB196	1.257	0.132	ND	ND	0.112	ND
ΣPCBs	25.569	21.793	20.888	21.516	19.289	15.655

ND stands for not detected; ∑PCBs stands for the total content of PCBs.

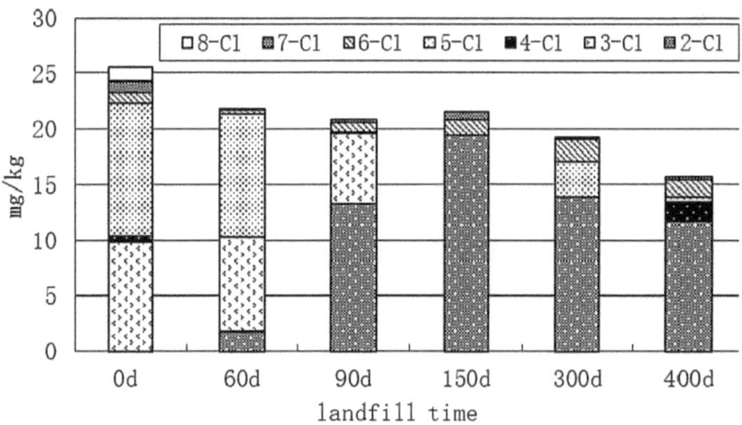

Figure 6.24 Distribution regularities of PCBs in the sludge obtained from different landfill stages.

and 5-Cl, whereas in the later period, more and more 2-Cl containing compounds were generated and accounted for most of the PCBs. Just like the PAHs, in the landfill period, the factors that can influence the contents of PCBs are also the evaporation, interception, and transformation. The analysis of related data showed that evaporation was not the main factor that influenced the content of PCBs, as the process of evaporation can pose the decrease of chlorinated compounds no matter they contained chlorinate more or less. It could be explained as the reduction of some compounds with more chlorinates via interception effect. The transformation of a substance can change the adsorption property and extract the feature of various pollutants, and the reaction of dechlorination in the transformation process can also explain the incensement of low chlorine compounds and the decrease of high chlorine compounds in landfill stages.

There are two different action modes in the degradation of PCBs: anaerobic dechlorination and aerobic biodegradation. The anaerobic environment could promote the dechlorination of PCBs. The anaerobic degradation rate of PCBs is in direct proportion to the chlorinated degree, and the rate is also determined by the number of replaced chlorine and replaced sites. By the dechlorination of PCBs, the environmental risk can be reduced through the following two aspects: (1) the PCBs' compounds can be degraded more easily by the aerobic bacteria if they can transform from high-chlorine compounds to low-chlorine compounds, (2) the dechlorination can decrease the bioaccumulation by transforming the high chlorine homologue to low chlorinated PCBs, which is not easy to be an enrichment to the food chain; meanwhile, it also reduces the carcinogenicity and the toxicity of dioxin analogues. In the aerobic condition, the PCBs would be degraded to hydroxyl-PCBs by aerobiotic reaction.

6.4.3 OCPs' Contents and Its Change in Sludge of Different Landfill Stages

Hexachloro-cyclohexane soprocide is a mixture consisted of 1, 2, 3, 4, 5, 6-hexachloride benzene. The isomer of six chlorine atoms is different in cyclohexane conversion of the space arrangement, and its main composition includes α-BHC, β-BHC, γ-BHC and δ-BHC, of which the γ-BHC (lindane) is the active ingredient. Hexachlorocyclohexane soprocide is now forbidden for using by the Chinese government, but the lindane is often used for pest controlling in the household, commerce, agriculture, forest, and so on. Now in China, although the pesticide named DDT has not been produced and used yet, it and its ramification as well as metabolin are rich in every trophic level of the biology as a result of the widespread distribution in the natural environment, long degradation period, and the hydrophobic and lipotropic characteristics.

It's indicated from Table 6.7 that the total contents of OCPs range from 1.78 to 2.37 μg/kg in sludge of different landfill stages. These data showed that the OCPs content was relatively lower in the sludge.

6.4.4 Source Apportionment of the Contents of PAHs, PCBs, and OCPs

There are many complex factors that can influence the contents of PAHs, PCBs, and OCPs. The kinds and contents of organic pollutants contained in sludge produced in different wastewater treatment plants or different periods of the same wastewater treatment plant usually have with various characteristics. These pollutants can be enriched in the solid sludge particles with an enrichment coefficient up to several orders of magnitude in the wastewater treatment process. For a variety of pollutants, their contents in the municipal sludge are determined by their contents in the wastewater; in other words, the source of the wastewater determines the contents of pollutants both in wastewater and sludge. Therefore, the key work of doing source apportionment of pollutants in sludge lies in ascertaining the main source of pollutants in wastewater.

Generally, the sludge mainly produced by treating domestic sewage has a low content of PAHs, while the sludge generated by processing industrial wastewater has a higher content of PAHs. The content of PAHs in municipal sludge also has some relationship with the wastewater treatment technology and the sludge type. Anaerobic conditions showed a higher

Table 6.7 Contents of OCPs in Sludge of Different Landfill Stages (μg/kg)

OCPs	0 d	60 d	90 d	150 d	300 d	400 d
α-BHC	ND	ND	ND	ND	ND	ND
β-BHC	ND	ND	ND	ND	ND	ND
γ-BHC	ND	ND	ND	ND	ND	ND
δ-BHC	ND	ND	ND	ND	ND	ND
p,p'-DDE	ND	ND	1.02	ND	0.04	ND
p,p'-DDD	2.26	2.13	0.87	2.23	1.99	1.78
o,p'-DDT	ND	0.24	ND	ND	ND	ND
p,p'-DDT	ND	ND	ND	ND	ND	ND
ΣOCPs	2.26	2.37	1.89	2.23	2.03	1.78

ND stands for not detected, $\sum$OCPs is the sum of the OCPs' content values.

efficiency than aerobic conditions when used to degrade organic pollutants in sewage treatment process. It's more efficient in degrading the POPs when combining the oxidation and reduction together.

There are two kinds of sources from which the PAHs in water environment come: (1) petroleum, may be produced from oil leakage, normal oil tanker emissions, urban ground flushing or other ways; and (2) burning, may be produced from vehicle emissions, coal, forest fires, biomass burning, smelting, etc. This kind of approach takes a significant proportion of PAHs with high molecular weight. The main source of PCBs and OCPs in sludge is the industrial sewage emitted from chemical industry (textile, printing and dyeing, and casting, etc.), wood processing, electrical industry, etc.

6.5 ABIOTIC ASSOCIATION OF PHTHALIC ACID ESTERS WITH HUMIC ACID IN A SLUDGE LANDFILL

Phthalic acid esters (PAEs) are pollutants ubiquitous in the aquatic and terrestrial environments. The hydrophobic PAEs with low solubility and high octanol—water partitioning coefficient are persistent against biodegradation in common wastewater treatment plants. The PAEs have been shown to migrate into the surface water and groundwater and continue to be a threat to the aquatic environment. In addition, sludge from sewage treatment is also a principal source of PAEs. Elevated concentrations at 11—144 mg/kg of PAEs are often found in dewatered sewage sludge, which exceed the limit of 100 mg/kg proposed by the Europe Union for land application.

There has been a growing concern regarding the potential health risk associated with PAEs in the soil system with sewage sludge application. The reduction of PAEs' content in sewage sludge is of great importance for further sludge landfill disposal. Biodegradation of PAEs by microorganisms under aerobic or anaerobic conditions is considered to be one of the major routes for PAEs removal. In addition to biotic degradation of PAEs in landfill, the abiotic association between HA and PAEs plays an important role in the fate and stability of PAEs. The association would decrease its mobility, and increase its bioavailability, and biodegradability resulting in the decreased toxicity. Therefore, in this section, the abiotic association of PAEs with HA as well as the subsequent influence on the fate of PAEs in sludge landfill were investigated via dialysis equilibrium technique and 14C-labeling.

6.5.1 Sludge Samples and Experimental Procedures

Dialysis equilibrium technique was used to study the association of ^{14}C-labeled PAEs with HA. The experiments were performed at various pH values in a polytetrafluoroethene chamber consisting of two half-cells with a volume of 10 mL, separated by a dialysis membrane (regenerated cellulose, cutoff 1000 Da, Union Carbide Company, USA). Diluted solutions of HA and a labeled PAEs solution were separately poured into each of the two half-cells. The pH values of the solutions were adjusted to values of 3, 7, or 9 using either NaOH or HCl before filling the cells. The labeled PAEs were allowed to equilibrate between the two half-cells at room temperature by rotating the cells at 10 r/min. Aliquots of homogeneous samples from both half-cells were taken at defined time intervals for an analysis of radioactivity by means of liquid scintillation counting. In addition, control experiments were carried out to investigate the distribution of ^{14}C-labeled PAEs when humic substances were absent from the system. The strength of the association of PAEs with humic substances was represented by partition coefficient (K_A) defined by the following equation:

$$K_A\left(kg^{-1}\right) = \frac{\text{Radioactivity associated with humic acids}}{\text{Free radioactivity}} \times \frac{1}{\{HA\}} = \frac{^{14}C_{HA} - {^{14}C_{H_2O}}}{^{14}C_{H_2O}} \times \frac{1}{\{HA\}} \tag{6.1}$$

where $\{HA\}$ is the concentration of HA in the medium (g/mL), $^{14}C_{HA}$ is the radioactivity in the half-cell containing HA, and $^{14}C_{H_2O}$ is the radioactivity in the HA-free half-cell.

6.5.2 Humic Acid (HA) Stabilization in Sludge Landfill

Fourier transform infrared spectrophotometer (FT-IR) is widely used for characterization of HA, and provides valuable information on the structural and functional properties of HA. Fig. 6.25 shows the FT-IR spectra of HA from different stabilization stages in the sludge landfill. There was an intensive absorbance band of asymmetric vibrations in the range of 2850–2950 cm^{-1} characterized as methyl and methylene groups, indicating the presence of aliphatic chains in HA and weak humification of the HA. This shape and intensive peak on IR spectra suggested a high degree of aliphaticity of HA, signifying that the sewage sludge was less mature. The strong peak at 1655 cm^{-1} was the structural vibrations of aromatic C=C or stretching of carboxylic groups, whereas a fairly intensive band at

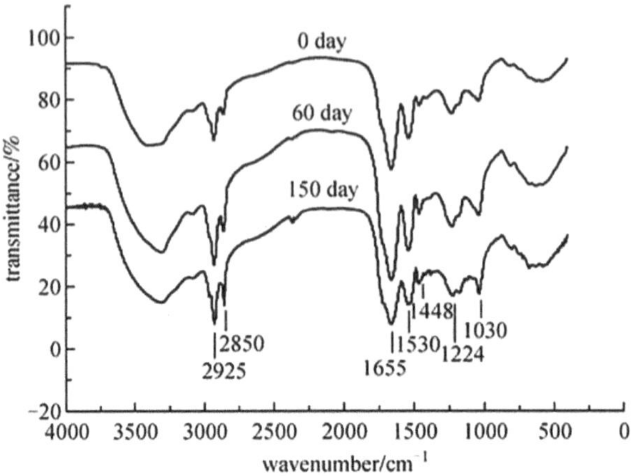

Figure 6.25 Fourier-transformed infrared spectra of humic acid (HA).

1530 cm^{-1} was a deformation of nitrogen functional groups NH. The absorption band of vibrations of carboxylic group C—O at 1224 cm^{-1} was also observed—another piece of evidence of the presence of carboxylic groups in HA. The strong absorbance of the absorption bands at 2925^{1} and 2850 cm^{-1} was due to stretching of aliphatic C—H and C═O of COOH, respectively. The absorbance of these bands decreased with the disposed time from 0 to 150 days, which indicated that aliphatic group of HA decreased during stabilization.

Table 6.8 shows the elemental contents of HA at different landfill times. The content of C generally increased with the landfill time while the contents of N showed an opposite trend. The content of N of HA was much higher than that of soil (2.45%), probably because the HA was from the humification of sludge with high content of protein.

To further investigate the characteristics of the HA, the atomic ratios of H/C, O/C, and N/C were calculated from the results of elemental analysis

Table 6.8 Elemental Composition and Rations of the HA From the Humification of Sludge

HA	N (%)	C (%)	H (%)	O (%)	C/N	O/C	H/C
Ha[a]	10.33	51.94	7.33	30.40	5.88	0.44	1.69
Hb	10.22	52.05	7.29	30.44	5.88	0.44	1.68
Hc	10.14	53.35	7.36	29.15	6.25	0.41	1.66

[a]Ha, Hb, Hc stand for HA extracted from the sludge at different landfill time of 0, 60, 150 days, respectively.

(Table 6.8). The H/C ratio normally represents the degree of maturity and aromatic condensation in HA, and can also be considered as a source indicator of OM. The H/C ratio greater than one in the HA samples used suggested a high fraction of aliphatic groups and a low content of aromatic rings in the structure. The decrease of H/C ratio of the HA in the sludge that had been landfilled for 150 days hinted that the degree of aromatization and maturity of HA increased with the landfill time. The changes of H/C ratio were consistent with the characteristic band of C—H aliphatic vibration observed on FT-IR spectra.

The O/C ratio of HA represents the degree of humification and the variation of oxygen functional groups. The O/C ratio of HA samples decreased with stabilization process, suggesting a decrease of the content of o-alkyl and carboxylic acid groups in the HA. The C/N ratio of the HA was increased from 5.88 to 6.25 during the 150-day sludge stabilization. The increase of C/N ratio with the landfill time might imply that HA only retained refractory forms of N during the stabilization process. The changes in ratios of H/C, O/C, and C/N were particularly visible at the later stage of stabilization process, suggesting that the most intensive course of humification probably occurred at the later stage of stabilization.

6.5.3 Association Intensity of Phthalic Acid Esters (PAEs) With Humic Acid (HA)

The association of ^{14}C-labeled diethyl phthalate (DEP) and dibutyl phthalate (DBP), with HA from the three different stabilization processes, was investigated. The associated forms of DEP and DBP in the half-cells with HA and their association coefficient values were calculated according to Eq. (6.1). The control experiment results showed that the percentages of ^{14}C-labeled PAEs in both sides of the membrane were 49.95% and 50.05%, respectively, after 24 h, which confirmed that the radioactivity was equally distributed on both sides of the membrane when HA were absent from the system.

DBPs have been demonstrated to have a strong association with HA, and the association appeared to be greatly dependent on the chemical structure of HA. Significant differences were observed in the association strength of HA from different sources. The total radioactivity in the half-cell containing the Hc was 64.9%, while that of Hb was 60.7%. The Hc showed a strong association with DBP ($K_A = 846$), and the association of Hb was relatively low ($K_A = 547$), while Ha did not show any association with DBP.

In general, HA showed a very limited association with DEP; the radioactivity was distributed almost equally on both sides of the cell. Specifically, the radioactivity in the cell with Ha was even a little lower than that in the other side, suggesting that Ha had no association with DEP and that the existence of Ha in the cell might have prevented DEP in the opposite side from diffusing into the cell. The results indicated that DBP had a stronger association with HA than DEP. This could be inferred from the high hydrophobicity and low water solubility of DBP ($\lg K_{ow} = 4.45$) in comparison with that of DEP ($\lg K_{ow} = 2.38$), implying that nonspecific hydrophobic interaction between PAEs and HA played an important role in the association. The association strength of Hb with DBP was lower than that of Hc, indicating that the association strength was dependent on chemical characteristics of HA. The high degree of aromatic condensation and maturity of Hc resulted in a stronger association with DBP, hinting a key role of aromatic core in binding of HA to DBP.

The degradation of PAEs in the sludge is related to the length of the alkyl chain of PAEs. As alkyl chain length increases, the degradation rate of PAEs decreases (corresponding half-life increases). Abiotic association between PAEs and HA affects the PAEs' biodegradation and bioaccumulation potential, which is also a determining factor controlling PAEs' distribution in landfill. For DBP, which is difficult to be completely biodegraded in sludge landfill, the great association strength with HA could decrease the mobility of DBP in the sludge landfill.

6.5.4 Effects of pH Value on the Association Strength

Table 6.9 summarizes $\lg K_A$ values of DBP at the three different pH values. Hc showed strong association in the pH range from 3 to 9 and the maximal association at acidic condition; more than 45.8% of DBP was associated with

Table 6.9 Association Intensity ($\lg K_A$) of ^{14}C-Labeled DBP (60 μg/L) With Humic Substances (0.1–1 mg/mL) at Various pH Values

	Association Intensity ($\lg K_A$)		
Humic Substances	**pH = 3**	**pH = 7**	**pH = 9**
Ha	—	—	—
Hb	3.00 ± 0.04	2.74 ± 0.02	2.10 ± 0.02
Hc	3.12 ± 0.03	2.93 ± 0.01	2.53 ± 0.02

K_A values (L/kg) are the means of three determinations at the equilibrium time.

Hc when pH was 3. It can be concluded that pH greatly affects the association strength of HA with DBP. For DBP, the association strength generally decreased with pH value. However, no association with Ha was detected at any of the experimental pH values.

The effect of pH was a result of changes in the aqueous-phase solute activity coefficients and a result from effects on DBP and HA interactions. The changes of physicochemical properties of HA with pH and its effect on sorption had been analyzed previously, based on an assumption that the organic domain becomes more hydrophobic and less solvated, when carboxyl groups in HA become protonated. However, as pH decreases, HA also becomes enriched in H-bond donor groups (−COOH). This can result in enhancement in H-binding between the proton donor groups of HA and the aromatic π system (which usually acts as an H-acceptor) of DBP. In addition, the decrease in association strength with increasing pH may be due to the effects of pH on the hydrolysis of DBP. The hydrolysis rate of esters depends largely on pH level of the aqueous environment. DBP undergoes base-catalyzed hydrolysis at an alkaline condition, and becomes monoethyl phthalate or monobutyl phthalate, which are more water soluble and therefore less likely to associate with HA.

As shown in Table 6.10, the association of Hb with a molecular weight greater than 10,000 Daltons is much stronger than that with a molecular weight greater than 1000 Da at pH values of 3, 7, and 9. This reflected that PAEs are readily adsorbed onto HA molecules with high molecular weights and containing many aromatic and peptide groups. The abundant micromolecule with high degradability in sludge results in a high degree of degradation of carbohydrate and simultaneously promotes a well development of aromatic core in HA. This also can be shown in the FT-IR spectra and results of elemental analysis. Based on these results, it can be speculated that aromatic rings play a main role in enhancing the molecular weight of HA.

Table 6.10 Association Intensity (lg K_A) of ^{14}C-Labeled DBP (60 μg/L) With Hb of Different Molecular Weight

	Association Intensity(lg K_A)		
Hb Molecular Weight	**pH = 3**	**pH = 7**	**pH = 9**
10,000 Dalton	3.05 ± 0.03	2.96 ± 0.01	2.53 ± 0.02
1000 Dalton	3.00 ± 0.04	2.74 ± 0.02	2.1 ± 0.02

K_A values (L/kg) are the means of three determinations at the equilibrium time.

6.6 CHEMICAL REDUCTION OF ODOR FOR SLUDGE IN THE PRESENCE OF FERRIC HYDROXIDE

The odorous compounds from sludge include reduced sulfur, nitrogen compounds, organic acids, aldehydes, and ketones. Among them, hydrogen sulfide (H_2S) is recognized as one of the predominant odorants and used to be the main indicator to assess nuisance odor. H_2S is produced as a by-product from anaerobic bacterial reduction of sulfate and decomposition of sulfur-containing organic constituents of sludge, and has an extremely low odor threshold relative to most other odorous compounds (around 0.5 ppb), along with high toxicity to mammals and aquatic species. When inhaled, H_2S inhibits cytochrome oxidase, which has an important role in mitochondrial respiration. It also causes eye irritation at concentrations as low as 50–100 ppm, while 300–500 ppm may result in severe poisoning, leading to unconsciousness or even death.

Chemical oxidation of H_2S and forming precipitation of insoluble sulfides with metals as well as a combination of both are common practices for minimizing H_2S emission in wastewater or sludge treatment processes. Ferric hydroxide (FH) has significant potential for odor control for dewatered sewage sludge. Made up of a short-range ordered arrangement of $Fe(O, OH, OH_2)_6$, FH is a very poorly crystallized hydrated oxyhydroxide, and can be dissolved by a range of reductants or by iron-reducing microorganisms in the absence of oxygen. Because of the special characteristics of reductive dissolution by sulfide, it possesses a high selectivity over other traditional oxidant alternatives, and provides a potential of long-term H_2S control for sewage sludge. The reductive dissolution of iron (hydr)oxides by sulfide was postulated based on the following sequence of reactions.

Reversible adsorption of HS^-:

$$> FeOH + HS^- \Leftrightarrow > FeS^- + H_2O \tag{6.2}$$

Reversible electron transfer:

$$> FeS^- \Leftrightarrow > Fe^{II}S \tag{6.3}$$

Reversible release of the oxidized product:

$$> Fe^{II}S + H_2O \Leftrightarrow > Fe^{II}OH_2^+ + S^{\cdot -} \tag{6.4}$$

Detachment of Fe^{2+}:

$$> Fe^{II}OH_2^+ + S^{\cdot -} \rightarrow \text{new surface site} + Fe^{2+} \tag{6.5}$$

and the following three equations are often used to describe the initial stage of iron–sulfide reactions for simplification. The ferrous iron supply stem from reductive dissolution of iron (hydr)oxides:

$$FeOOH + e^- + 3H_2O + HS^- \rightarrow Fe^{2+} + 5H_2O \quad (6.6)$$

Ferric iron is reduced by sulfide together with the precipitation of colloidal sulfur, and ferrous iron reacts with S^{2-} to precipitate as insoluble FeS.

$$2Fe^{3+} + HS^- \rightarrow 2Fe^{2+} + S^0 + H^+ \quad (6.7)$$

$$Fe^{2+} + S^{2-} \rightarrow FeS \quad K_{sp} = 3.7 \times 10^{-19} \quad (6.8)$$

Despite the frequent use of Eqs. (6.6)–(6.8), a variety of different oxidation products have been reported for reactions of sulfide with iron (hydr)oxides, depending significantly on both the type of iron (hydr)oxides and environment conditions. Various other end products, such as pyrrhotite varying in composition from FeS to Fe_4S_5, ferric sulfide Fe_2S_3, smythite Fe_3S_4, and pyrite and marcasite have also been cited as the end products of iron–sulfide reactions. The environmental conditions such as dissolved oxygen level, pH and the presence of ions such as $SO_4{}^{2-}$, Mg^{2+} and Ca^{2+} can affect the reaction rate.

In this section, the effectiveness of different dosages of FH to reduce odor emission from dewatered sewage sludge was investigated. The odor reduction was evaluated with respect to sensual odor using an electronic nose and odorous compounds (H_2S, NH_3, and VFAs). Electronic nose is complex system with a human nose-like structure, which comprises an array of electronic chemical sensors with partial specificity and an appropriate pattern-recognition system. The concentration of H_2S and odor was correlated using Steven's law.

6.6.1 Experimental Procedure

Sludge samples were collected from a domestic WWTP (Shanghai, China). The experimental procedure is described schematically in Fig. 6.26. Hundred grams of dewatered sludge samples were homogenized with 0, 0.05, 0.10, and 0.25 g FH (<0.125 mm), respectively, and were transferred into 500-mL glass flasks with stoppers (6 repetitions per dosage). After being flushed with N_2 to exclude oxygen, the flasks were then sealed with glass caps and placed in a biochemical incubator at 25°C. Odor measurements were performed by an electronic nose and the results were expressed in

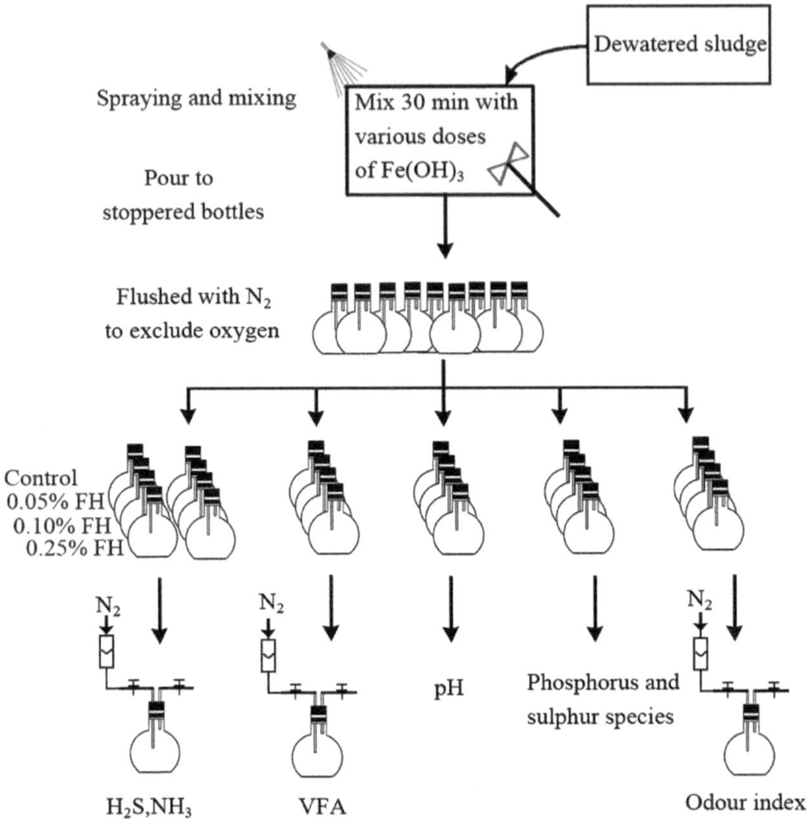

Figure 6.26 Schematic description of the laboratory experimental procedure for chemical reduction of odor for sludge in the presence of ferric (Fe) hydroxide.

odor index. Gas samples were injected in polyester bag, diluted with odorless air, and measured with a portable digital odor sensor (XP-329IIIR, Cosmos, Japan). The odor index value is dimensionless and can be converted as odor concentration [dilution factor to detection threshold (DT) or European odor units (OU)] by Eq. (6.9).

$$\text{Odourindex} = 10 \times \log_{10}(\text{DT}) \tag{6.9}$$

6.6.2 Effect of FH on H_2S, NH_3, VFAs, and Odor Abatement

H_2S can be eliminated very effectively by FH for dewatered sludge (Fig. 6.27A). Without FH, H_2S generation rate was first increased from 495 to 2132 μg and then decreased gradually to 613 μg after 32 days. After a 32-day incubation, the removal efficiencies of H_2S were 81.3% (73.4–87.8%), 93.7% (84.9–96.4%), and 97.5% (92.8–98.8%) at 0.05%,

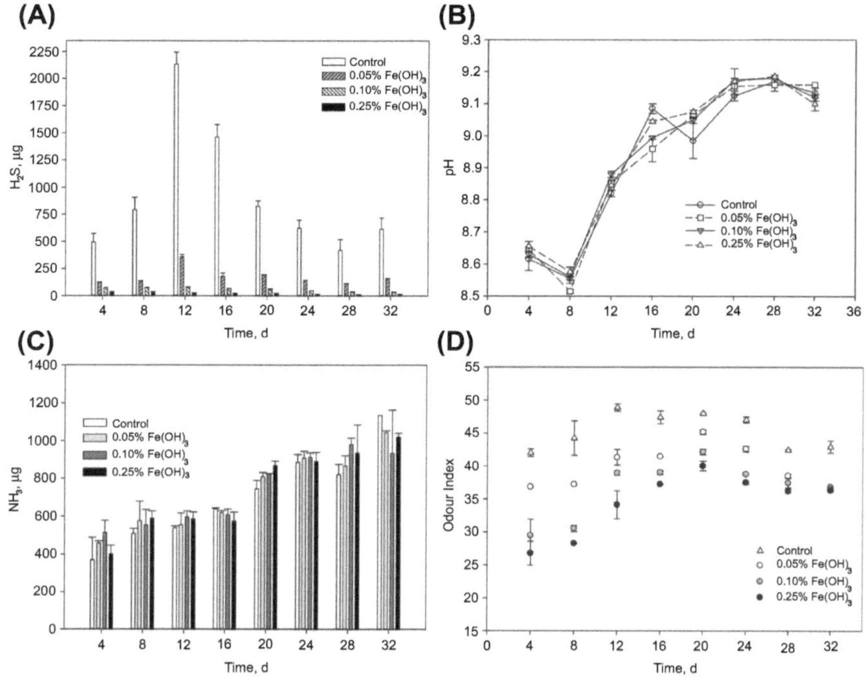

Figure 6.27 Effect of FH on odorous compounds, odor index and pH during anaerobic incubation of sludge.

0.10%, and 0.25 wt% of FH, respectively. The addition of FH reduced H_2S production dramatically. Repeated measures indicated that the corresponding removal rates were statistically significant ($p < .001$). The removal efficiencies were positively correlated with the FH dosage.

FH did not raise pH in sludge samples during the anaerobic incubation (Fig. 6.27B). The general trend was that pH was slightly decreased from about 8.64 to about 8.55 for the first 8 days. This might be contributed by the formation of organic acids during the decomposition of OM. When the organic acids further succeed the biological decomposition, the pH gradually rose to about 9.10 and then remained constant. The pH values in all sludge samples were about 9.10 at the end of the experiment. The production of NH_3 was not affected in the presence of FH. The release rate of NH_3 in dewatered sludge in the first 4-day period was at 369 μg, and then increased gradually to 1135 μg over the test time (Fig. 6.27C). The variation in NH_3 release rate over time varied significantly ($p < .001$). This was probably contributed by the increasing pH in sludge samples. The total amount

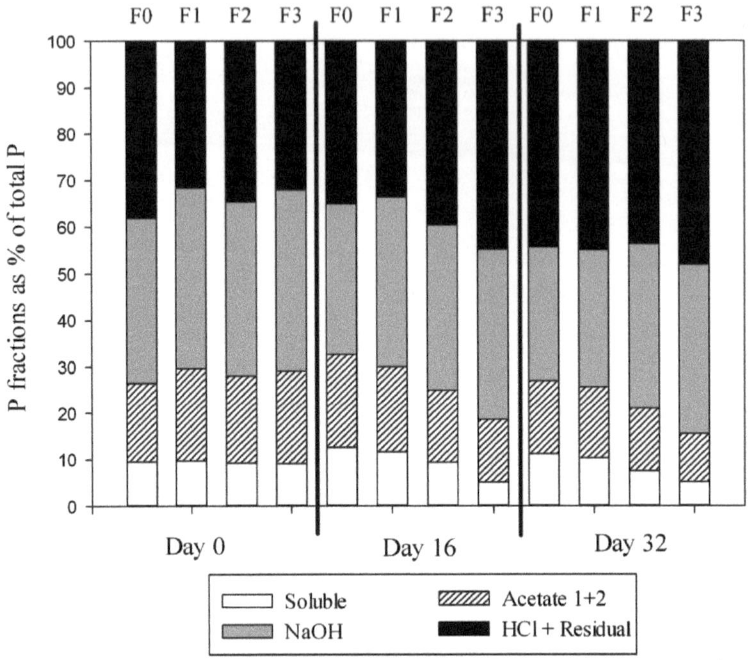

Figure 6.28 Fraction distribution of phosphorus before and after incubation of sludge at various rates of FH. F0: control, F1: 0.05% FH, F2: 0.10% FH, F3: 0.25% FH.

of NH_3 released were 5.64, 5.83, 5.92, and 5.86 mg, and the average NH_3 release rates were 176, 182, 185, and 183 μg per day in dewatered sludge at 0, 0.05, 0.10, and 0.25 g FH doses, respectively. The production of NH_3 between treatments and the control sample did not vary significantly. No evidence showed the trend of decreasing NH_3 release rate after the 32-day incubation. To certain degree, increasing doses of FH decreased VFAs' production, which was dominated by acetic, propionic, isobutyric, and isovaleric acids (Table 6.11). The removal efficiencies of these acids were respectively 72.0%, 73.2%, 75.0%, and 76.3% in the presence of 0.25% FH. Hydroxyl radical might be produced during the reductive dissolution of FH by sulfide. Apart from HS^-, it would be neutralized by organic acids including VFAs formed during the decomposition of OM in sludge. This could explain the changes of pH and the reduction of VFAs.

Odor indexes of the dewatered sludge were lessened by the addition of FH (Fig. 6.27D). The general trend of odor index was similar to H_2S emission; without FH, it increased from 42.1 to 49.0 during the first 12 days and then gradually decreased to 43.0 at the end of the experiment. The

Table 6.11 Effect of FH on VFAs' Production During a 32-day Anaerobic Incubation for Sludge

VFAs (μg)	Acetic	Propionic	n-Butyric	Isobutyric	n-Valeric	iso-Valeric
Control	155.13	45.90	2.21	20.63	3.85	15.25
0.05% FH	76.50	20.84	n.d.[a]	10.16	n.d.	7.14
0.10% FH	46.84	13.44	n.d.	5.17	n.d.	3.69
0.25% FH	43.55	12.30	n.d.	5.17	n.d.	3.61

[a]*n.d.*, not detected.

reduction for odor indexes ranged from 2.8 to 7.5, 5.1 to 13.8, and 5.3 to 16.0, accounted for the removal efficiencies (converted by Eq. 6.9) of 69% (48–83%), 84% (68–96%) and 89% (76–98%), respectively, in the FH doses of 0.05%, 0.10%, and 0.25%. The variation in odor indexes between the treatments and the control sample varied significantly ($p < .01$). It also should be noted that the discrepancy of odor index among different doses of FH was decreased from 10.0 to 0.5, and that the discrepancy between undosed and dosed sludge was also reduced. This phenomenon could be explained by the decline of H_2S emission rate over time.

6.6.3 Sulfur and Phosphorus Species Analysis

The sulfur species analysis before and after anaerobic incubation is shown in Table 6.12. Without FH, the incubation changed the concentrations of SO_4^{2-}—S, acid-volatile sulfide (AVS), Cr(II)-reducible sulfide (CRS), and elemental sulfur (ES) in the dewatered sludge from 197.6, 364.5, 46.1, and 1.6 μg/g in the beginning of incubation to, respectively, 71.2, 323.4, 50.9, and 2.6 μg/g at the end of 32-day incubation. The decrease of SO_4^{2-}—S concentration was owing to a sulfate-reducing by sulfate-reducing bacteria (SRB). The potential sources of AVS included dissolved S(-II) species (H_2S, HS^-, $FeHS^+$), FeS clusters, iron sulfide nanoparticles, mackinawite (FeS), pyrite (FeS_2), and greigite (Fe_3S_4). Pyrite (FeS_2) can be recovered using the AVS analysis procedure used. Thus, the decline of AVS concentration in sludge without FH was probably due to the emission of dissolved S(-II) species. It also found that the amounts of CRS and ES increased slightly over time. Chemical phosphorus removal (CPR) generally involves the addition of iron as coagulant, and is often adopted for phosphorus (P) removal in wastewater treatment, resulting in different iron species including Fe bound to organic materials, FHs and Fe phosphates in sludge. Thus, the increase of CRS and ES concentrations after incubation is probably owing to the reaction between FHs and sulfide.

Table 6.12 Changes of Sulfur Species Among Fractions During Aanaerobic Incubation of Sludge

Sulfur Species		Before Incubation	After 32-d Incubation	Difference
Sulfate sulfur (μg/g)	Control	197.6 (±14.8)[a]	71.2 (±0)	−126.4 (±14.8)
	0.05% FH	187.3 (±51.8)	90.8 (±1.8)	−96.5 (±51.9)
	0.10% FH	163.1 (±24.1)	63.1 (±2.7)	−100.0 (±24.2)
	0.25% FH	166.8 (±6.2)	71.2 (±5.4)	−95.6 (±8.2)
AVS (μg/g)	Control	364.5 (±1.1)	323.4 (±9.2)	−41.2 (±9.2)
	0.05% FH	339.7 (±27.5)	356.1 (±4.8)	16.4 (±27.9)
	0.10% FH	334.0 (±5.0)	412.3 (±35.8)	78.3 (±36.1)
	0.25% FH	318.1 (±2.1)	346.6 (±12.9)	28.5 (±13.1)
CRS (μg/g)	Control	46.1 (±2.4)	50.9 (±0.7)	4.5 (±2.5)
	0.05% FH	43.1 (±2.4)	57.7 (±4.5)	13.8 (±5.1)
	0.10% FH	42.7 (±2.7)	72.4 (±3.7)	28.0 (±4.6)
	0.25% FH	49.9 (±4.3)	75.8 (±0.7)	24.4 (±4.3)
ES (μg/g)	Control	1.6 (±0.2)	2.6 (±0.8)	1.1 (±0.8)
	0.05% FH	1.3 (±0.1)	2.5 (±0.7)	1.2 (±0.7)
	0.10% FH	1.4 (±0.1)	10.5 (±1.1)	9.1 (±1.1)
	0.25% FH	1.6 (±0.3)	5.9 (±2.3)	4.3 (±2.4)

[a]Standard deviation.

The changes of sulfur species among different fractions hinted the formation of FeS, FeS_2, and S^0. After 32-day incubation, the presence of FH at 0.05%, 0.10%, and 0.25% enhanced concentrations of AVS by 16.4, 78.3, and 28.5 μg/g sludge, CRS by 13.8, 28.0, and 24.4 μg/g sludge, and ES by 1.2, 9.1, and 4.3 μg/g sludge, respectively. A Chi-square test revealed significant differences in sulfate ($p < .01$), AVS ($p < .001$), CRS ($p < .001$), and ES ($p < .001$) between the treatments and the control sample. The increase of CRS and ES indicates the FeS_2 and S^0 are products of the oxidation reactions of sulfide with FH in dewatered sludge. The increase of AVS-S also implies that dissolved S(-II) formed during sulfate reduction by SRB can form insoluble iron (II) sulfide (FeS) during the reductive dissolution of FH. As a result, the formation of FeS_2, S^0 and FeS reduces H_2S emission from the dewatered sludge (Fig. 6.27A).

In the dewatered sludge sample without FH treatment, the percentages of H_2O-P, acetate-P, NaOH-P, and HCl/residual-P were 9.6%, 16.9%, 35.4%, and 38.1% before the incubation, as compared to 11.2%, 15.7%, 28.8%, and 44.3% after the 32-day incubation (Fig. 6.8). The NaOH-P fraction is considered to be soluble reactive P (SRP) from Fe(III)—hydroxy—P and organic P, the acetate-P is $MgNH_4PO_4$ (struvite)-P adsorbed to $CaCO_3$ and some P from amorphous Ca−P precipitates, and the HCl/residual-P is calcium phosphates. The decline of NaOH-P and the increase of HCl/residual-P were probably due to the degradation of organic P and the formation of calcium phosphates. The addition of FH changed the P fractions. After the 32-day incubation, the percentages of H_2O-P and acetate-P were 10.4% and 15.2%, 7.6% and 13.4%, and 5.2% and 10.3%, while the percentages of NaOH-P and HCl/residual-P were 29.5% and 44.9%, 35.4% and 43.6%, and 36.4% and 48.0% at 0.05%, 0.10%, and 0.25% of FH, respectively. The percentage of H_2O-P varied significantly between treatments and the control sample ($p = .001$). The above results indicated that soluble and loosely bound P were immobilized by the addition of FHs, forming calcium phosphates. The changes of the P fraction distribution would eventually affect the removal of by FH because phosphates could inhibit the reaction between FH and H_2S by binding Fe on the particle surface. However, the effect of phosphates on odor abatement in the presence of FH is not yet quantified, and requires further study.

6.6.4 Odor-H_2S Correlation

The correlation between odor index and H_2S concentrations was described in Fig. 6.29. Odor level correlates with H_2S concentration in a modified Steven's law:

$$C_{(OU)} = m[H_2S]^n \tag{6.10}$$

where $C_{(OU)}$ is the odor concentration and $[H_2S]$ is the H_2S concentration. The values of m and n vary with the composition of odor. The data from dewatered sludge fitted well to Eq. (6.10), with m = 543.27, n = 0.7118, and $R^2 = 0.8049$ (line A, *solid line*). Moreover, it fitted nicely when the data were divided into two phases based on incubation time, line B (*dotted line*, day 0−16 and day 24−32) with m = 104.15, n = 0.9475 and $R^2 = 0.9703$, and line C (*dotted line*, day 16−24) with m = 20,117.99, n = 0.5624, and $R^2 = 0.9692$. The phenomenon can be explained by the variation of odor compositions during the incubation, especially the emission of unknown volatile substances from dewatered sludge from day 16 to day 24. These results also imply that H_2S can be adopted by WWTP authorities as a measure of the odor level of ambient air at dewatered sludge disposal processes using a modified Steven's law when H_2S dominates the resulting odor emanating from sludge.

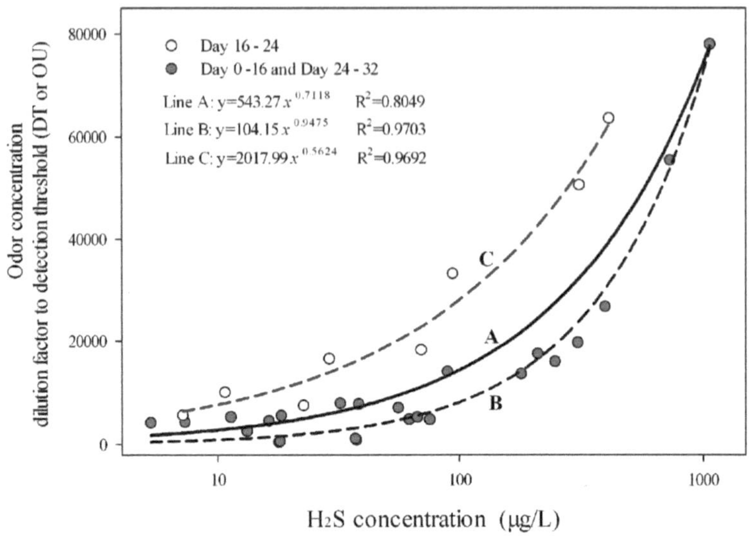

Figure 6.29 Correlation between odor and H_2S concentrations using a modified Steven's law relationship for sludge. Line A: fits for all data, Line B and C: fit for specific data divided based on incubation time (Line B: day 0–16 and day 24–32; Line C: day 16–24).

6.6.5 Mechanisms of Pyrite Formation

The mechanism of pyrite formation in sewage sludge is important to elucidate the process of sulfide removal by FH. The formation of pyrite was coupled to a process in which free sulfide was oxidized to form polysulfides, which again reacted with FeS to form pyrite (pathway 1). In pathway 1, elemental sulfur (ES) was the oxidant. However, the fact that only a small portion of ES was produced implied that pyrite was not formed by the pathway 1. Another mechanism (pathway 2) in which ferric iron (hydr)oxides are the oxidants to form polysulfides and subsequently pyrite.

$$8FeOOH + 5H_2S \rightarrow 8Fe^{2+} + S_5^{2-} + 2H_2O + 14OH^- \tag{6.11}$$

$$4FeS + S_5^{2-} + 2H^+ \rightarrow 4FeS_2 + H_2S \tag{6.12}$$

In summary,

$$8FeOOH + 8H_2S \rightarrow 4FeS_2 + 4Fe^{2+} + 8H_2O + 8OH^- \tag{6.13}$$

Moreover, pathway 1 should predominate at the oxygen-sulfide interface of very productive environments of high OM supply; pathway 2 might occur in those environments where decomposition rates are not high enough for the development of such a redox microstructure. Although there was high OM content, sewage sludge was incubated under strictly anaerobic conditions and the oxygen-sulfide interface did not exist, and thus pyrite was probably formed through oxidation of sulfide by FH. The fact that none of the FeS was produced in Eq. (6.12) was due to excess amount of ferric oxide. Furthermore, because the rate of reoxidation of ferrous iron to sulfide-reactive ferric oxide is too low, FeS could accumulate during our experiment. Thus pyrite formation is suggested by the following equation:

$$2FeOOH + 3H_2S \rightarrow FeS_2 + FeS + 4H_2O \qquad (6.14)$$

6.6.6 Implications for Use of Fe Hydroxide to Eliminate Sludge Odor

These results clearly demonstrate that FH can mitigate the odor in sludge over a long time. As H_2S is mainly responsible for the odor, its reaction with FH to form nontoxic FeS, FeS_2, and S^0 has more benefits than drawbacks. The reaction with FH does not destroy the sludge environments such as oxidation-reduction potential (ORP), pH, and biological activity of sludge bacterium compared to other conventional oxidants. This suggests that FH may be used in sludge disposal process including composting and anaerobic digestion in addition to dewatering and land application. The operating cost of this potential approach will be a key factor. Although the low dosage and long-lasting effectiveness may allow us to use pure FH (99.8%) for sludge odor control in sewage treatment plants, reuse of FH-containing waste is more desirable and cost-effective. In the reactions of Fenton and Fenton-like, which is employed for oxidative treatment of industrial wastewater, a large amount of FH sludge is generated, and needs to be properly disposed at a high cost. Thus, the application of FH sludge, rather than pure FH, for sludge odor control may be more economically and environmentally feasible.

6.7 STABILIZATION OF SEWAGE SLUDGE USING NANOSCALE ZERO-VALENT IRON (NZVI) FOR AN ABATEMENT OF ODOR AND IMPROVEMENT OF BIOGAS PRODUCTION

Nanoscale zero-valent iron (nZVI) received a lot of attention, especially its capability to reduce chlorinated benzenes and hexavalent chromium. The microscopic and spectroscopic studies suggested that nZVI in aqueous environment consists mainly of metallic iron and a surface layer of iron oxide. The presence of two nano-constituents in the core—shell: the metallic iron acts as the electron source and exerts a reducing character, while the oxide shell facilitates sorption and surface complexation, and meanwhile, permits electron passage from the metal core. This special core—shell structure of nZVI may have potential to control H_2S for a long term and improve methane production during anaerobic digestion.

Moreover, microscale or coarser ZVI has been tested as an electron-donor for methanogenesis in anaerobic sludge. As ZVI is believed to be readily utilized as a slow-release electron donor for methanogenesis in anaerobic sludge, increased surface area of nanoscale ZVI should have a positive effect on the methane and biogas production during sludge anaerobic digestion.

In this section, nZVI is investigated as a potentially alternative agent for the abatement of odor to improve biogas production during anaerobic digestion. Two commercial grade microscale iron powders are adopted for comparison. The odor reduction was evaluated by measuring the amounts of odorous compounds (H_2S and NH_3).

6.7.1 Effect of ZVI on H_2S and NH_3 Concentration in Biogas

Nanoscale zero-valent iron (nZVI) (20 nm) was obtained from Aladdin Chemistry Co., Ltd. (China). Two microscale zero-valent iron (mZVI) products were also utilized to compare with nZVI. One of them had particle diameter smaller than 74 μm and Fe content greater than 99.0% (Sinopharm Chemical Reagents Co., China, denoted as mZVI-1) and the other had particle diameter smaller than 19 μm and Fe content greater than 99.3% (Shanghai CNPC Powder Material Co., China, denoted as mZVI-2) (Fig. 6.30).

The result showed that nZVI could eliminate H_2S very effectively during sludge anaerobic incubation. Without ZVI, H_2S concentration in biogas was first increased from 1.16 to 7.90 μg/mL and then decreased gradually to 4.67 μg/mL (Fig. 6.31A). The addition of nZVI significantly reduced H_2S formation. During the 17-day incubation, the concentrations of H_2S in biogas were decreased by 98.0% (96.2—98.9%) at 0.10 wt% of nZVI, compared with 47.0% (33.8—60.6%) at 0.10 wt% of superfine iron particle (mZVI-2) and no change in the presence of mZVI-1, in which the average H_2S concentration in biogas was increased from 4.26 to 5.39 μg/mL during anaerobic incubation.

The addition of ZVI had little effect on NH_3 formation during anaerobic incubation. In general, the NH_3 concentration in biogas was at 3.73 μg/mL initially, and then increased to 6.97 μg/mL during the incubation period in

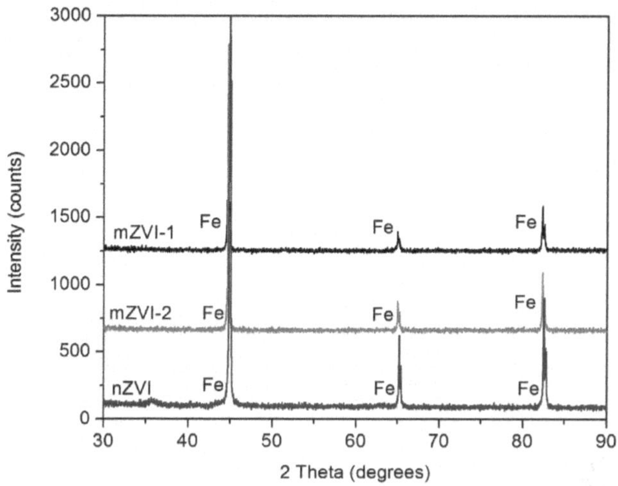

Figure 6.30 XRD spectrum of mZVI and nZVI used in this experiment. For clarity, mZVI-1 is moved 1200 units up and mZVI-2600 units in X-ray diffractograms.

sludge without ZVI (Fig. 6.31B). This phenomenon was possibly at least partly contributed by the increasing pH in sludge samples during anaerobic incubation. The average NH_3 concentrations in biogas were 5.6 μg/mL in the sludge without ZVI, compared to 5.5, 5.6, and 6.3 μg/mL in the sludge with mZVI-1, mZVI-2, and nZVI, respectively, suggesting no significant change in NH_3 release after 17-day incubation.

Compared with microscale ZVI, nZVI had demonstrated significantly high H_2S removal efficiency, probably due to its core–shell structure of nZVI. FeS may be formed when sulfide reacts directly with ZVI, as shown in Eq. (6.15).

$$Fe^0 + H_2S = FeS + H_2(g) \tag{6.15}$$

However, Eq. (6.15) would not be the main pathway for the abatement of H_2S during sludge anaerobic digestion. Previous studies indicate that nZVI can form a surface layer of iron oxide when situated in aqueous environment, and therefore the reaction may be simplified as Eq. (6.16).

$$Fe^0 + 2H_2O = FeOOH + 1.5H_2(g) \tag{6.16}$$

Because of its special structure and characteristics, the neoformed surface layer possesses a high selectivity for sulfides over other traditional oxidant alternatives, and provides a potential of long-term H_2S control during sludge anaerobic incubation. The chemical composition of the surface layer consists of a mixed Fe(II)/Fe(III) phase in proximity to the metallic core and a predominantly Fe(III) oxide phase at the surface of the nanoparticles. TGA-DTG

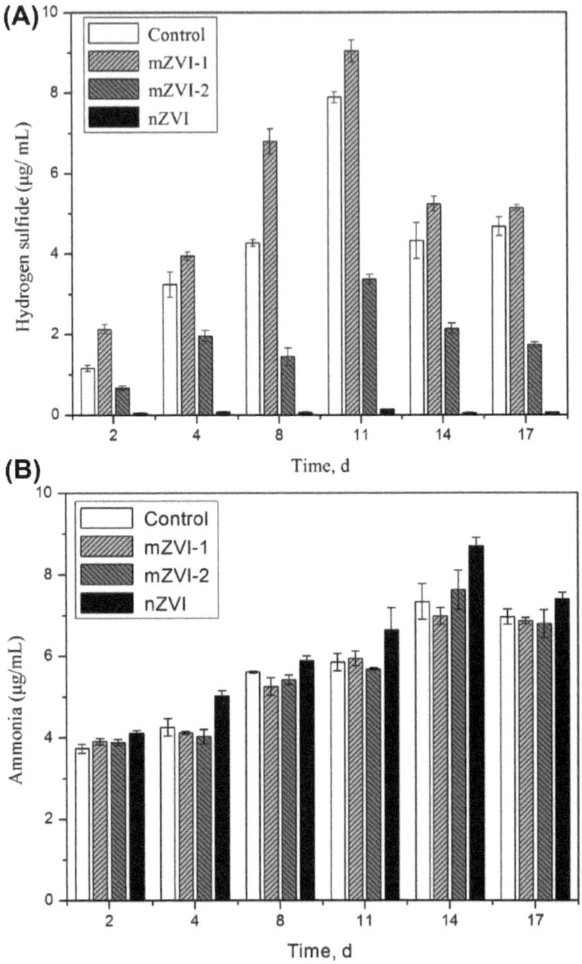

Figure 6.31 Effect of ZVIs on the H_2S (A) and NH_3 (B) concentration in biogas during sludge anaerobic incubation.

analysis further proved the presence of bound water at this layer when nZVI was dosed in sludge (Fig. 6.32), in which 0.1 wt% nZVI was added to sludge [TS (total solids) = 10 g/L] and anaerobic incubated for 20 days. The weight loss at the temperature of 100–150°C could be attributed to the water loss from the layer of nZVI particles.

The H_2S produced from the reduction of sulfate by SRB can effectively react with the hydrous Fe(II)/Fe(III) oxides layer of the nanoparticles via Eq. (6.17), in which S^{2-} was immobilized as FeS or oxidized as S_8:

$$2FeOOH + 3H_2S = 2FeS + 1/8S_8 + 4H_2O \tag{6.17}$$

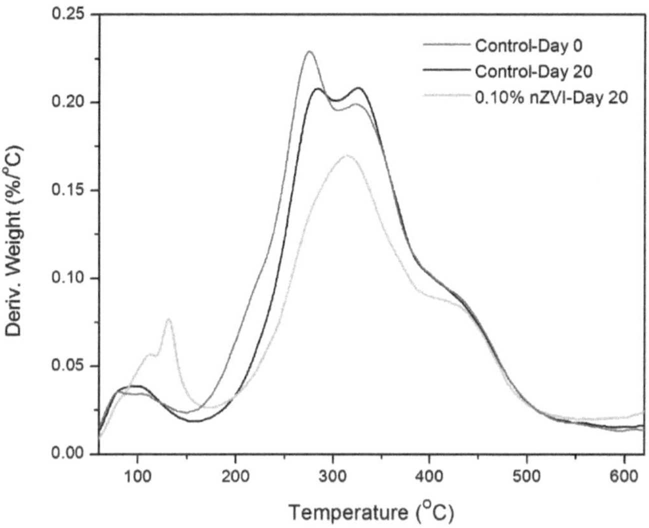

Figure 6.32 TGA/DTG analysis implicated the presence of bound water at the iron oxides layer in nZVI particle when situated in sludge.

Despite the frequent use of Eq. (6.17), oxidation product pyrite (FeS_2) has also been cited for reaction of sulfide, as in Eq. (6.18).

$$2FeOOH + 3H_2S = FeS_2 + FeS + 4H_2O \tag{6.18}$$

H_2S sequestration may take place predominately via pathway in Eqs. (6.4)—(6.5), rather than the pathway involving metallic iron (Eq. 6.15), in which nZVI is situated in a saturated H_2S solution. Sulfide formed by SRB may be immobilized according to Eqs. (6.15)—(6.18) to form FeS, S_8, and FeS_2, resulting in remarkable reduction of H_2S concentration in biogas.

Moreover, nZVI could kill SRB or inhibit SRB activity by coating the cells and therefore inhibiting cell proliferation. The deactivation of SRB by nZVI would thus decrease the production of H_2S. However, compared with the chemical sulfide immobilization, the deactivation of SRB probably contributed only a minor part for H_2S abatement because of the low dosage (0.1 wt%) of nZVI and high TS content in the sludge in this experiment. Perhaps, the different H_2S removal efficiencies of mZVI and nZVI can be the best explained by the relatively higher reactivity of nZVI with water than that of mZVI in forming hydrous Fe (II)/Fe (III) oxides layer, rather than the effect of deactivating SRB.

6.7.2 Effect of ZVI on Biogas Composition and Methane Production

Biogas upgrading (i.e., increasing the CH_4 content in biogas) was also observed in sewage sludge with ZVI during anaerobic digestion. Fig. 6.33 shows the biogas compositions from sludge anaerobic incubation treated with different ZVIs. The concentration of CH_4 in biogas gradually increased from 42.8% to 64.2% without ZVI, but from 48.2% to 70.6% in the presence of nZVI during incubation, accounting for 5.1−13.2% increase of CH_4 content in the presence of 0.10% nZVI. The addition of mZVI also had ability to improve CH_4 content in biogas, but at a lower degree. The average CH_4 concentrations in biogas were increased from 53.84% (without ZVI) to 54.56%, 56.89%, and 62.52% in the presence of mZVI-1, mZVI-2, and nZVI, respectively, during anaerobic digestion. The improvement of CH_4 concentration in biogas by ZVI is considered beneficial to the overall quality and usefulness of biogas.

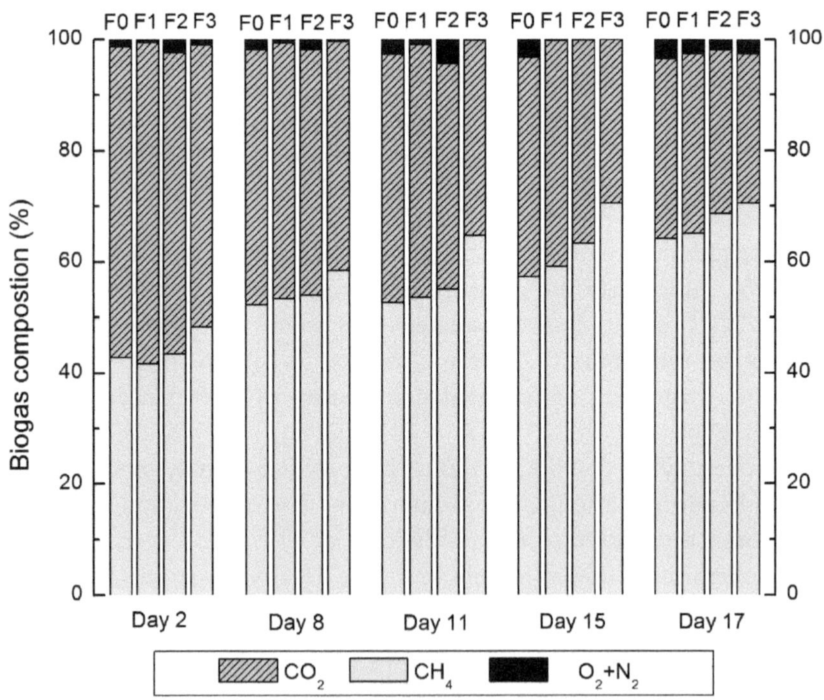

Figure 6.33 Biogas composition analysis during anaerobic digestion of sludge in the presence of ZVIs. F0: control; F1: 0.10% mZVI-1; F2: 0.10% mZVI-2; F3: 0.10% nZVI.

Fig. 6.34 shows data on the final cumulative volumes of methane and biogas produced during the anaerobic digestion. Without ZVI, the cumulative volumes of biogas and methane were 15.46 and 8.16 L. Adding 0.10% nZVI remarkably increased those respective volumes to 22.20 and 13.70 L, accounting for 30.4% and 40.4% enhancement. However, adding 0.10% mZVI-1 only slightly improved the respective volumes to 16.27 and 8.60 L, whereas mZVI-2 decreased the respective volumes to 12.36 and 6.77 L although the methane content in biogas was increased.

The conversion of ZVI and CO_2 to methane has been found with different species of hydrogenotrophic methanogens via Eq. (6.19), which would explain the general upgrade of biogas in the presence of ZVIs.

$$\begin{gathered} 8H^+ + 4Fe^0 + CO_2 \rightarrow CH_4 + 4Fe^{2+} + 2H_2O \\ \Delta G^{0\prime} = -150.5\ \text{kJ/mol}\ CH_4 \end{gathered} \tag{6.19}$$

The result of biogas and methane production in this study indicated that nZVI provided a much higher methane production compared to mZVI-1, suggesting that nZVI can act as an effective electron donor for methane production. Field studies had revealed that the microbial density of in the ZVI zone can be one to three orders of magnitudes higher than its surrounding

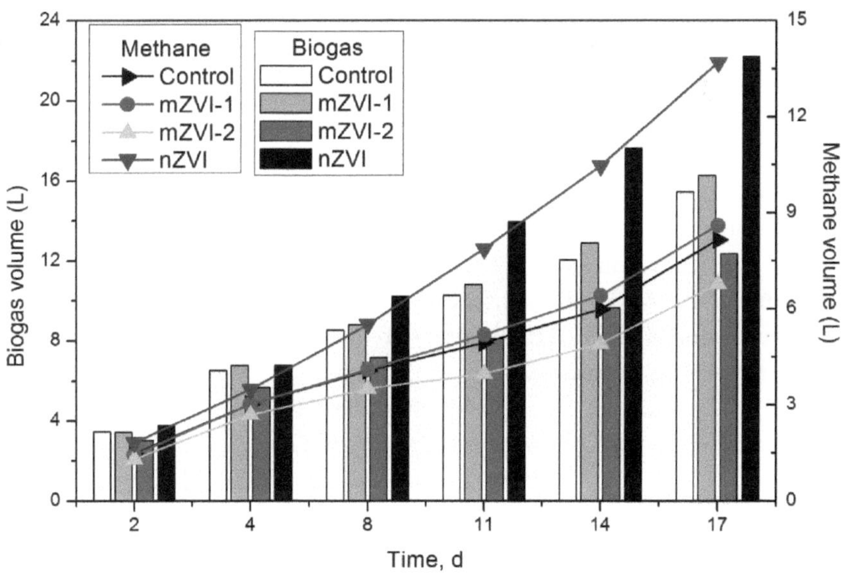

Figure 6.34 Final cumulative methane and biogas production during anaerobic digestion of sludge in the presence of ZVIs.

area. Therefore, the increase of biogas and methane production during anaerobic digestion may be attributed by the increase of methane-producing bacteria (MPB) activities and the augment of MPB biomass. Moreover, sulfide can severely impair methanogenesis (e.g., 60 mg S/L), and thus the immobilization of sulfide by nZVI can reduce the toxicity of sulfide to the MPB, and benefit CH_4 production. It appears that mZVI-2 does not provide evident benefit to the methane production. The effects of mZVI-2 on the SRB activity and the competition of SRB and MPB are not clear from our result, and should be further studied.

6.7.3 Effect of ZVI on Phosphorus Species' Distribution

Table 6.13 shows the P fractions according to the analysis of model compounds. In general, the proportion of soluble and acetate-extractable P (considered to be accessible by the microorganisms) was decreased in the presence of ZVI, while the proportion of NaOH-extractable P was elevated. The fractions of P extracted by ultrapure water (UPW) (represented as soluble or weakly bound P) and by 0.1 M acetate buffer decreased significantly from 778.7 to 1132.0 mg/kg in the control sludge to 490.9 and 1044.5 mg/kg for mZVI-1 treatment, 457.8 and 944.4 mg/kg for mZVI-2 treatment, and 449.0 and 713.3 mg/kg for nZVI treatment, respectively. The corresponding percentages of bioavailable fractions P of the total concentration were decreased from 76.8% to 66.9%, 62.3%, and 52.5%. Meanwhile, the proportions of NaOH-extractable P [containing P from Fe (III)-hydroxy-P and iron phosphate mainly] were elevated from 449.3 mg/kg (18.0%) to 662.8 (28.5%), 714.1 (31.8%) and 894.4 mg/kg (40.4%) for mZVI-1, mZVI-2, and nZVI treatments. The change of P species in the sludge with nZVI was similar to that in the sludge with hydrous ferric oxide.

The results of distribution of P species showed that the presence of ZVIs in the sludge immobilized bioavailable fraction P by forming Fe (III)-hydroxy-P or iron phosphates. Furthermore, the rates of P immobilization appeared to be positively correlated with the surface area of ZVIs. However, because of low level of iron phosphate in sludge, XRD analysis was not able to detect iron phosphates directly. To further clarify the main product of P immobilization, 0.1 wt% of nZVI was added into sludge with TS of 10 g/L and anaerobic incubated for 20 days. The dried sludge containing nZVI was analyzed by XRD, and the result (Fig. 6.35) showed clear increase in intensity of vivianite [$Fe_3(PO_4)_2$] in sludge with nZVI as the intensity of quartz (SiO_2) decreased, demonstrating that iron phosphate was formed to immobilize soluble P during anaerobic digestion. Decreased

Table 6.13 Effect of ZVI-Dosing on the Concentration (mg/kg) and Distribution of Phosphorus in Sludge after Anaerobic Incubation

	UPW	Acetate 1	Acetate 2	NaOH	HCl	Residual	Bioavailable Fractions[a]	Total Concentration
Undosed	778.72 (SD 5.53)	1041.89 (SD 13.34)	90.07 (SD 6.15)	449.26 [18.0%] (SD 6.04)	91.91 (SD 10.03)	37.42 (SD 2.68)	1910.68 [76.8%] (SD 15.70)	2489.27 (SD 19.76)
mZVI-1	490.90 (SD 7.82)	942.94 (SD 10.87)	101.55 (SD 0.03)	662.75 [28.5%] (SD 14.15)	81.21 (SD 5.34)	45.38 (SD 3.19)	1535.39 [66.0%] (SD 13.39)	2324.73 (SD 20.45)
mZVI-2	457.81 (SD 9.41)	654.82 (SD 9.09)	289.58 (SD 4.47)	714.10 [31.8%] (SD 7.37)	86.49 (SD 19.81)	46.24 (SD 3.26)	1402.21 [62.3%] (SD 13.83)	2249.04 (SD 25.47)
nZVI	448.99 (SD 16.44)	623.77 (SD 4.67)	89.50 (SD 4.65)	894.41 [40.4] (SD 18.30)	72.05 (SD 2.35)	83.50 (SD 5.50)	1162.26 [52.5%] (SD 17.71)	2212.22 (SD 17.71)
P species[b]	Soluble P, P weakly bound to particles	Struvite, P adsorbed to $CaCO_3$, some P from amorphous Ca-P precipitates		Soluble reactive P from Fe(III)-hydroxy-P, iron phosphate	Calcium phosphates	Tailing from calcium phosphates		

[a]Made up of the first three fractions for P (UPW, acetate 1 and acetate 2), and considered to be accessible by the microorganisms.
[b]Interpretation of phosphorus species.

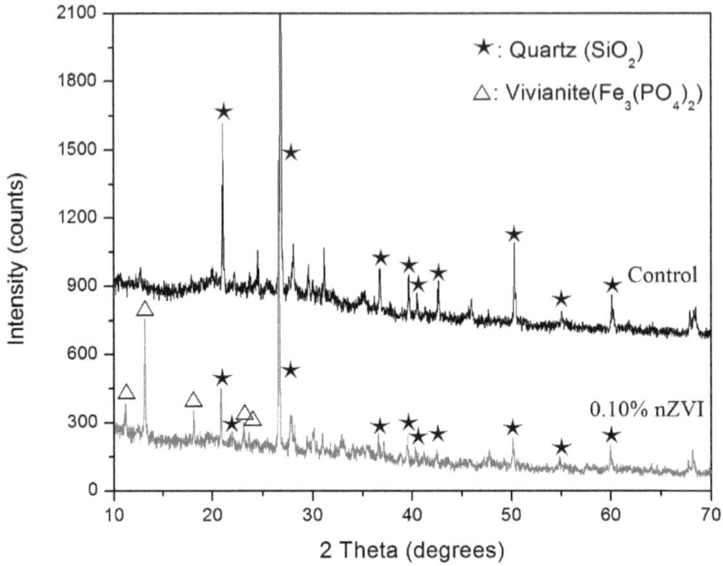

Figure 6.35 Formation of vivianite in sludge treated with nZVI. For clarity, control is moved 600 units up.

concentrations of bioavailable P in digesters could be a possible cause of impaired anaerobic digestion. However, the percentages of the bioavailable fractions P in the total concentration of P were still relatively high, and therefore it could be speculated that bioavailable P had little impact, if any, on sludge anaerobic digestion. However, many different parameters including the doses of ZVI, particle diameters (surface areas) of ZVI, sludge characteristics (moisture content and initial P concentration) may affect the bioavailable concentration of P, and therefore impact the anaerobic digestion process eventually.

6.7.4 Roles of nZVI in Sludge Anaerobic Digestion

Based on these findings, the roles of nZVI in sludge anaerobic digestion can be showed in Fig. 6.36 schematically: (1) NZVI reacted with H_2O to form a hydrous Fe(II)/Fe(III) oxides layer (Eq. 6.3) or reacted with previously accumulated sulfide in sludge directly to form FeS (Eq. 6.2); (2) the neoformed hydrous Fe (II)/Fe (III) oxides layer at the surface of the nanoparticles possesses a high reactivity toward sulfides, resulting in the abatement of H_2S in biogas during anaerobic digestion; (3) the hydrous Fe (II)/Fe (III) oxides layer at the surface of nZVI immobilized the soluble or weakly bound P as vivianite, and the lowered bioavailable P fraction

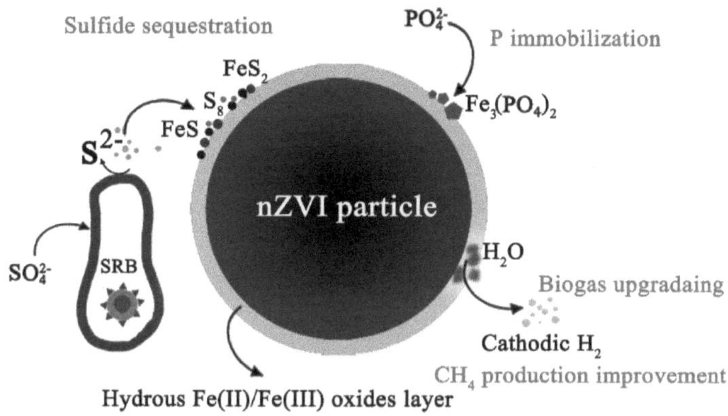

Figure 6.36 Roles of nZVI in sludge anaerobic digestion.

further impaired the anaerobic digestion process possibility; and (4) H_2 generated by the chemical reaction between H_2O and nZVI under anaerobic conditions upgraded the biogas and elevate the methane production during anaerobic digestion. In summary, abatement of odor and improvement of biogas production could be achieved by applying nZVI during sludge anaerobic digestion.

6.8 USE OF CORE–SHELL ZERO-VALENT IRON NANOPARTICLES FOR REMOVAL OF SULFIDE IN LONG-TERM SLUDGE ANAEROBIC DIGESTION

Nanoscale zero-valent iron (nZVI) particles have been widely used for treating hazardous and toxic wastes due to their small size, large specific surface area, and elevated reactivity for rapid contaminant transformation. During the reactions with water (surface hydroxylation), a thin surface layer (thickness of ca. 2–4 nm) mainly consisting of iron oxide (or amorphous oxide) will be formed at the agglomerate surface and between the individual particles. Two nanoconstituents are present in the core–shell structure. Metallic iron acts as the electron source and reductant, and the oxide shell facilitates sorption and surface complexation while allowing electrons to pass into the metal core. The core–shell structure provides nZVI with manifold functional properties. Several lines of studies have indicated that oxide shells possess the sportive properties of sulfide, and the long-term effectiveness of hydrogen sulfide removal in sludge anaerobic digestion remains unknown. The objectives of this section are to: (1) evaluate the long-term potential of nZVI for removing sulfide in biogas from biodigesters; (2) investigate the effects of nZVI on VS degradation and methane production; and

(3) clarify the profiles of phosphorus, iron, and reduced inorganic sulfur (RIS) in nZVI-dosed sludge.

6.8.1 Effect of nZVI on pH

Fig. 6.17A presents the pH trends of sludge in the presence of nZVI during anaerobic digestion. The nZVI-dosed sludge showed a steep increase in pH during the first 2 days. The pH increased from 6.5 in the control sludge to 7.0, 7.2, and 7.5 in the sludge treated with 0.05%, 0.10%, and 0.20% nZVI, respectively. In the subsequent period of 18 days, the pH continued to increase at a much slower rate and then gradually stabilized at the end of the experiment. The initial pH surge was attributed to rapid oxidation when nZVI entered the aqueous phase, as shown in Eq. (6.20).

$$Fe^0(s) + 2H_2O \rightarrow Fe^{2+} + H_2(g) + 2OH^- \qquad (6.20)$$

Next, as hydroxyl ions accumulated on the surfaces of the nZVI particles, the formation of a passivating oxyhydroxide coating (core–shell structure) slowed down the oxidation rate and increased the pH. The changes in pH during anaerobic digestion could be divided into three steps based on the different pH values with and without nZVI treatment (Fig. 6.37B). Step 1 included the formation of the core–shell structure of nZVI in the sludge within 2 days. Step 2 implied that the pH was buffered by organic acid anions, such as acetate derived from the decomposition of OM. Next, the changes in pH values between the undosed and nZVI-dosed sludges were gradually stabilized.

6.8.2 Effect of nZVI on Hydrogen Sulfide Concentrations in Biogas

The sludge without nZVI treatment produced biogas with an average hydrogen sulfide concentration of 300 mg/Nm^3 during 20 d of anaerobic digestion. Because of the consumption of sulfate by SRB over time, the concentration of hydrogen sulfide in the biogas gradually decreased from 663 to 84 mg/Nm^3. nZVI doping significantly ($p < .005$) reduced the average hydrogen sulfide concentration in the biogas to 6.1, 0.9, and 0.5 mg/Nm3 in the treatments with nZVI concentrations of 0.05%, 0.10%, and 0.20% (Fig. 6.38A), respectively. The corresponding removal efficiencies were 98.0% (96.8–100%), 99.7% (98.3–100%), and 99.8% (98.3–100%), respectively.

In the sulfate-rich sludge without nZVI, the introduction of 1000 mg/L of sodium sulfate increased the average hydrogen sulfide concentration from 312 to 3620 mg/Nm^3 during anaerobic digestion (Fig. 6.38B). With the

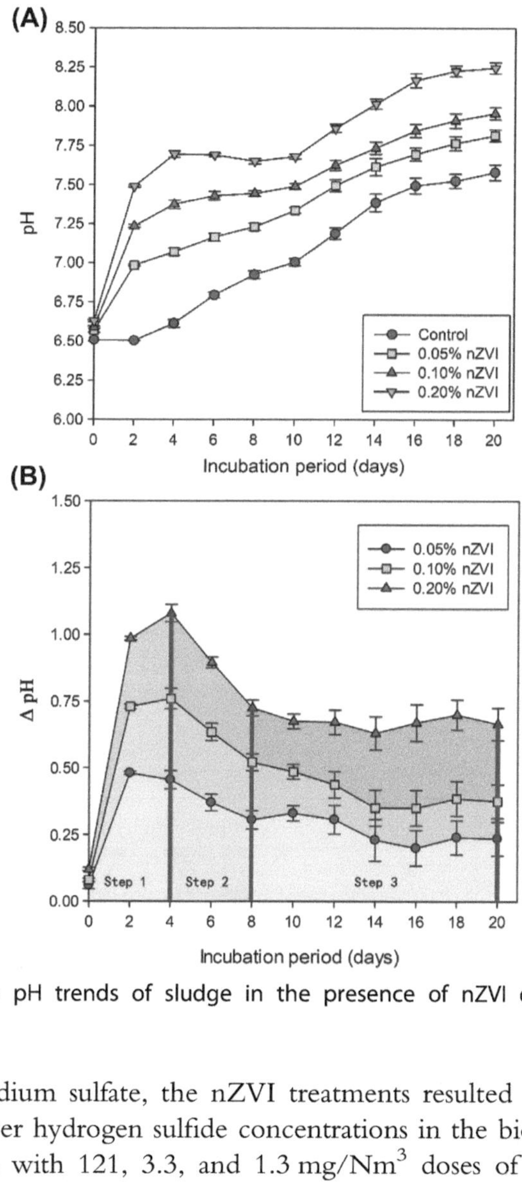

Figure 6.37 The pH trends of sludge in the presence of nZVI during anaerobic digestion.

addition of sodium sulfate, the nZVI treatments resulted in significantly ($p < .001$) lower hydrogen sulfide concentrations in the biogas relative to the treatments with 121, 3.3, and 1.3 mg/Nm^3 doses of nZVI (0.05%, 0.10%, and 0.20%, respectively). The corresponding removal efficiencies reached 97.2% (91.7–99.6%), 99.9% (99.7–100%), and nearly 100%, and the removal efficiencies were positively correlated with the nZVI dosages. Generally, the addition of 0.05% nZVI results in an average hydrogen sulfide concentration of less than 10 mg/Nm^3 (or 150 mg/Nm^3 of sulfate-rich sludge), which was lower than most of the manufacturers' specifications

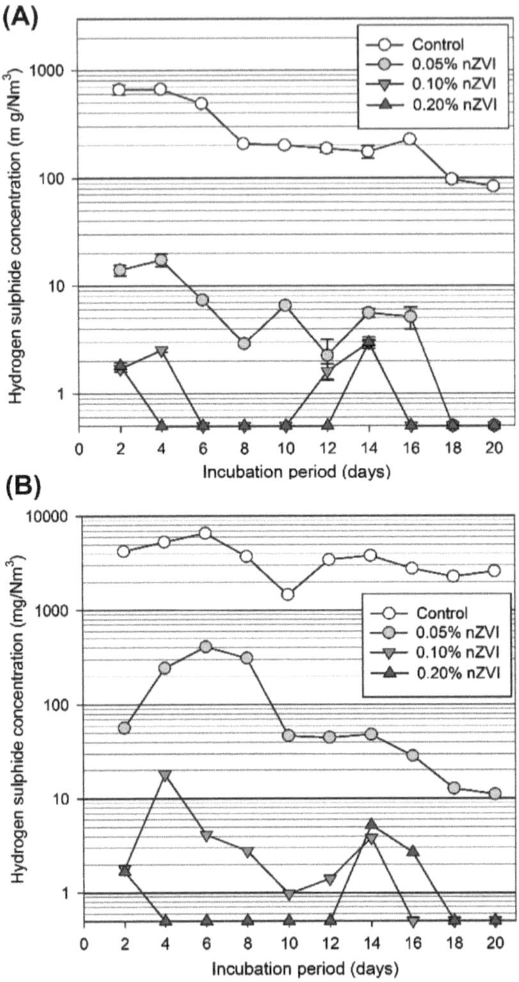

Figure 6.38 Effect of nZVI doping on hydrogen sulfide concentration in biogas (A: without sulfate addition, B: with the addition of 1000 mg/L sodium sulfate).

for power generation. Furthermore, a dose of 0.10% NZVI can decrease the average hydrogen sulfide concentration to less than 1 mg/Nm^3 (or 5 mg/Nm^3 of sulfate-rich sludge), which could meet the requirements of other uses (e.g., fuel for vehicles).

When situated in aqueous environments, an oxide layer will be formed on the surface of the nZVI. The chemical composition results indicated that the oxide layer consisted of two phases, a mixed phase of Fe(II)/Fe(III) near the metallic core and a phase of predominantly Fe(III) oxide near the oxide/water interface. The surface of the oxide invariantly contains hydroxide groups after exposure to water, resulting in an apparent surface stoichiometry that is similar to that of FeOOH. The characterization of the oxide layer was similar to that of FH (FeOOH), which was an effective and long-lasting reagent for removing hydrogen sulfide from sewage sludge in anaerobic environments according to our previous study.

6.8.3 Effect of nZVI on Methane Production and Volatile Solid Degradation

Fig. 6.39 shows the cumulative volumes of methane produced during 20 d of anaerobic digestion. The volume of methane is a useful measure of sludge treatment because it is directly related to the combustion capability in electric power plants or CHP plants. The methane production in the presence of different doses of nZVI varied. With respect to methane production, nZVI dosages of 0.05% and 0.10% increased the methane production by 9.8% and 4.6%. However, the methane production decreased by 8.8% at the elevated nZVI dose (0.20%). ZVI can produce hydrogen gas and stimulate sulfate

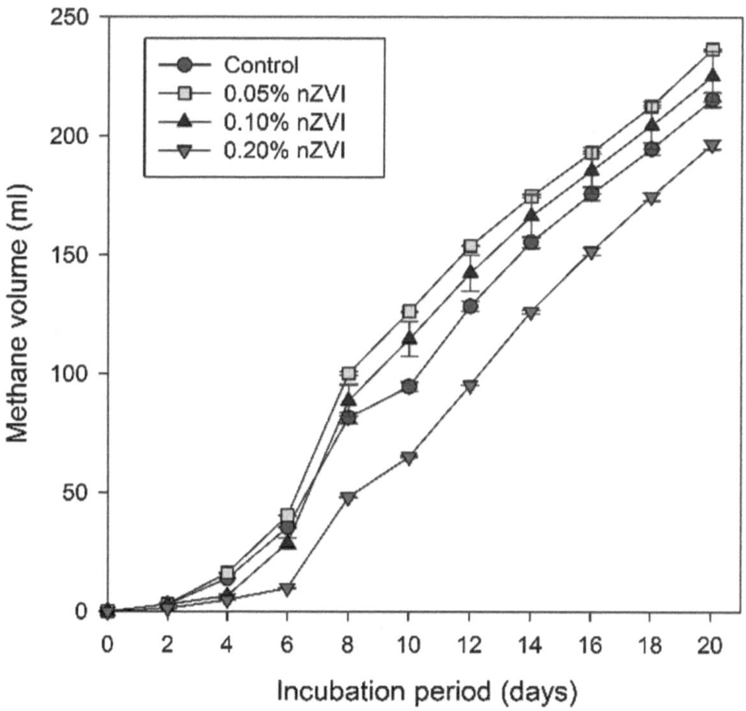

Figure 6.39 The final cumulative volumes of methane produced during 20-d anaerobic digestion for sludge.

reducers and methanogenic populations. The metallic iron core could act as a slow-release electron donor. Enhancement of the proportion of acetate conversion, acidogenesis, under anaerobic conditions by adding iron powder has been reported. However, the biotoxicity of nZVI was also recorded due to the cell membrane disruption when small nanoparticles enter the cell. The effects of each factor on methane yield were not quantified, and further studies are needed to distinguish between these potential mechanisms.

Fig. 6.40 shows the differential thermal gravimetric analysis (DTG) data of sludge samples after 20 d of anaerobic digestion. As suggested by a peak in the DTG curves, the weight loss occurred at approximately 100−150°C in the nZVI-dosed sludge samples, which could be attributed to the evaporation of bound water in the core−shell structure of the nZVI when situated in the sludge. The largest sludge sample peaks appeared at 175−550°C, mainly as a result of OM pyrolysis. The introduction of nZVI and the biodegradation of OM could both result in decreased peak intensities at approximately 175−550°C in the nZVI-dosed sludge samples.

Due to the formation of the core−shell structure of nZVI in the sludge, the difference of LOI550-LOI105 cannot reflect the OM content in the sludge. Weight loss of the sludge samples at 175−550°C by TG analysis

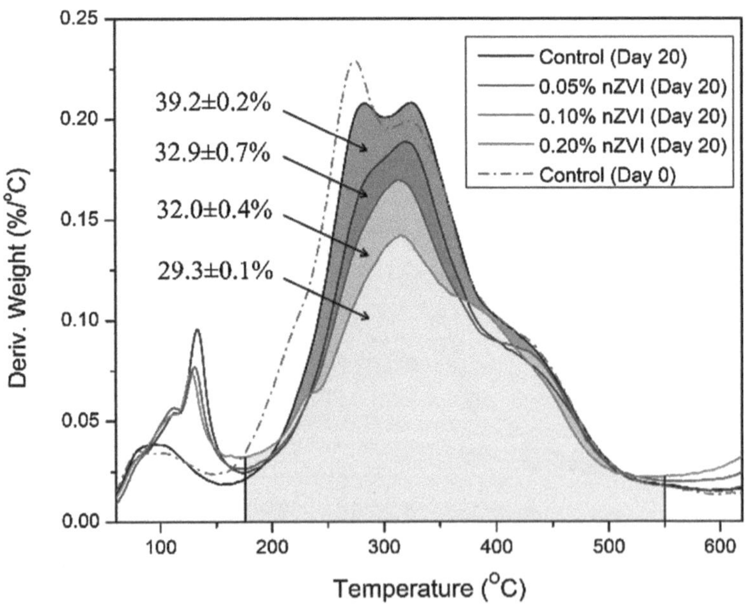

Figure 6.40 TG-DSC curves of sludge after 20 d anaerobic digestion.

was attributed to the OM content in the sludge samples. Based on this assumption, the VS contents (%, dry basis) decreased from 39.2% to 32.9%, 32.0%, and 29.3% in the treatments with 0.05%, 0.10%, and 0.20% nZVI after 20 d of anaerobic digestion. Because the amended Fe would change the composition of the solids, which would change the fraction of VS generated per g of total solids, the values of VS in g/L were calculated according to Eq. (6.21). The VS contents (g/L, wet basis) of the sludge decreased from 3.38 ± 0.02 g/L to 3.14 ± 0.07 and 3.24 ± 0.04 g/L in the presence of 0.05% and 0.10% nZVI, and were not different from those in the 0.20% nZVI-dosed sludge (3.39 ± 0.01 g/L). The methane production and VS degradation implied that nZVI doses of 0.5−0.10% could accelerate sludge stabilization during anaerobic digestion. Methane production could be inhibited at elevated NZVI doses (e.g., 0.20%) due in part to limited soluble P availability.

$$\mathrm{VS(g/L, wet\ basis)} = \mathrm{VS(\%, dry\ basis)} \times \mathrm{TS(g/L)} \tag{6.21}$$

6.8.4 Effect of nZVI on the Distributions of Phosphorus, Iron, and RIS Species

Table 6.14 demonstrates the fractionation profile of P for sludge after anaerobic digestion. The addition of nZVI reduced the proportion of H_2O-P and acetate-P and elevated the proportion of NaOH-P. A chi-square test revealed significant differences in the P species distributions between the treatments and the control ($p < .001$). The percentages of the bioavailable fractions of P (H_2O-P + acetate-P) relative to the total concentrations were significantly decreased ($p < .001$) from 58.1% to 5.32%, 0.78%, and 0.38% in the treatments with 0.05%, 0.10% and 0.20% NZVI, respectively. Among the fractions accessible by the microorganisms, the fraction of H_2O-P significantly decreased from 79.4 mg/L in the control to 0.4, 0.2, and 0.2 mg/L for the three nZVI dosages. Meanwhile, the P in Fe(III)-hydroxy-P and iron phosphate (mainly extracted by NaOH) increased from 73.2 mg/L to 200, 209, and 206 mg/L for the nZVI treatments of 0.05%, 0.10%, and 0.20%.

The P species distribution results indicated that soluble and loosely bound P were immobilized by the addition of nZVI due to formation of Fe (III)-hydroxy-P or iron phosphates. Further XRD analysis (Fig. 6.41) confirmed a clear increase in the intensity of vivianite [$Fe_3(PO_4)_2$] in sludge treated with nZVI. The lower concentrations of bioavailable P (especially soluble P) in the digest reactors could impair digestibility by limiting soluble

Table 6.14 Effects of nZVI-Dosing on the Concentration and Distributions of Phosphorus in the Sludge After Anaerobic Digestion

	H_2O-P		Acetate-P					
Treatments	**Soluble**	**UPW**[a]	**Acetate 1**	**Acetate 2**	**NaOH**	**Residual**	**Bioavailable Fractions**[b]	**Total Concentration**
Undosed	66.05 ± 0.04	13.38 ± 0.05	39.91 ± 0.08	11.54 ± 0.29	73.17 ± 0.21	21.16 ± 0.61	130.88 ± 0.31 58.12%	225.21 ± 0.71
0.05% nZVI	0.10 ± 0.01	0.30 ± 0.04	3.36 ± 0.01	8.57 ± 0.39	200.10 ± 0.30	19.55 ± 0.76	12.33 ± 0.39 5.32%	231.98 ± 0.91
0.10% nZVI	0.08 ± 0.00	0.14 ± 0.00	0.40 ± 0.00	1.17 ± 0.01	208.66 ± 6.62	18.33 ± 0.03	1.79 ± 0.01 0.78%	228.78 ± 6.62
0.20% nZVI	0.11 ± 0.01	0.12 ± 0.00	0.36 ± 0.01	0.27 ± 0.02	205.70 ± 8.54	19.95 ± 0.32	0.86 ± 0.02 0.38%	226.51 ± 8.85
P species[c]	Soluble P species	P Weakly bound to sludge particles	Struvite; P adsorbed to $CaCO_3$; P from amorphous Ca-P precipitates	Soluble reactive P from Al, or Fe. Organic P	Iron phosphate			

[a]Ultrapure water (18 MΩ cm).
[b]Made up of the first four fractions for P (soluble, UPW, acetate 1, acetate 2) and considered to be accessible by the microorganisms.
[c]Interpretation of phosphorus species.

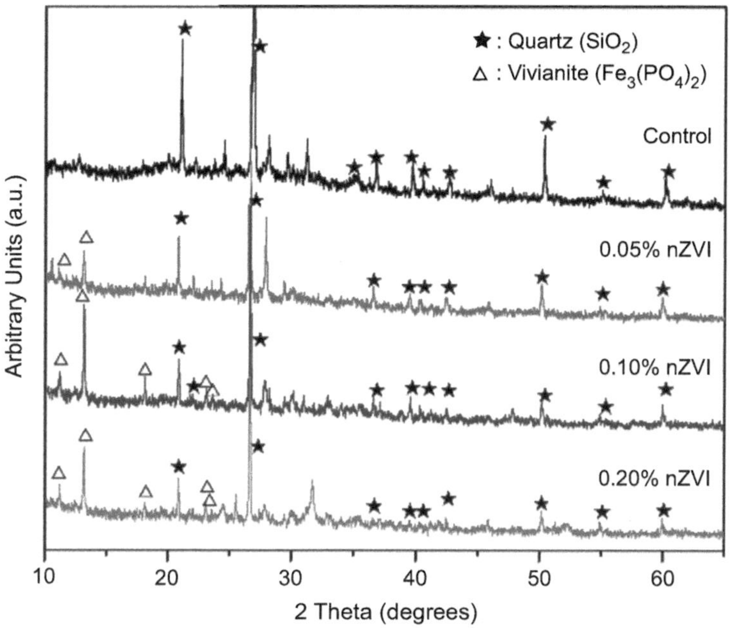

Figure 6.41 XRD spectra of the crystalline phases in sludge after anaerobic digestion.

P, which would further decrease biogas production and decelerate sludge stabilization. Another interesting point is that the immobilization of P by the addition of nZVI could help recover P in phosphate-containing wastewater, especially in sludge when using alkali extraction technology.

The characteristics of Fe in the sludge are shown in Fig. 6.42. Understandably, the addition of nZVI markedly increased the total Fe content in the sludge from 148.6 mg/L for the control to 617.8, 1155, and 1905 mg/L when nZVI was added at rates of 0.05%, 0.10%, and 0.20%, respectively. The concentrations of bioavailable Fe extracted in the first three fractions (soluble + UPW + KNO_3 + KF) were not vastly different in the presence of nZVI. Our results indicated that adding nZVI resulted in the conversion of most of the Fe to the fraction of Fe hydroxides or phosphates during anaerobic digestion after 20 days of anaerobic digestion. The formation of Fe hydroxides and Fe phosphate most likely resulted from the H_2O-nZVI reaction and resulted in immobilization of the bioavailable P fraction. Moreover, the determination of Fe-ASC indicated the presence of a boundary of reactive iron oxides in the nZVI-dosed sludge after anaerobic digestion. The amorphous Fe concentration increased from 69 mg/L to 466, 760, and 1644 mg/L when nZVI was added at 0.05%, 0.10%, and

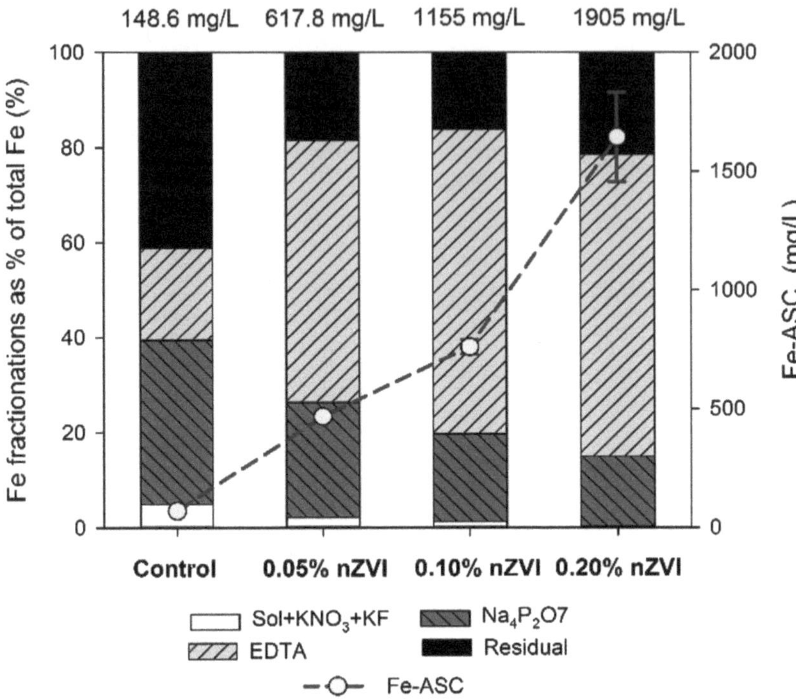

Figure 6.42 Effects of nZVI-dosing on the distribution of iron in sludge after anaerobic digestion. Interpretation of iron species, $Na_4P_2O_7$-Fe: Organically bound Fe; EDTA-Fe: iron hydroxides and Fe-hydroxy-phosphates; and Residual Fe: tailings from Fe—P, Fe phosphate and FeS.

0.20%, respectively, which implied that reusing nZVI-dosed sludge to eliminate hydrogen sulfide could be conducted to reduce operation costs.

The distribution of RIS species was determined to elucidate the effects of nZVI in sulfide reaction, as shown in Fig. 6.43. After 20 days of incubation, the presence of nZVI at 0.05%, 0.10%, and 0.20% enhanced the acid-volatile sulfide (AVS) concentrations by 3.85, 4.42, and 7.49 mg/L, the Cr(II)-reducible sulfide (CRS) concentrations by 0.12, 0.15, and 0.14 mg/L and the ES concentrations by 0.08, 0.11, and 0.11 mg/L, respectively. Repeated ANOVA tests revealed significant differences in AVS ($p < .05$) and CRS ($p < .05$) between the treatments and the control. Potential sources of AVS include dissolved S(-II) species (H_2S, HS^-, $FeHS^+$), FeS clusters, iron sulfide nanoparticles, mackinawite (FeS), pyrite, and greigite (Fe_3S_4). Therefore, the increase of AVS-S in sludge treated with nZVI suggested that the dissolved S(-II) produced during sulfate reduction was immediately immobilized in the form of insoluble FeS, potentially due to

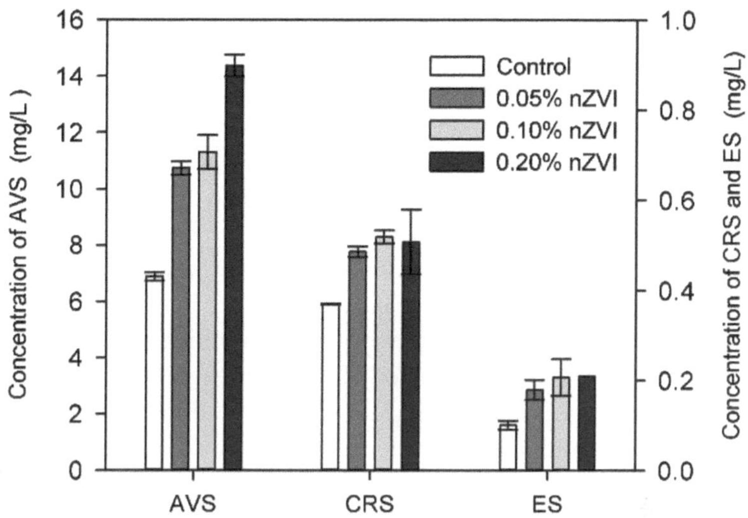

Figure 6.43 Effects of nZVI-dosing on the distribution of reduced inorganic sulfur (RIS) species in sludge after anaerobic digestion.

the direct reaction between sulfide accumulated in the sludge and ZVI, as shown in Eq. (6.22).

$$Fe^0 + H_2S = FeS + H_2(g) \tag{6.22}$$

The variations in CRS-S (pyrite) and ES between the treatment and control suggested that FeS_2 and S^0 were by-products of the reactions between hydrogen sulfide and nZVI during sludge anaerobic digestion. nZVI can form a surface layer of iron oxide when situated in the aqueous environment, as shown in Eq. (6.23) (for simplification).

$$Fe^0 + 2H_2O = FeOOH + 1.5H_2(g) \tag{6.23}$$

Apart from Eq. (6.22), the hydrogen sulfide produced from the reduction of sulfate by SRB can effectively react with the hydrous Fe(II)/Fe(III) oxide layers of the nanoparticles via Eqs. (6.24)—(6.25), in which S_2- was immobilized as FeS or FeS_2 or oxidized as S_8:

$$2FeOOH + 3H_2S = 2FeS + 1/8S_8 + 4H_2O \tag{6.24}$$

$$2FeOOH + 3H_2S = FeS_2 + FeS + 4H_2O \tag{6.25}$$

The distribution of RIS indicated that most of the hydrogen sulfide produced was immobilized as iron (II) sulfide and that a minor portion of the RIS was oxidized to pyrite and elemental sulfur on the surface of nZVI.

According to the above-mentioned results, the nZVI presented multiple physical-chemical properties during sludge anaerobic digestion that were closely correlated with the core–shell structure. The hydrogen and hydroxyl ions generated by the Fe^0-H_2O reaction under anaerobic conditions favored an increase in pH. After the core–shell structure was formed when nZVI was added to the aqueous environment, the nanoscale surface oxide layer significantly decreased the hydrogen sulfide concentration in the biogas by forming iron sulfide (mostly), pyrite and elemental sulfur, which improved the quality of the biogas and was beneficial for energy use and odor abatement. The methane yield was enhanced at a low dose of nZVI (0.05–0.10%) but decreased as the dose was elevated (0.20%). Methane production and VS degradation analyses implied that doses of 0.5–0.10% nZVI could accelerate sludge stabilization in anaerobic digestion. However, the immobilization of soluble or weakly bound P through the formation of vivianite with nanoscale oxide layers at the surface of nZVI reduced the bioavailable P fraction and could further inhibit methane production processes at elevated nZVI doses (e.g., 0.20%).

6.9 TREATMENT OF AGED-LANDFILL-LEACHATE USING AGED-SLUDGE-BASED BIOREACTOR

Landfill is the main disposal method for sludge generated in municipal wastewater treatment plant. The mineralization extents of the aged sludge are certainly dependent on the age of sludge placed in landfills. The resultant stabilized sludge is defined as "aged sludge" with properties of about 10% in OM and less than 50% in moisture content. The aged sludge contains a great number of microorganisms, which have strong decomposition capability and are very resistant to suboptimal environmental conditions. The permeability of the sludge is greatly improved as a result of the formation of a large quantity of microporous structure. When the leachate leaches through the bioreactor, due to the strong adsorption capability of aged sludge, a lot of organic and in OMs are adsorbed. The microorganisms present in the aged sludge then degrade the absorbed contaminants and lead to a decrease of COD and $NH_4^+ - N$.

6.9.1 Collection of Aged Sludge and Landfill Leachate

Aged sludge was excavated from a column, which had stabilized for 5-year aiming at simulation of the stabilization process of the sludge at landfill, and was used as the feeding materials of the bioreactor. Externally, the aged sludge looks like soil and smells almost inodorous. Three parallel samples, taken from the top, middle, and bottom of the column respectively, were mixed together and used for analyzing the characteristics. The parameters typically used for the analysis of soil were chosen to determine the properties of aged sludge, including WC, pH, OM, cation exchange capacity (CEC), total nitrogen (TN), total phosphorous (TP), and some microorganism parameters, etc (Table 6.15). It should be noticed that aged sludge used for bacteria determination was obtained from a bioreactor (aged sludge: coal ash, 10: 0) after one month of leachate treatment. The amount of nitrobacter and nitrosomonas was much higher than that of denitrobactor, indicating a strong capability for converting ammonia nitrogen to nitrate or nitrite, but a weak capability for transforming nitrate or nitrite to nitrogen.

Aged leachate was sampled from Shanghai Refuse Landfill, the largest landfill in China (Table 6.16). The biodegradation of contaminants in leachate was assessed on the basis of reduction in the concentration denoted by COD, BOD, TOC, and total carbon (TC) and changes in TN, $NH_4^+ - N$ and TP. Lower BOD/COD and higher TN/COD means that the leachate was extremely difficult to be biologically treated.

6.9.2 Aged-Sludge-Based Bioreactor

Coal ash (48% of SiO_2 and 35% of Al_2O_3) taken from a power plant in Shanghai was added to the bioreactor to increase the permeability of filling

Table 6.15 Chemical Parameters and Bacteria Population in the 5-Year Old Aged Sludge

Parameters	WC (%)	OM (g/kg)	CEC (Milliequivalent/ 100 g Dried Aged Sludge	SC (μS/cm)	TN (g/kg)	TP (g/ kg)	pH
	50.3	95.6	96.7	222	7.39	5.12	7.38
Type of bacteria	*Coliform*	*Nitrobacter*	*Nitrosomonas*	*Denitrobacter*	Total bacteria		
Number/g dry sludge	4.21×10^3	2.54×10^6	3.56×10^5	3.32×10	2.87×10^7		

Table 6.16 Typical Parameters of the Aged Leachate Used (mg/L, Except pH)

COD	$NH_4^+ - N$	TN	TOC	TC	BOD	pH
581–2618	317–1569	1158–1314	161–257	1150–1689	50–250	8.1–8.5

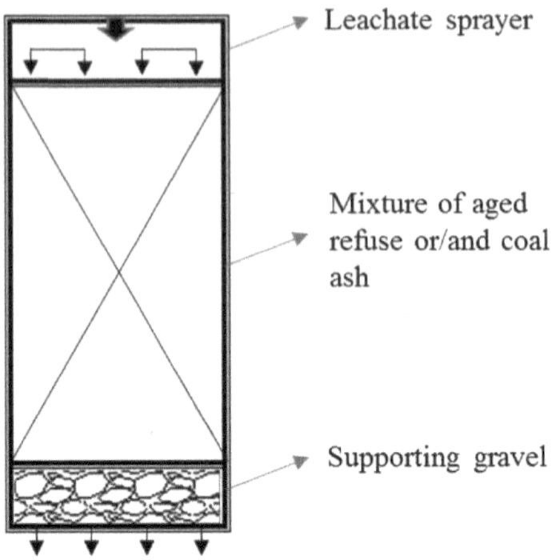

Figure 6.44 Schematic of aged-sludge-based bioreactor.

mass. Four cylindrical bioreactors (denoted as C1, C2, C3, and C4), with effective inner diameter of 15 cm, were filled with the mixture of 3 kg (in dry basis) of 5-year old aged sludge and coal ash with ratios of 10: 0, 10: 1, 10: 2, 10: 3 (by weight on dry basis), respectively (Fig. 6.44). As a result, the mixture height was 50 cm, and the initial moistures of the mixtures were around 48−50%, depending on the coal ash ratios. Stones (<1 cm in diameter) were used as supporting materials installed at the bottom of bioreactors.

Leachate was sprayed over the aged-sludge surface through sprayer at two to four intervals a day and 0.25−4 h per interval. The flow rate was 2.1 mL/min. Effluent was collected from the collection pipe. COD, BOD, TC, IC, TOC, TN, and $NH_4^+ - N$ concentrations, and pH in inlet and outlet leachate were monitored by conventional methods, and described as percent removal.

6.9.3 Effect of Domestication Process

Leachate was introduced into the bioreactors continuously for one month with a typical flow rate of 16.7 mL/min and spraying time of 30 min/d (i.e., 0.5 L/d for each bioreactor). This process can be called domestication process. The effluent discharged from the bottom of the bioreactor was sampled for analysis. Both influent and effluent characteristics at domestication period were illustrated in Fig. 6.45. It can be seen that the effluent COD increased with the increase of the influent while the effluent $NH_4^+ - N$ was nearly independent on the influent. Effluent COD

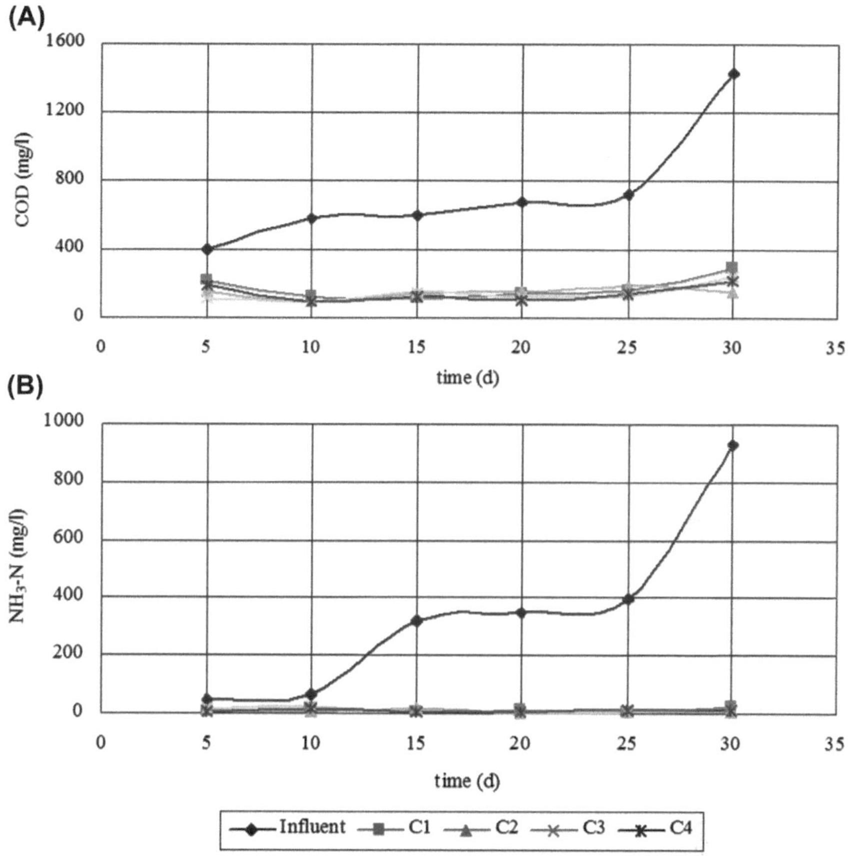

Figure 6.45 Removal of COD and NH_3—N in domestication process at hydraulic load of 70.8 L/m^3/d using the aged-sludge bioreactor.

and $NH_4^+ - N$ were below 250 mg/L and 15 mg/L when the corresponding influent COD and $NH_4^+ - N$ were 1429 mg/L and 928 mg/L, respectively, with a removal percentage of over 80% and 98% in this period.

C1 bioreactor (without coal ash) was blocked and the treatment efficiency deteriorated, but can be recovered when the surface aged sludge was overturned intermittently. It was also observed that the addition of coal ash in the aged sludge could improve the permeability of the mixture while the treatment efficiency can be kept unchanged when aged sludge/coal ash ratio was below 10: 3 (by weight on dry basis). Higher coal ash ratios would lead to a deterioration of effluent.

6.9.4 Effect of Hydraulic Loading

As shown in Fig. 6.46, higher hydraulic loading led to a lower removal efficiency of contaminants. The highest removal of COD and $NH_4^+ - N$ was 83% and 99.6% at a hydraulic load of 17.4 L/m^3/d, compared with 62% and 94% removals at 70.8 L/m^3/d. It was also found that the treatment efficiencies of C1, C2, C3, and C4 were quite consistent, regardless of the presence of coal ash, though C1 bioreactor may work slightly better than the other three bioreactors. The COD and $NH_4^+ - N$ concentrations in effluent were always observed to be below 200–250 mg/L and 2–10 mg/L at the influent concentrations.

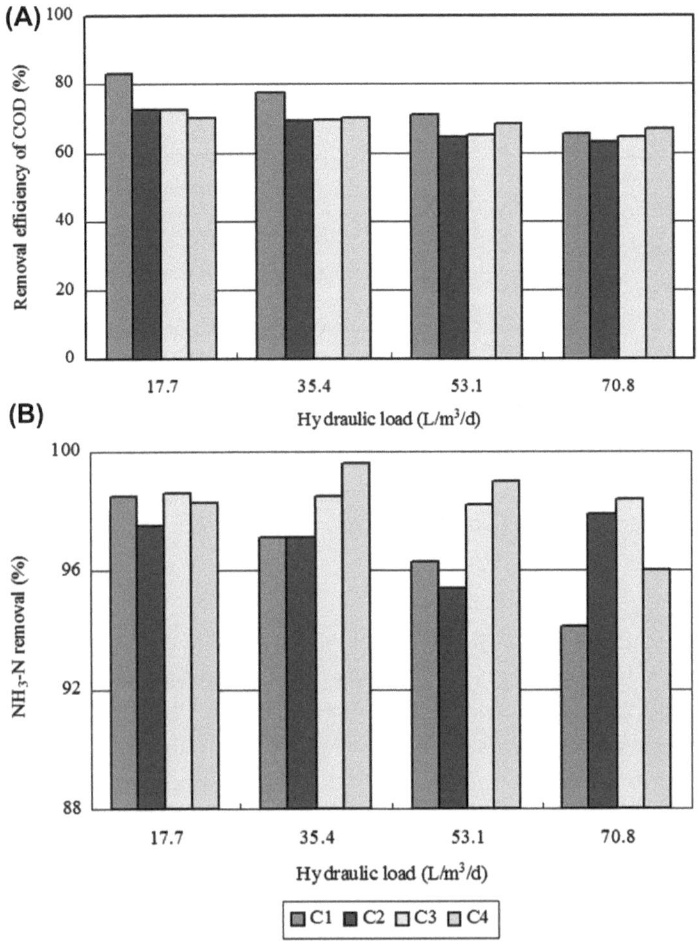

Figure 6.46 Removal of COD and NH_3–N in different hydraulic loads using the aged-sludge bioreactor.

6.9.5 Effect of Leachate Introduction Intervals

The aged-sludge-based bioreactor is considered to be somewhat similar to a land treatment system. The influent was introduced into the bioreactor for a definite time (e.g., 0.5 h) and then stopped for a while (e.g., 2.5 h). The ratios of influent introduction time (wet time) to stopping time (dry time) are referred to as "wet/dry" ratios in this work. The pollutants in the leachate will be absorbed onto the surface of the aged sludge during the "wet time" and then biodegraded during the "dry time". Typically, a rate of 2–8 mL/min was used, with a total leachate volume of 62.5–250 mL for each introduction, with 1–4 times of introduction within 12 h of daytime (8 a.m.–8 p.m.).

Table 6.17 shows the relationships among wet/dry ratios, interval time of introduction, and removal of COD $NH_4^+ - N$ and IC. The optimum conditions for maximum COD removal were found to be 81.7% at wet/dry ratios of 1: 11 and a spraying interval of 3 h. While $NH_4^+ - N$ removals kept unchanged and IC decreased slightly as the spraying intervals increased.

6.9.6 Removals of COD and BOD over Operation Time

The influent and effluent concentrations of COD, BOD, $NH_4^+ - N$, and TN were monitored for a long period and the typical results of C1 are illustrated in Figs. 6.47–6.48. The effluent qualities of the other three bioreactors were nearly consistent with those of C1 bioreactor. 78–84% of COD can be removed at an influent COD of 500–2600 mg/L. In the first 100 days of operation, the aged-sludge-based bioreactor produced an

Table 6.17 The Treatment Effect of Leachate in C1 in Different Conditions (Hydraulic Load 35.4 L/m^3/d) Using the Aged-Sludge Bioreactor

Intervals (h) 8: 00 a.m.–8: 00 p.m.	Wet/Dry Ratios	Spraying Time (h)	COD Removal (%)	NH_3–N Removal (%)	IC Removal (%)
3	1:5	0.5	79.1	99.5	95.3
	1:11	0.25	81.7	99.5	96.0
	1:23	0.12	80.0	99.5	95.2
6	5:1	1	80.9	99.6	95.5
	1:11	0.5	79.2	99.6	95.0
	1:23	0.25	76.0	99.6	94.8
12	1:5	2	80.6	99.4	92.1
	1:11	1	80.3	99.3	91.8
	1:23	0.5	76.7	99.3	89.7

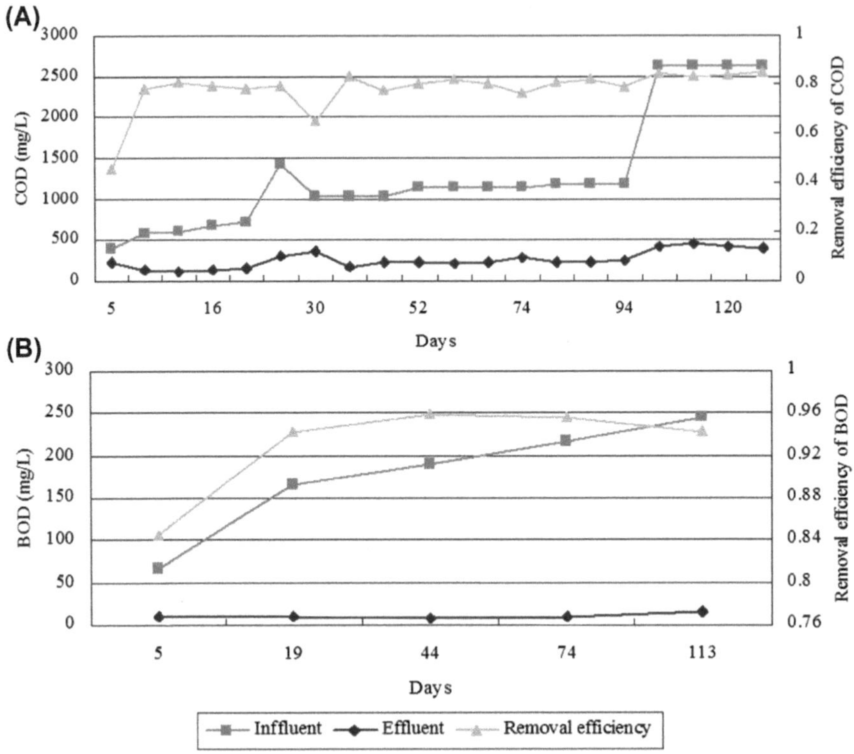

Figure 6.47 Removal of COD and BOD over operation time at a hydraulic load of 35.4 L/m^3/d using the aged-sludge bioreactor.

effluent (COD < 300 mg/L) at influent COD of below 1200 mg/L and the effluent COD increased to 390−450 mg/L with the increase of influent COD to over 2500 mg/L (Fig. 6.47A). The influent and effluent BODs were tested at intervals of 25−30 days, and the effluent values were always lower than 10−15 mg/L with initial concentrations of 65−246 mg/L (Fig. 6.47B).

6.9.7 Removals of $NH_4^+ - N$ and TN over Operation Time

The TN in the raw leachate used was over 1200 mg/L, in which ammonia nitrogen predominated, with much lower concentrations of organic nitrogen, nitrate, and nitrite. As shown in Fig. 6.48A, 98% of ammonia nitrogen was removed, with 20−60% removal of TN, illustrating the stronger capability of converting ammonia nitrogen to nitrate or nitrite and a weaker capability of transforming nitrate and nitrite to nitrogen, which was consistent with the bacteria distribution.

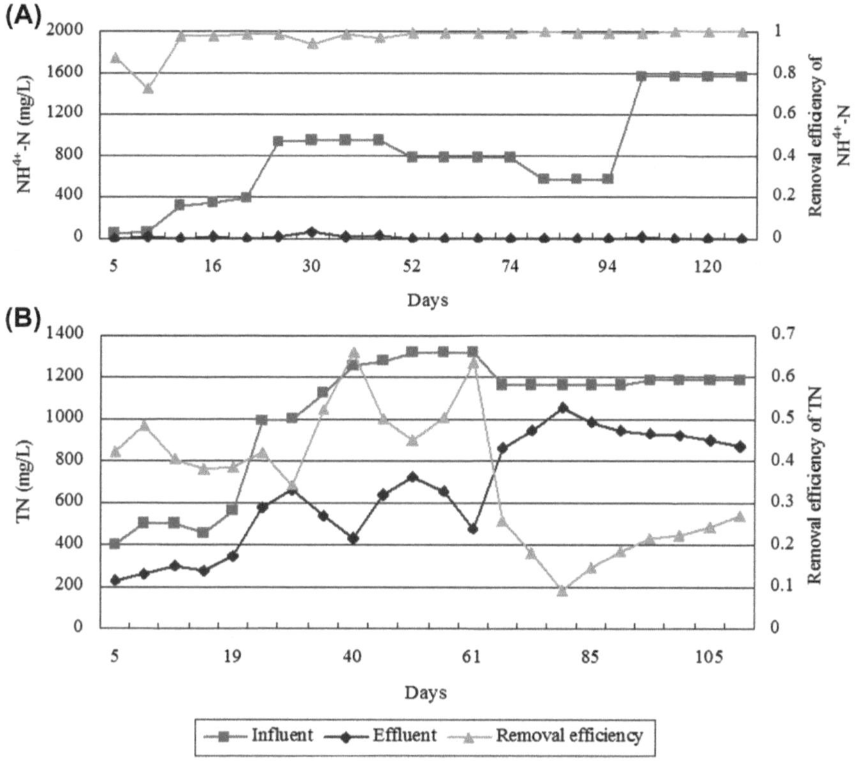

Figure 6.48 Removal of $NH_4^+ - N$ and TN over operation time at a hydraulic load of 35.4 $L/m^3/d$ using the aged sludge bioreactor.

There was a drop of TN removal efficiency after two months of treatment (Fig. 6.48B), which was mainly due to the following two reasons: First of all, the dominant bacteria species in aged sludge were gradually evolved from anaerobe to aerobic bacteria; second, the physical structures of aged sludge in bioreactor were relatively steady, which led to a definite treatment capacity of contaminants, especially for the rate of nitrate transformation by anaerobe. Hence, if the nitrogen concentration in the influent increased, the corresponding removal efficiency decreased.

The dissolved oxygen concentrations in the effluent and influent leachates were intermittently tested and the results were approximately 3 mg/L in the effluent while nearly 0 mg/L in the influent, which mainly due to the spraying process of leachate into the superficial area of aged sludge. In the distribution process, the anaerobic leachate was sprayed into little drips that were exposed to atmosphere around and reoxygenated simultaneously. The dissolved oxygen in the leachate greatly contributed to the removal of $NH_4^+ - N$ while took a negative function on the transformation of nitrate and nitrite to nitrogen gas.

6.9.8 Removal of TP Over Operation Time

Contrasted with the TN, TP in landfill leachate used was much less typically below 30 mg/L. It was found that the TP was almost completely removed by the aged-sludge-based bioreactor.

6.9.9 Outlooks of Aged-Sludge-Based Bioreactor for Landfill Leachate Treatment

The temperature effect was neglected in this study, as it's impossible to control in practical application. This study was conducted from Autumn throughout Spring of the year, the coldest season in Shanghai, China, typically 0–20°C. Under this adverse condition, the contaminants' removal efficiency still remained at an acceptable level. Effluent quality will be improved in summer.

The permeation of the aged sludge increased by air drying for several days. The height of the bioreactor used in this work was 50 cm, which could provide a perfect condition for nitrification and an adverse condition for denitrification. However, both nitrobacter and denitrobacter existed in the aged-sludge-based bioreactor, leading to a higher removal of $NH_4^+ - N$ and lower removal of TN. This phenomenon indicates that the microbes in aged sludge have a much stronger adaptability after years of anaerobic treatment such as landfilling. The surface of the aged sludge in the bioreactors may settle slightly at the starting period and then remain the level after 4 weeks of operation, as proved in the testing period of 4 months. As for leachate with higher concentrations of pollutants, such as COD $>$ 2000 mg/L, two- or three-stage aged-sludge-based bioreactor may be needed or the height of the bioreactor may be increased to ensure an acceptable effluent quality.

The volume load of the aged-sludge-based bioreactor may be lower than the traditional methods. But the bioreactor produced negligible residuals and the operating and performance are simple and could be easily controlled by working staffs. Meanwhile, the process flowsheet is compact and the maintenance of facilities used is less costly.

As one of the potential resources, the aged sludge is abundant and far-flung. The aged-sludge bioreactor not only makes full use of the sludge generated in the municipal wastewater treatment plant, but also achieves a better treatment effect for leachate. The whole process can be regarded as "using waste to treat waste", which is consistent with "sustainable landfilling" technology.

6.10 LANDFILLING AND STABILIZATION PROCESS IN GENERAL FOR SLUDGE SANITARY LANDFILL

It is considered that a recycle of "landfill + sludge landfilling + sludge biodegradation + aged sludge excavation + relandfilling of fresh sludge" should be possible. However, the operation of landfilling of sludge was found quite difficult if it contained over 60% moisture with extremely low strength and permeability. The geotechnical properties of sludge can be improved by mixing additives such as aged refuse excavated from long-term closed refuse landfill, coal ash, demolition wastes, and soil.

In this section, both chemical sludge from flocculation process alone and biological sludge from biological process in a sewage treatment plant were tested. It was found that the moisture had a significant effect on geotechnical properties of sludge. The geotechnical properties of sludge for landfill operation should meet the requirements of the following specifications: bearing capacity over 50 kPa, vane shear strength over 25 kPa, permeability of over the magnitude of 10^{-6}—10^{-5} cm/s, and moderate odor. Side slope of 12.6 degrees may be required for preventing sliding of sludge piles in landfill. To meet these conditions, the sludge with 80% moistures should be mixed with the additives and ratios of additives to sludge should be over 6: 10, 9: 10, 7: 10 and 9: 10 for coal fly ash, aged refuse, demolition wastes, and soil, respectively.

The annual production rate of leachate from fresh sludge landfill with a moisture of 80% was about 44 L/t. The biodegradability of leachate decreased and annual production of leachate increased to 287 L/t in the presence of 33% aged refuse in the mixtures. Meanwhile, it was observed that the presence of the additives can accelerate the stabilization of sludge. Pollutants' concentrations in leachate from mixtures with additives were always lower than those of the fresh sludge.

A field lysimeter with 36 t of fresh sludge was established and its long-term stabilization process was monitored. It was found that pollutants' concentrations in the leachate from the field-test lysimeter were significantly lower than those of the laboratory lysimeters. Moreover, it seemed that the degradation in the field lysimeter was faster. After 498 days' biodegradation, the OMs decreased by 67.1%, 61.6%, 30.5%, and 51.4% for the laboratory-scale lysimeters of biosludge, biosludge + aged refuse, chemical sludge, and field test lysimeter of chemical sludge, respectively. The biodegradation of the biosludge was observed to be faster than the chemical sludge.

The humidification of sludge in the lysimeters was studied. It was found that humus, humic percentum, and humic index slightly increased during the stabilization process. Molecular weight of HA decreased in early days revealing its mineralization process, and increased with time in metaphase and anaphase, showing its maturation process. Fulvic acid decomposed in prophase and composed in metaphase and anaphase. Humic percentum decreased in the sequence of the laboratory-scale lysimeters of sludge + aged refuse, biosludge, field test lysimeter of chemical sludge, and chemical sludge, while humic index decreased in the sequence of the laboratory-scale lysimeters of biosludge, biosludge + aged refuse, field test lysimeter of chemical sludge, and biosludge.

Biological toxicity faded during the stabilization of sludge, slow in prophase and rapid in metaphase and anaphase, according with logarithm formula. The germinating index of biosuldge, aged refuse + sludge, chemical sludge, and field-test chemical sludge can reach to 80% at 612, 626, 650, and 670 days of placement, respectively. Mixing of aged refuse could reduce the toxicity of the biosludge and increased the germinating index.

Settlement of simulated landfills could be classified as: the initial stress settlement occurred mainly in the first year; and secondary degradation settlement in the following years. The predicted ultimate settlement of the chemical sludge in the field-test lysimeter, chemical sludge, biosludge, and aged refuse + biosludge in the laboratory scale lysimeters were evaluated to be 12.33%, 20.54%, 13.77%, and 26.57% of the initial sludge height, and the corresponding settlements in first year were 98.9%, 94.4%, 97.4%, and 98.9% of the ultimate settlement, respectively. 99.99% of the ultimate settlement may reach at 870, 1170, 995, and 762 days of the placement, respectively. Consolidation coefficient of the field-test lysimeter was calculated to be 0.00249 cm^2/s.

The equations among the settlements, pollutants concentrations in leachate (COD, NH_3—N, etc.), chemical compositions in the sludge (OMs, BDM, VS, TS, moisture, etc.) and the placement time, were established. Taking the leachate discharge standard and the BDM in soil as reference, it was estimated that resultant biodegraded sludge can be regarded as the stabilized sludge or aged sludge which can be excavated and recycled after around 5.82 years placement, when the BDM in the sludge decreased from around 30—60% in sludge to 4.76% in soil and COD to 100 mg/L in the leachate. If the aged sludge was to be used for planting, 1.68—1.84 years at least would be needed.

REFERENCES

[1] Adav Sunil S, Lee DJ, Tay JH. Extracellular polymeric substances and structural stability of aerobic granule. Water Res. 2008;42:1644–50.

[2] Anipsitakis GP, Dionysiou DD. Radical generation by the interaction of transition metals with common oxidants. Environ. Sci. Technol. 2004;38(13):3705–12.

[3] Andreoli CV, von Sperling M, Fernandes F. Biological wastewater treatment series-sludge treatment and disposal. London (UK): IWA Publishing, Alliance House; 2007.

[4] Appels L, Houtmeyers S, Degreve J, Van Impe J, Dewil R. Influence of microwave pre-treatment on sludge solubilization and pilot scale semi-continuous anaerobic digestion. Bioresour. Technol. 2013;128:598–603.

[5] Appels L, Lauwers J, Gins G, Degreve J, Van Impe J, Dewil R. Parameter identification and modeling of the biochemical methane potential of waste activated sludge. Environ. Sci. Technol. 2011;45:4173–8.

[6] Bala Subramanian S, Yan S, Tyagi RD, Surampalli RY. Extracellular polymeric substances (EPS) producing bacterial strains of municipal wastewater sludge: isolation, molecular identification, EPS characterization and performance for sludge settling and dewatering. Water Res. 2010;44:2253–66.

[7] Bolzonella D, Pavan P, Zanette M, Cecchi F. Two-phase anaerobic digestion of waste activated sludge: effect of an extreme thermophilic prefermentation. Ind. Eng. Chem. Res. 2007;46:6650–5.

[8] Bougrier C, Delgenès JP, Carrère H. Effects of thermal treatments on five different waste activated sludge samples solubilisation, physical properties and anaerobic digestion. Chem. Eng. J. 2008;139(2):236–44.

[9] Brown S, Beecher N, Carpenter A. Calculator tool for determining greenhouse gas emissions for biosolids processing and end use. Environ. Sci. Technol. 2010;44:9509–15.

[10] Cantrell KJ, Yabusaki SB, Engelhard MH, Mitroshkov AV, Thornton EC. Oxidation of H_2S by iron oxides in unsaturated conditions. Environ. Sci. Technol. 2003;37:2192–9.

[11] Carrere H, Bougrier C, Castets D, Delgenes JP. Impact of initial biodegradability on sludge anaerobic digestion enhancement by thermal pretreatment. J. Environ. Sci. Health A 2008;43:1551–5.

[12] Chen JJ. Biomass to renewable energy processes. Boca Raton (London, New York): CRC Press, Taylor & Francis Group; 2010.

[13] Chen W, Westerhoff P, Leenheer JA, Booksh K. Fluorescence excitation–emission matrix regional integration to quantify spectra for dissolved organic matter. Environ. Sci. Technol. 2003;37:5701–10.

[14] Chai XL, Hao Y, Zhao X, Liu G, Zhu Y, Ji R, Wu J, Tong H, Zhao YC. Abiotic association of phthalic acid esters with humic acid of a sludge landfill. Front. Environ. Sci. Eng. 2012;6(6):778–83.

[15] Chiavola A, Ridolfi A, D'Amato E, Bongirolami S, Cima E, Sirini P, et al. Sludge reduction in a small wastewater treatment plant by electro-kinetic disintegration. Water Sci. Technol. 2015;72:364–70.

[16] Chu LB, Yan ST, Xing XH, Yu AF, Sun XL, Jurcik B. Enhanced sludge solubilization by microbubble ozonation. Chemosphere 2008;72:205–12.

[17] Climent M, Ferrer I, Baeza MD, Artola A, Vazquez F, Font X. Effects of thermal and mechanical pretreatments of secondary sludge on biogas production under thermophilic conditions. Chem. Eng. J. 2007;133:335–42.

[18] Coelho NMG, Droste RL, Kennedy KJ. Evaluation of continuous mesophilic, thermophilic and temperature phased anaerobic digestion of microwaved activated sludge. Water Res. 2011;45:2822–34.

[19] Daniels L, Belay N, Rajagopal BS, Weimer PJ. Bacterial methanogenesis and growth from CO_2 with elemental iron as the sole source of electrons. Nature 1987;237:509–11.
[20] Eskicioglu C, Terzian N, Kennedy KJ, Droste RL, Hamoda M. Athermal microwave effects for enhancing digestibility of waste activated sludge. Water Res. 2007;41:2457–66.
[21] Frolund B, Palmgren R, Keiding K, Nielsen PH. Extraction of extracellular polymers from activated sludge using a cation exchange resin. Water Res. 1996;30:1749–58.
[22] Gassara F, Brar SK, Pelletier F, Verma M, Godbout S, Tyagi RD. Pomace waste management scenarios in Quebec-Impact on greenhouse gas emissions. J. Hazard. Mater. 2011;192:1178–85.
[23] Ge HQ, Jensen PD, Batstone DJ. Increased temperature in the thermophilic stage in temperature phased anaerobic digestion (TPAD) improves degradability of waste activated sludge. J. Hazard. Mater. 2011;187:355–61.
[24] Gil O, Vale C. Evidence for polychlorinated biphenyls dechlorination in the sediments of Sado Estuary, Portugal. Mar. Pollut. Bull. 2001;42:453–61.
[25] Grady CPL, Daigger GT, Lim HC. Biological wastewater treatment. New York: Marcel Dekker, Inc.; 1999.
[26] Guo X, Liu J, Xiao B. Bioelectrochemical enhancement of hydrogen and methane production from the anaerobic digestion of sewage sludge in single-chamber membrane-free microbial electrolysis cells. Int. J. Hydrogen Energy 2013;38:1342–7.
[27] Hospido A, Moreira MT, Martin M, Rigola M, Feijoo G. Environmental evaluation of different treatment processes for sludge from urban wastewater treatments: anaerobic digestion versus thermal processes. Int. J. Life Cycle Assess. 2005;10:336–45.
[28] Houghton JI, Quarmby J, Stephenson T. Municipal wastewater sludge dewaterability and the presence of microbial extracellular polymer. Water Sci. Technol. 2001;44(2–3):373–9.
[29] Jin B, Wilén BM, Lant P. Impacts of morphological, physical and chemical properties of sludge flocs on dewaterability of activated sludge. Chem. Eng. J. 2004;98:115–26.
[30] Karri S, Sierra-Alvarez R, Field JA. Zero valent iron as an electron-donor for methanogenesis and sulfate reduction in anaerobic sludge. Biotechnol. Bioeng. 2005;92:810–9.
[31] Kim DH, Jeong E, Oh SE, Shin HS. Combined (alkaline + ultrasonic) pretreatment effect on sewage sludge disintegration. Water Res. 2010;44:3093–100.
[32] Kuglarz M, Karakashev D, Angelidaki I. Microwave and thermal pretreatment as methods for increasing the biogas potential of secondary sludge from municipal wastewater treatment plants. Bioresour. Technol. 2013;134:290–7.
[33] Lachemi M, Sahmaran M, Hossain KMA, Lotfy A, Shehata M. Properties of controlled low-strength materials incorporating cement kiln dust and slag. Cem. Concr. Compos. 2010;32:623–9.
[34] Li XY, Yang SF. Influence of loosely bound extracellular polymeric substances (EPS) on the flocculation, sedimentation and dewaterability of activated sludge. Water Res. 2007;41(5):1022–30.
[35] Li M, Zhao YC, Guo Q, Qian X, Niu D. Bio-hydrogen production from food waste and sewage sludge in the presence of aged refuse excavated from refuse landfill. Renew. Energy 2008;33:2573–9.
[36] Li YY, Noike T. Upgrading of anaerobic-digestion of waste activated-sludge by thermal pretreatment. Water Sci. Technol. 1992;26:857–66.
[37] Lin M, Zhao YC. Filtration improvement of zinc sludge by using unconventional alkalization sequence. J. Environ. Sci. 2000;12(1):63–7.
[38] Long JH, Bullard CM. Waste activated sludge pretreatment to boost volatile solids reduction and digester gas production: market and technology assessment. Fla. Water Resour. J. 2014:44–50.

[39] Lu XQ, Zhen GY, Liu Y, Hojo T, Adriana LE, Li YY. Long-term effect of the antibiotic cefalexin on methane production during waste activated sludge anaerobic digestion. Bioresour. Technol. 2014;169:644–51.

[40] Lubken M, Wichern M, Schlattmann M, Gronauer A, Horn H. Modelling the energy balance of an anaerobic digester fed with cattle manure and renewable energy crops. Water Res. 2007;41:4085–96.

[41] Ma JL, Zhao YC, Wang JM, Wang L. Effect of magnesium oxychloride cement on stabilization/solidification of sewage sludge. Constr. Build. Mater. 2010;24(1):79–83.

[42] Ma JL, Wang JM, Zhao YC, Liu T, Wang L. Stabilization of heavy metals in sewage sludge using Sorel cement. In: International conference on chemical, biological and environmental engineering (CBEE 2009, Singapore, 2009/10/9–2009/10/11); 2010. p. 69–75.

[43] McCarty PL, Smith DP. Anaerobic wastewater treatment. Environ. Sci. Technol. 1986;20:1200–6.

[44] Mehdizadeh SN, Eskicioglu C, Bobowski J, Johnson T. Conductive heating and microwave hydrolysis under identical heating profiles for advanced anaerobic digestion of municipal sludge. Water Res. 2013;47:5040–51.

[45] Metcalf E. Wastewater engineering: treatment, disposal, reuse. New York: Metcalf & Eddy Inc., McGraw-Hill; 2003.

[46] Nielsen PH, Jahn A. Extraction of EPS. In: Wingender J, Neu TR, Flemming HC, editors. Microbial extracellular polymeric substances: characterization, structure and function. Berlin Heidelberg: Springer-Verlag; 1999.

[47] Naganathan S, Razak HA, Hamid SNA. Effect of kaolin addition on the performance of controlled low-strength material using industrial waste incineration bottom ash. Waste Manage. Res. 2010;28:848–60.

[48] Neyens E, Baeyens J, Dewil R, De heyder B. Advanced sludge treatment affects extracellular polymeric substances to improve activated sludge dewatering. J. Hazard. Mater. 2004;106(2–3):83–92.

[49] Onyeche TI. Economic benefits of low pressure sludge homogenization for wastewater treatment plants. In: IWA specialist conferences moving forward wastewater biosolids sustainability, Moncton, New Brunswick, Canada; 2007. p. 417–22.

[50] Pilli S, Bhunia P, Yan S, LeBlanc RJ, Tyagi RD, Surampalli RY. Ultrasonic pretreatment of sludge: a review. Ultrason. Sonochem. 2011;18:1–18.

[51] Pilli S, More TT, Yan S, Tyagi RD, Surampalli RY. Fenton pre-treatment of secondary sludge to enhance anaerobic digestion: energy balance and greenhouse gas emissions. Chem. Eng. J. 2016;283:285–92.

[52] Pilli S, More T, Yan S, Tyagi RD, Surampalli RY. Anaerobic digestion of thermal pre-treated sludge at different solids concentrations-computation of mass-energy balance and greenhouse gas emissions. J. Environ. Manage. 2015;157:250–61.

[53] Poxon TL, Darby JL. Extracellular polyanions in digested sludge: measurement and relationship to sludge dewaterability. Water Res. 1997;31:749–58.

[54] Rabinowitz B, Stephenson R. Full-scale demonstration of waste activated sludge homogenization at the Los Angeles County Joint Water Pollution Control Plant. In: Proceedings of the water environment federation. Residuals and biosolids management; 2006. p. 445–59.

[55] Ren NQ, Chua H, Chan SY, Tsang YF, Wang YJ, Sin N. Assessing optimal fermentation type for bio-hydrogen production in continuous-flow acidogenic reactors. Bioresour. Technol. 2007;98:1774–80.

[56] Riau V, De la Rubia MA, Perez M. Temperature-phased anaerobic digestion (TPAD) to obtain class A biosolids: a semi-continuous study. Bioresour. Technol. 2010;101:2706–12.

[57] Rittmann BE, Lee HS, Zhang HS, Alder J, Banaszak JE, Lopez R. Full-scale application of focused-pulsed pre-treatment for improving biosolids digestion and conversion to methane. Water Sci. Technol. 2008;58:1895—901.
[58] Sahinkaya S, Kalipci E, Aras S. Disintegration of waste activated sludge by different applications of Fenton process. Process Saf. Environ. Prot. 2015;93:274—81.
[59] Sasaki D, Sasaki K, Watanabe A, Morita M, Matsumoto N, Igarashi Y, et al. Operation of a cylindrical bioelectrochemical reactor containing carbon fiber fabric for efficient methane fermentation from thickened sewage sludge. Bioresour. Technol. 2013;129:366—73.
[60] Su L, Zhao YC. Chemical reduction of odour in fresh sewage sludge in the presence of ferric hydroxide. Environ. Technol. 2013;34(1—4):165—72.
[61] Su L, Shi X, Guo G, Zhao A, Zhao YC. Stabilization of sewage sludge in the presence of nanoscale zero-valent iron (nZVI): abatement of odor and improvement of biogas production. J. Mater. Cycles Waste Manage. 2013;15(4):461—8.
[62] Su L, Zhen GY, Zhang L, Zhao YC, Niu D, Chai XL. The use of the core—shell structure of zero-valent iron nanoparticles (NZVI) for long-term removal of sulphide in sludge during anaerobic digestion. Environ. Sci. Process. Impacts 2015;17:2013—21.
[63] State Environmental Protection Administration of China. Discharge standard of pollutants for municipal wastewater treatment plant. Beijing: Standard Press of China; 2002GB18918.
[64] Taricska JR, Long DA, Chen JP, Hung Y, Zou S. Anaerobic digestion. In: Wang LK, Pereira NC, Hung Y, editors. Biological treatment processes. New York: Humana Press; 2009. p. 589—634.
[65] Thauer RK, Jungermann K, Decker K. Energy conservation in chemotrophic anaerobic bacteria. Bacteriol. Rev. 1977;41:100—80.
[66] Tervahauta T, Bryant IM, Leal LH, Buisman CJN, Zeeman G. Improved energy recovery by anaerobic grey water sludge treatment with black water. Water Air Soil Poll. 2014;6:2436—48.
[67] Turovskiy IS, Mathai PK. Wastewater sludge processing. Hoboken (New Jersey): John Wiley & Sons, Inc.; 2006.
[68] Tyagi VK, Lo SL. Microwave irradiation: a sustainable way for sludge treatment and resource recovery. Renew. Sustain. Energy Rev. 2013;18:288—305.
[69] Urbain V, Block JC, Manem J. Bioflocculation in activated-sludge, an analytic approach. Water Sci. Technol. 1992;25:441—3.
[70] Wilén BM, Jin B, Lant P. The infiuence of key chemical constituents in activated sludge on surface and flocculating properties. Water Res. 2003;37:2127—39.
[71] Wahidunnabi AK, Eskicioglu C. High pressure homogenization and two-phased anaerobic digestion for enhanced biogas conversion from municipal waste sludge. Water Res. 2014;63:430—46.
[72] Wang LC, Li X, Zhao YC, Cao WH. Sludge treatment and resource series-sludge drying and incineration (Chinese edition). China: Metallurgical Industry Press; 2010.
[73] Wang LK, Shammas NK, Hung YT. Biosolids engineering and management. New York (USA): Humana Press; 2008.
[74] Wu B, Chai XL, Zhao YC. Enhanced dewatering of waste-activated sludge by composite hydrolysis enzymes. Bioprocess Biosyst. Eng. 2016;39:627—39.
[75] Xin D, Chai XL, Zhao W. Hybrid cement-assisted dewatering, solidification and stabilization of sewage sludge with high organic content. J. Mater. Cycles Waste Manag. 2016;18:356—65.
[76] Zhen GY, Lu XQ, Wang BY, Zhao YC, Chai XL, Niu DJ, Zhao TT. Enhanced dewatering characteristics of waste activated sludge with Fenton pretreatment: effectiveness and statistical optimization. Front. Environ. Sci. Eng. 2014;8(2):267—76.

[77] Zhen GY, Lu XQ, Zhao YC, Chai XL, Niu DJ. Enhanced dewaterability of sewage sludge in the presence of Fe(II)-activated persulfate oxidation. Bioresour. Technol. 2012;116:259−65.
[78] Zhen GY, Lu XQ, Li YY, Zhao YC, Wang BY, Song Y, Chai XL, Niu DJ, Cao XY. Novel insights into enhanced dewaterability of waste activated sludge by Fe(II)-activated persulfate oxidation. Bioresour. Technol. 2012;119:7−14.
[79] Zhen GY, Lu XQ, Wang BY, Zhao YC, Chai XL, Niu DJ, Li YY, Song Y, Cao XY. Synergetic pretreatment of waste activated sludge by Fe(II)-activated persulfate oxidation under mild temperature for enhanced dewaterability. Bioresour. Technol. 2012;124:29−36.
[80] Zhen GY, Lu XQ, Li YY, Zhao YC. Innovative combination of electrolysis and Fe(II)-activated persulfate oxidation for improving the dewaterability of waste activated sludge. Bioresour. Technol. 2013;136:663−4.
[81] Zhen GY, Yan XF, Zhou HY, Chen H, Zhao TT, Zhao YC. Effects of calcined aluminum salts on the advanced dewatering and solidification/stabilization of sewage sludge. J. Environ. Sci. 2011;23(7):1225−32.
[82] Zhen GY, Lu XQ, Cheng XB, Chen H, Yan XF, Zhao YC. Hydration process of the aluminate $12CaO \cdot 7Al_2O_3$-assisted Portland cement-based solidification/stabilization of sewage sludge. Constr. Build. Mater. 2012;5(30):675−81.
[83] Zhen GY, Zhou HY, Zhao TT, Zhao YC. Performance appraisal of controlled low-strength material (CLSM) using sewage sludge and refuse incineration bottom ash. Chin. J. Chem. Eng. 2012;20(1):80−8.
[84] Zhen GY, Lu XQ, Zhao YC, Niu J, Chai XL, Su LH, Li YY, Liu Y, Du JR, Hojo T, Hu Y. Characterization of controlled low-strength material (CLSM) obtained from dewatered sludge and MSWI bottom ash: mechanical and microstructural perspectives. J. Environ. Manage. 2013;129:183−9.
[85] Zhen GY, Lu XQ, Li YY, Zhao YC. Combined electrical-alkali pretreatment to increase the anaerobic hydrolysis rate of waste activated sludge during anaerobic digestion. Appl. Energy 2014;128:93−102.
[86] Zhen GY, Lu XQ, Li YY, Liu Y, Zhao YC. Influence of zero valent scrap iron (ZVSI) supply on methane production from waste activated sludge. Chem. Eng. J. 2015;263:461−70.
[87] Zhen GY, Lu X, Kobayashi T, Li YY, Xu KQ, Zhao YC. Mesophilic anaerobic co-digestion of waste activated sludge and *Egeria densa*: performance assessment and kinetic analysis. Appl. Energy 2015;148:78−86.
[88] Zhen GY, Lu XQ, Kobayashi T, Kumar G, Xu KQ. Promoted electromethanosynthesis in a two-chamber microbial electrolysis cells (MECs) containing a hybrid biocathode covered with graphite felt (GF). Chem. Eng. J. 2016;284:1146−55.
[89] Zhen GY, Kobayashi T, Lu XQ, Xu KQ. Understanding methane bioelectrosynthesis from carbon dioxide in a two-chamber microbial electrolysis cells (MECs) containing a carbon biocathode. Bioresour. Technol. 2015;186:141−8.
[90] Zhen GY, Kobayashi T, Lu XQ, Kumar G, Xu KQ. Biomethane recovery from *Egeria densa* in a microbial electrolysis cells (MEC)-assisted anaerobic system: performance and stability assessment. Chemosphere 2016;149:121−9.
[91] Zhu Y, Zhang H, Zhao YC. Sludge treatment and resource series-sludge recycling sanitary landfill technology (Chinese edition). China: Metallurgical Industry Press; 2010.
[92] Zhu Y, Zhao YC. Stabilization process within a sewage sludge landfill determined through both particle size distribution and content of humic substances as well as by FT-IR analysis. Waste Manage. Res. 2010;29(4):379−85.

[93] Zhu Y, Chai XL, Li H, Zhao YC. Combination of combustion with pyrolysis for studying the stabilization process of sludge in landfill. Thermochim. Acta 2007;464:59–64.

[94] Zhu Y, Han Z, Liu X, Li J, Liu F, Feng S. Study on the effect and mechanism of hydrothermal pretreatment of dewatered sewage sludge cake for dewaterability. J. Air Waste Manage. Assoc. 2013;63(8):997–1002.

[95] Zhu Y, Zhang M, Han Z, Feng SP. Variation of PAHs, PCBs and OCPs contents and influencing factors in sludge landfill process. Appl. Mech. Mater. 2014;522–524:797–805.

INDEX

'*Note*: Page numbers followed by "f" indicate figures, "t" indicate tables.'

D

E

F

G

H

I

K

L

M

N

O

X

Z

FSC
www.fsc.org
MIX
Papier | Fördert
gute Waldnutzung
FSC® C083411

Kolibri 360 GmbH
Ferdinand-Jühlke-Straße 7
99095 Erfurt, Deutschland
produktsicherheit@kolibri360.de